Food Proteins

Edited by

John E. Kinsella
Cornell University
Ithaca, New York

William G. Soucie
Kraft, Inc.
Research and Development
Glenview, IL

The American Oil Chemists' Society
Champaign, IL

Mention of firm names of trade products does not imply endorsement or recommendation by the editors or contributors over other firms or similar products not mentioned.

Copyright © 1989 by the American Oil Chemists' Society. All rights reserved. No part of this book may be reproduced or transmitted in any form or by any means without written permission of the publisher.

Library of Congress Cataloging-in-Publication Data

Food proteins.

    Proceedings of the Protein and Co-products Symposium, held in the spring of 1988 during the National AOCS Meeting in Phoenix; sponsored by the American Oil Chemists' Society, Protein and Co-products Division.
    Includes bibliographical references.
    1. Proteins—Congresses. I. Soucie, William G. II. Kinsella, John E., 1938- . III. Protein and Co-products Symposium (1988 : Phoenix, Ariz.) IV. American Oil Chemists' Society. Protein and Co-products Division.
TP453.P7F662   1989   664   89-18159
ISBN 0-935315-26-8

Printed in the United States of America

Preface

Burgeoning developments in the food industry and rapid changes and innovations in development of consumer food products have resulted in an increasing demand for versatile ingredients possessing appropriate functional properties. In this regard, proteins represent an important class of functional ingredients, particularly because they possess a wide range of dynamic functional properties, show versatility during processing and possess the capacity to form network structures and stabilize emulsions and foams in addition to providing essential amino acids. The toxic factors associated with food proteins are also very important.

Because of the heterogenous nature of most food proteins, a knowledge of their molecular structure and the relationship between structure and functional attributes in food systems is relatively limited. Furthermore, processing conditions and environmental factors greatly affect the behavior of proteins in various applications. With the advent of protein engineering, an understanding of the molecular structure and molecular interactions between proteins and other food components, including toxins, are clearly needed to enable the food processor to select the most appropriate functional and healthy ingredients for specific applications. The American Oil Chemists' Society, Protein and Co-Products Division, organized a symposium (held in the Spring of 1988 during the national AOCS meeting in Pheonix) to discuss the above topics. The emphasis was on 1) protein structure and function, 2) the functional attributes required for use in food, and 3) the safety of different food proteins and associated toxins.

Perusal of the Table of Contents will reveal that the information contained in this book represents the current thinking of prominent researchers in the various scientific fields represented herein. This book should be a valuable addition to the libraries of academic and industrial research laboratories. Scientists and students alike should find it a useful resource for current information regarding protein functional relationships and associated toxic compounds in food systems.

This text is comprised of a collection of chapters representing the latest reviews and research regarding protein structure, structural-functional relationships of food proteins and mechanisms of protein:protein interactions in gels, emulsions and foams. It also contains papers that discuss the most recent developments regarding toxic compounds associated with food proteins. The effects of these toxic compounds must be understood when the harboring plants or plant products are used in food, or when the toxins are bound to the proteins themselves. Much has been learned about these toxins, especially regarding their nutritional impact and procedures for controlling their potential deleterious effects.

The authors of the chapters are to be complimented for their cooperation and their expeditious preparation of manuscripts that allowed for the timely publication of this book.

It should also be noted that funds to support the Protein and Co-Products Symposium were received from Kraft, Inc., Nabisco Brands, Inc., The Quaker Oats Co., Central Soya and The Campbell Institute for Research and Technology. Without this corporate support, the symposium would not have been possible.

William G. Soucie
John E. Kinsella

Contents

Preface iii

Chapter 1 **The Role of Dynamics and Solvation in
Protein Structure and Function**
—*J.W. Brady* 1

Chapter 2 **Protein Structure in Solution**
—*I.D. Kuntz* 19

Chapter 3 **Interrelationship of Molecular and Functional
Properties of Food Proteins**
—*Srinivasan Damodaran* 21

Chapter 4 **Structure: Function Relationships in Food Proteins,
Film and Foaming Behavior**
—*J.E. Kinsella and L.G. Phillips* 52

Chapter 5 **Film Properties of Modified Proteins**
—*Srinivasan Damodaran* 78

Chapter 6 **Glycosylation of β—Lactoglobulin and
Surface Active Properties**
—*Ralph D. Waniska and John E. Kinsella* 100

Chapter 7 **Molecular Properties of Proteins Important in Foams**
—*J.B. German and L.G. Phillips* 132

Chapter 8 **Lipid—Protein—Emulsifier—Water Interactions in
Whippable Emulsions**
—*N.M. Barfod, N. Krog, and W. Buchheim* 144

Chapter 9 **Molecular Properties and Functionality of Proteins
in Food Emulsions: Liquid Food Systems**
—*M.E. Mangino* 159

Chapter 10 **Are Comminuted Meat Products Emulsions
or a Gel Matrix**
—*Joe M. Regenstein* 178

Chapter 11 **Molecular Properties and Functionality of
Proteins in Food Gels**
—*E. Allen Foegeding* 185

Chapter 12 **Functional Roles of Heat Induced Protein
Gelation in Processed Meat**
—*James C. Acton and Rhoda L. Dick* 195

Chapter 13 **Effects of Medium Composition, Preheating,
and Chemical Modification upon Thermal Behavior
of Oat Globulin and β—Lactoglobulin**
—*V.R. Harwalkar and C.-Y. Ma* 210

Chapter 14 **Effect of Molecular Changes (SH Groups and
Hydrophobicity) of Food Proteins
on Their Functionality**
—*E. Li-Chan and S. Nakai* 232

Chapter 15 **Relationship of SH Groups to Functionality
of Ovalbumin**
—*Etsushiro Doi, Naofumi Kitabatake, Hajime Hatta,
and Taihei Koseki* 252

Chapter 16 **Use of Radio—Labeled Proteins to Study the
Thiol—Disulfide Exchange Reaction in Heated Milk**
—*Bong Soo Noh and Tom Richardson* 267

Chapter 17 **Genetic Modification of Milk Proteins**
—*Lawrence Creamer, Sang Suk Oh, Robert McKnight, Rafael
Jimenez—Flores, and Tom Richardson* 277

Chapter 18 **Inactivation and Analysis of Soybean Inhibitors
of Digestive Enzymes**
—*Mendel Friedman, Michael R. Gumbmann,
David L. Brandon, and Anne H. Bates* 296

Chapter 19 **The Nutritional Significance of Lectins**
—*Irvin E. Liener* 329

Chapter 20 **α—Amylase Inhibitors of Higher Plants and
Microorganisms**
—*John R. Whitaker* 354

Chapter 21 **Toxic Compounds in Plant Foodstuffs: Cyanogens**
—*Jonathan E. Poulton* 381

Chapter 22 **New Perspectives on the Antinutritional Effects
of Tannins**
—*Larry G. Butler* 402

Chapter 23 **Nutritional and Physiological Effects of Phytic Acid**
—*Lilian U. Thompson*....................................410

Chapter One
The Role of Dynamics and Solvation in Protein Structure and Function

J.W. Brady

Department of Food Science
Cornell University
Ithaca, New York 14853

The techniques of genetic engineering make it possible for the primary sequence of proteins to be altered almost at will. This ability opens up the possibility of making specific mutants of proteins with the intention of improving or modifying the structural or functional properties of the molecules, such as enzyme activity, specificity, or thermal stability (1), and even of designing *de novo* for specific purposes completely novel proteins with desired functions. Unfortunately, it is often quite difficult to guess which potential mutations will produce the desired changes in functional properties. Since random alterations are not likely to be successful, guidance from a theoretical understanding of protein folding and the determinants of protein conformation and function would be extremely useful. The development of practical theoretical methods for predicting the possible functional properties of hypothetical mutant proteins would limit the need to actually produce mutants. A smaller number, with a higher probability of success could be produced. By combining the basic structural information concerning wild types gained from experimental studies with various forms of molecular modeling, it is now becoming possible to make improved evaluations of the functional consequences of specific mutations (2).

In recent years an overall understanding of the general factors which determine protein conformation and function has begun to emerge (3-5). These discoveries have been based on the structural information provided by a wide variety of sources. Globular proteins are generally presumed to fold up into compact shapes in order to remove as much of their nonpolar sidechain surface area as possible from contact with solvent, due to the high entropic cost of solvating these non-hydrogen-bonding groups in the unfolded state. In this view, the free energy gain arising from the diminished "structuring" of the aqueous solvent as nonpolar groups are removed to the protein interior, combined with a more favorable potential energy (lower enthalpy) in the native state, outweighs the lower configurational entropy that the chain itself has in the more restricted folded state, producing a negative free energy for the folding process at physiological temperatures. Aqueous solvation thus plays a critical, determining role in protein folding and in processes involving the physical properties of protein systems.

It is also becoming increasingly clear that atomic and molecular motions are often quite important, not only for protein structure, but in determining function as well (6). Proteins are known to undergo a wide variety of motions, many of which are involved in such processes as substrate binding or catalytic activity. For example, it has been demonstrated by analysis of the crystal structure of myoglobin that if the protein were completely rigid, it would not be not possible for oxygen ligands to diffuse into the interior pocket of this protein to the heme binding site, or back out again once there (7). However, fluctuations involving small changes in the positions of certain buried sidechains produce a feasible pathway which is consistent with the known binding kinetics. Because of the importance of aqueous solvation and molecular motions in determining the properties of biopolymers, it may often be necessary to include these factors in simulation methods which attempt to model proteins realistically.

The most powerful theoretical techniques for the study of protein conformation generally fall under the overall heading of molecular mechanics calculations. Molecular mechanics studies (8-11), which include such things as conformational energy minimization, Monte Carlo studies, molecular dynamics simulations, and normal mode analysis, attempt to numerically simulate the physical properties and behavior of a system on a high speed computer, based upon an assumed understanding of the variation of the energy of the system with molecular coordinates. Molecular mechanics calculations have long been used to analyze polypeptide conformation and structure/function relationships in biopolymers (12-20), as well as to study the microscopic physical properties of a variety of non-biological systems, including solids, liquids, and aqueous solutions (21-23). Theoretical studies are capable of revealing information about structure and dynamics on the molecular level which is difficult or impossible to obtain by other means. This paper will discuss how molecular motions and aqueous solvation can be included in theoretical models of proteins through the use of molecular dynamics (MD) simulations.

Molecular Mechanics Calculations

Molecular mechanics studies exploit a knowledge of molecular energies as a function of atomic coordinates to theoretically simulate the properties of molecular systems. Since it is not possible to accurately calculate the quantum mechanical energy for macromolecules, analytic, semi-empirical energy functions with theoretically reasonable functional forms which have been parameterized to the results of experiments and simple calculations are commonly used. A typical example of such a function of the internal coordinates q might be (17):

$$V(\underset{\sim}{q}) = \Sigma\, k_b\, (b-b_o)^2 + \Sigma\, k_\theta(\theta-\theta_o)^2 + \Sigma\, k_\phi[1+\cos(n\phi-\delta)] + \underset{i>j}{\Sigma}(A_{ij}/r_{ij}^{12} - B_{ij}/r_{ij}^6 + q_iq_j/r_{ij}) \quad (1)$$

which includes harmonic terms for bond stretching and bending with force constants k_b and k_ϕ, a periodic hindered torsional term in the torsional angles ϕ, with force constant k_ϕ, periodicity n, and phase factor δ, Lennard-Jones-type nonbonded interactions, and a Coulombic term between the atomic partial charges q_i and q_j. Such a function permits the evaluation of the variation in energy with the position of every atom in the system. In order for a computer model of a system to be useful, it is necessary that the potential energy functions be sufficiently realistic to adequately reproduce physical behavior. For this reason, considerable effort goes into the development of the various parameters (8,9,17) which appear in energy functions such as equation (1).

The most common types of molecular mechanics calculations are conformational energy studies (12-14). Conformational energy calculations involve the search — usually through energy minimization — for the molecular geometry with the lowest potential energy. Minimization calculations are based upon the implicit assumption that the structure with the lowest mechanical energy, as calculated from the semiempirical energy functions, will be the form found in nature. In this approach, modeling physical behavior for proteins or other biopolymers involves varying the atomic coordinates of the protein until the lowest energy conformation is located. Predictions of structure-dependent physical properties of the protein are then made from this lowest-energy conformer.

Model systems are often employed in conformational energy studies of polypeptides. The alanine "dipeptide", N-methylalanylacetamide, has long been used to model the general backbone conformational energy of amino acid residues whose sidechains are not branched at the β-carbon (12,13). This molecule,

$$CH_3 - \underset{\underset{O}{\parallel}}{C} - \underset{\underset{H}{|}}{N} - \underset{\underset{H}{|}}{\overset{\overset{CH_3}{|}}{C^\alpha}} - \underset{}{\overset{\overset{O}{\parallel}}{C}} - \underset{\underset{H}{|}}{N} - CH_3$$

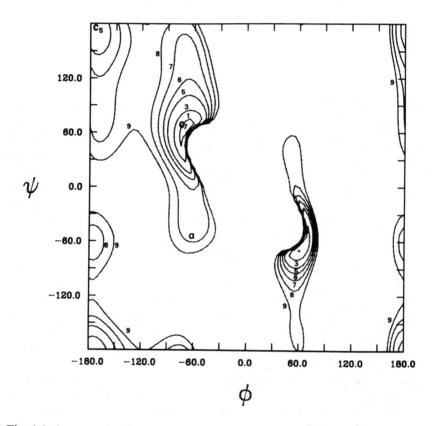

Fig. 1.1. An example of a conformational energy map for the alanine dipeptide. Energy contours calculated from the potential energy functions of Rossky, *et al.* (24) are plotted as a function of the backbone torsional angles ϕ and ψ. The contour intervals are indicated in kcal/mol above the global minimum, C_7^{eq}, at (-70,60), and this conformation as well as the C_5 and α-helical conformations are indicated by symbols. The unlabeled minimum at approximately (60,-60) corresponds to the C_7^{ax} conformation.

is called a dipeptide because while it has only one asymmetric α-carbon with an alanine sidechain, it contains two peptide bonds on either side of this group, and is thus like a segment excised from the middle of an extended polypeptide chain. By evaluating the energy of this dipeptide molecule for all combinations of the backbone torsional angles ϕ and ψ, a map of the conformational energy of the system as a function of these angles can be prepared. An example of such a conformational energy map is illustrated in Figure 1. The energy function used in the preparation of Figure 1 (24), which has been superseded by more refined functions (17,18), has several minimum energy conformations, including the so-called C_7^{ax}, C_7^{eq}, and C_5 forms (24). The α-helical and β-sheet conformations are not local minima for single residues on this surface, and the lowest energy conformation for this function is the C_7^{eq} form; in newer energy parameterizations, C_7^{eq} is not a minimum on the energy surface and the α-helix and β-sheet forms are at a minimum (17,18). However, this function (as is true for most other such semi-empirical functions) has reasonably low energies for the known possible low-energy polypep-

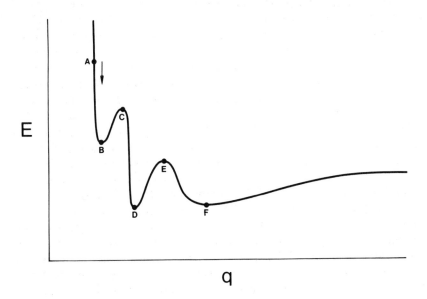

Fig. 1.2. A schematic illustration of the variation of the energy of a system in one dimension as a function of a generalized coordinate q, showing a global energy minimum at D and two higher energies, local minima at B and F.

tide conformations, with most other forms ruled out by extremely high energies resulting from repulsive overlaps of atomic cores (12,13).

A major difficulty with energy minimization studies is that due to the immense complexity of protein molecules, the structure with the lowest potential energy is difficult to identify. There will generally be a very large number of higher-energy local minima on the energy surface for the molecule in the high-dimensional hyperspace of the protein's many internal coordinates. Since there is no general solution to the problem of identifying which of these local minima has the lowest energy, most energy minimization procedures will usually relax the system to the nearest local minimum energy structure not separated from the starting structure by any barriers on the potential energy surface (11). This problem is schematically illustrated for a one-dimensional system in Figure 2, where a typical minimization algorithm beginning with the starting geometry A would simply follow the gradient down to position B, and be unable to cross the barrier at C to the global minimum energy structure at D. This difficulty is often referred to as the multiple-minimum problem. Thus, an energy minimization of the alanine dipeptide molecule starting from a structure close to the C_5 geometry would predict the C_5 geometry to be the lowest energy conformer, even though it is approximately 5 kcal/mole higher in energy with this energy function than the C_7^{eq} conformer. Because of the complexity of this problem, it is not currently possible to completely predict the tertiary conformation of a protein solely from primary sequence, even if there were no other complicating factors. However, other problems, including the neglect of structural fluctuations and solvent, also limit the usefulness of simple energy minimization calculations.

Minimization studies are based on the assumption that polypeptides exist in a single well-defined conformation. Globular proteins, however, are not static (with a single fixed structure) in character, but are continuously fluctuating between a very large number of closely related conformations which all have the same overall folding pattern but which differ in specific local details (6,25). The protein conformations determined by x-ray diffraction studies of crystals are actually approximate time-averages of these various constantly interconverting, related conformations, in spite of the impression of frozen rigidity conveyed by pictures and models of these structures (6). The assumption of a single lowest-energy conformation is even less valid for most small oligopeptides, or for non-globular proteins such as the caseins, which in many cases may be interconverting between a large number of quite unrelated, essentially random conformations. Such multiple conformations could frustrate attempts to model physical behavior based solely upon minimum energy geometry even if the lowest energy structure could be identified. Physical properties must be calculated as an appropriately weighted average over all of the dynamically accessible states.

Another implicit assumption of the energy minimization approach to predicting polypeptide conformations is that the mechanical energy alone determines the physically observed conformation. Actual folding, of course, is determined by the system free energy, with the relevant energy quantities being the free energy differences between the various possible globular states, as well as the many possible unfolded states. These free energy differences contain both enthalpic and entropic contributions. Such free energies will contain contributions from any change in the configurational entropy of the chain and in the entropy of the solvent. The importance of water "structuring" in protein folding has already been discussed. In addition, solvent water molecules provide important hydrogen bond partners for protein dipoles and act as a heat source and sink, thermalizing the solute protein molecule. Unfortunately, conformational energy calculations must typically ignore the role of solvent in determining polymer structure due to the difficulty and expense of including solvation effects, although attempts have been made to incorporate solvation into minimization studies (26).

Because of the importance of flexibility and solvation in biological systems, it would clearly be desirable in many cases to be able to include such effects in molecular mechanics studies. While it is generally not practical to include these factors in conformational energy minimization calculations, other molecular mechanics techniques are available which can more readily incorporate these effects. Molecular dynamics simulations (21) and Monte Carlo calculations (27) can be used to calculate free energies (although this is neither simple nor inexpensive) and can explicitly include the role of solvent water molecules. Molecular dynamics simulations can also be used to study protein motions, to examine structural fluctuations and process rates, and to allow systems to cross over the energy barriers which so often frustrate energy minimization studies.

Molecular Dynamics Simulations

Molecular dynamics studies are a specific type of molecular mechanics calculation in which the actual atomic motions of a system (such as a protein molecule) are directly simulated on a large digital computer by numerically solving the classical equations of motion of every atom in the system responding to the forces arising from a given force field (8,17,21). MD simulations were originally developed to model the behavior of simple physical systems such as argon and nitrogen (21), but they have been used to study a wide variety of more complex systems, including liquid water and aqueous solutions (16,20,22,24). Following the pioneering work of Karplus and coworkers (15), molecular dynamics simulations have also now been extensively applied to the study of the dynamics of proteins, nucleic acids, and more recently, simple carbohydrates (18,28-31). With the advent of modern computers, it is now possible to simulate the motions of biological solutes in aqueous solution.

The potential energy functions used in MD simulations to describe the variation of the total molecular potential energy as a function of atomic coordinates are generally the same as those used in other types of molecular mechanics calculations, such as energy minimization studies. Since the negative gradient, or derivative, of the potential energy with respect to atomic position is the force acting on that atom,

$$F_i = -\nabla_i V \qquad (2)$$

these derivatives can be used to numerically integrate Newton's equations of motion,

$$F_i = m_i a_i \qquad (3)$$

in principal producing a complete description of the position of every atom in the molecule as a function of time, $x(x_o, v_o, t)$, given the initial positions x_o and velocities v_o at $t=0$. Because of the general conservation priniciples of classical physics, the energy of any given trajectory of this type is conserved (remains constant), while the instantaneous temperature fluctuates. A conventional molecular dynamics simulation thus represents a statistical mechanical microcanonical ensemble (9,10). MD simulations of other ensembles, such as the canonical (N,V,T) ensemble, are also possible (9). If numerical simulations are propagated for a sufficiently long period of time, properties averaged over the entire period of the simulation will converge to the measured values of that property. Such a complete description of the way in which a molecular system evolves with time allows the examination of structure and dynamics at a level of detail unavailable by any other means. Because of their inherently dynamic character, MD simulations are an excellent way to examine rates for various processes, and for studying structural fluctuations and transitions.

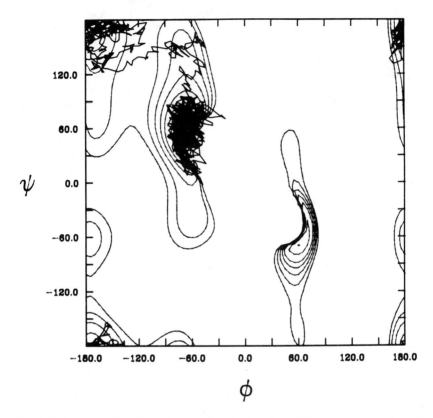

Fig. 1.3. An example of a trajectory for alanine dipeptide, superimposed on the conformational energy map of Figure 2. This trajectory began in the C_5 well in the upper left and underwent a transition to the C_7^{eq} well.

As a concrete example of the utility of this type of calculation, one might consider the case of the alanine dipeptide. Figure 3 illustrates a typical trajectory for the alanine molecule through the Ramachandran (ϕ,ψ) space as calculated from an MD simulation that began with a starting conformation close to the C_5 form. As can be seen, the molecule is in continual motion, initially fluctuating in the C_5 well around the lowest energy structure for that well, but only rarely having exactly the lowest energy in the C_5 structure. The time that the molecule spends in each region of the conformation space is determined by the Boltzmann-weighted probability of that conformational energy relative to the lowest energy. The physical and structural properties of the molecule must be averaged over the entire trajectory, rather than be calculated from the minimum energy geometries, to give the best approximation to physical values.

Approximately one third of the way through the illustrated trajectory, the molecule experienced a conformational fluctuation sufficiently large as to carry it over the barrier separating the C_5 conformations from the C_7^{eq} well, which is the lowest energy geometry with this potential energy function in vacuum (see above). Following this transition, the molecule fluctuated in the C_7^{eq} well for the remainder of the simulation. While the example illustrated here is for a small dipeptide, MD simulations have been conducted for a number of proteins (15,28,32), which have been found to be in continual motion, with many transient and local conformational changes. These conformational changes often include transitions in backbone torsional angles, which generally occur as anti-correlated pairs in order to conserve the overall direction of the main chain. An accurate calculation of molecular structural properties requires that they be averaged from trajectory simulations over all such conformations.

Solvation

The trajectory illustrated in Figure 3 is for a dipeptide molecule fluctuating in vacuum. In biological systems proteins are frequently in contact with water, and the way in which the presence of solvent affects motions of the type illustrated in Figure 3 can be quite important for both protein structure and function. The role of water in biological systems can also be directly simulated using molecular dynamics or Monte Carlo calculations by explicitly including the solvent molecules in the system which is being modeled (16,20,24). In such a calculation, the water molecules are treated like the atoms of the protein, with new terms in the total system potential energy to represent the water-protein interactions as well as the water-water interactions.

Biological simulations of solutions raise special problems. In order to simulate dilute systems, a model would need to be studied that contained enormous quantities of water in order to avoid a system dominated by edge effects at its boundaries. Numerical simulations of systems of this size are not possible due to the computational cost. Because of this a cubic "box" containing the solute molecule and a few hundred to a few thousand water molecules is taken as the primary system, and this box is then surrounded with exact images of itself, so that water molecules at the edge experience interactions with image neighbors rather than vacuum, and the equations of motion for these image molecules do not need to be separately integrated since they can be generated by symmetry operations upon the coordinates of the molecules in the primary box (27).

The effect of the presence of solvent molecules upon the conformational and dynamical behavior of proteins can be quite profound. As already noted, the potential energy function illustrated in Figure 1 is probably somewhat unrealistic in that it has the C_7^{eq} conformation, which is almost never observed in proteins, as the global minimum energy structure, with the α-helix not being a stable minimum energy form. When a model of alanine dipeptide using this potential energy function is put into aqueous solution, however, the free energy surface of the total system is apparently quite different. In a series of extensive simulations of this molecule in aqueous solution (J. Brady and M. Karplus, unpublished results) the molecule was found to undergo spontaneous transitions to other conformations, most notably the α-helical conformation. The reason for this preference is apparently the improved hydrogen-bonding to solvent molecules which is possible in the α-helical conformation (see Figure 4) as well as the greater configurational entropy of the α-helical structure. (It should be remembered that for a dipeptide, the "α-helical" conformation merely refers to the values of ϕ and ψ; because there is no extended polypeptide chain, there is no Pauling helix with its characteristic hydrogen bond reducing the chain entropy.) It is interesting that this conformational preference is found in spite of the bias of the potential energy function for C_7^{eq}.

The extraordinarily detailed information available from molecular dynamics simulations allows a direct examination of the microscopic details of water behavior around biopolymers (33). For example, such simulations can be used to investigate the so-called water binding by biopolymers. This controversial subject has long been a matter of debate, with the amount of water classified as bound being dependent upon the experimental technique used to probe the system. From MD simulations, one can directly examine the interaction of water molecules with a solute. As might be expected, a continuum of behaviors can be observed. For example, Figure 5 illustrates the atomic pair distribution function for water molecules around a hydroxyl oxygen atom in an MD simulation of a carbohydrate. This function is basically the probability of finding a solvent molecule a given distance from the solute oxygen atom. The sharp, narrow first peak represents the nearest neighbor, hydrogen-bonded water molecules. The sharpness of this peak is characteristic of hydrogen bonding, as is the deep first minimum in this

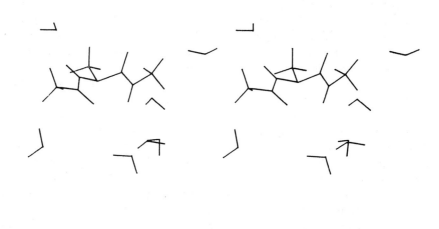

A

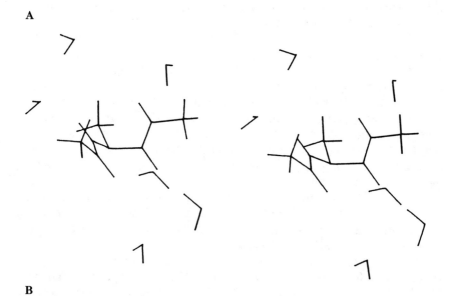

B

Fig. 1.4. Stereo drawings of a "snapshot" from molecular dynamics simulations of the alanine dipeptide in A) the C_7^{eq} and B) α-helical conformations in aqueous solution. The positions of the water molecules closest to the dipeptide dipoles are also indicated.

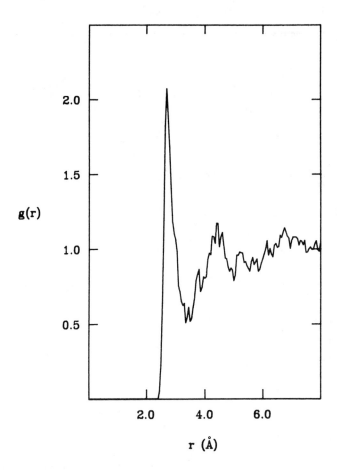

Fig. 1.5. The atomic pair distribution function for water molecules about a hydroxyl group of glucose, as calculated from a molecular dynamics simulation.

function, indicating the rather strong spatial localization of these water molecules. However, similar pair distribution functions for water molecules around other groups, such as nonpolar methylene carbon atoms or the amide proton of peptide groups, can be quite different. In general, the simulations find aqueous solvation to be a complex phenomenon.

Although hydrogen-bonded water molecules are well localized while they are near the solute, they are not bound in the conventional sense, since they have not lost their kinetic identity (i.e., they are not translating with the solute) and are continually exchanging with other water

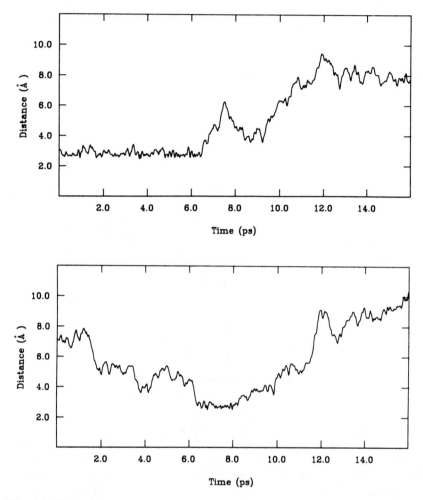

Fig. 1.6. Plots of the distance, as a funciton of time, for the oxygen atoms of two different water molecules from the same oxygen atom of a glucose molecule, as calculated from a molecular dynamics simulation. Note that as the water molecule illustrated in A escapes from the first hydration shell of the hydroxyl, it is simultaneously replaced by the molecule illustrated in B, which also subsequently escapes and is replaced by a third.

molecules in the bulk of the solution. This process is illustrated in Figure 6, which plots the distance of two water molecules from the solute hydroxyl oxygen as a function of time during the dynamics simulation.

One is initially hydrogen-bonded to the hydroxyl, and remains quite close to the hydroxyl oxygen atom. After approximately six ps, however, this molecule escapes into the bulk of the solution and is temporarily replaced by the other molecule in a concerted exchange. This second water molecule also subsequently escapes and is replaced by a third. Typical residence times in peptide and carbohydrate systems were found to be a few picoseconds. Water molecules adjacent to the solute functional groups, while kinetically independent, have a reduced translational diffusion coefficient relative to the bulk water molecules, and their mobility varies slightly depending upon the nature of the functional groups such as alanine methyl groups and peptide NH and CO dipoles. This reduction in mobility found in dynamics simulations for water molecules near surfaces is consistent with experimental measurements (33–35). The rotational freedom of water molecules immediately adjacent to solute atoms is also reduced, with those close to nonpolar groups being more affected than those next to backbone dipoles. All of these properties can be measured directly from MD simulations and related to the structure and properties of the individual solute functional groups. From studies such as these it may eventually be possible to develop general rules to help rationalize the folding behavior and solution properties of more complicated solutes, such as long polypeptides and proteins.

Free energies can, in principle, be extracted from MD simulations or Monte Carlo calculations by directly calculating the configuration integral, and thus the partition function, from the simulations (9). Unfortunately this procedure is not practical, since the integrals contain contributions from high energy regions where the trajectories spend only brief periods, providing inadequate averaging statistics. Nonetheless, the entropic contributions to molecular processes are present in the simulations, and provided the simulations are of sufficient length, various techniques are available for approximating free energy changes for molecular processes. One new technique, thermodynamic perturbation calculations (9), has shown particular promise in calculating practical free energy quantities such as the free energy difference between protein mutants binding with substrates in solution. This type of simulation has already been successfully applied to practical problems, such as the design of enzymatic inhibitors (36), predicting the effects of point mutations upon protein stability (37), and upon enzymatic binding and catalysis (38).

Conclusions

Protein engineering, the deliberate design of novel mutant proteins with desired properties, has already become a realistic goal with the advent of modern recombinant DNA techniques. The ability to modify the activity of proteins has been demonstrated for several systems of practical importance, and the development of a biotechnology industry based upon the production of biological products with engineered functional properties could become the "next industrial revolution". However, the efficient use of genetic engineering techniques to produce improved proteins will require considerable guidance from chemical theory in order to select those mutants which will be most likely to exhibit the desired functional properties.

Although the greatest potential of rational molecular design perhaps lies in the pharmaceutical industry, numerous applications can also be expected in the food industry. Those food applications which are most likely to occur in the immediate future will involve the manipulation of the properties of globular food proteins, such as increasing the thermal stability or binding specificity of the whey protein β-lactoglobulin (39,40), or in modifying the flavor properties of such molecules as the intensely sweet proteins monellin and thaumatin (41). Many important food proteins (e.g., the caseins and collagen) are not globular, and many other food proteins exhibit their greatest utility (e.g., in foams or gels) in a non-globular, denatured state. The properties of these food molecules will be very difficult to simulate using molecular mechanics since their physical behavior is an average over a great many disordered states which interconvert slowly on the molecular dynamics timescale. The large size of many important food proteins will also limit the application of modeling calculations to these molecules with current computers, and initial studies of the non-globular food proteins will necessarily begin with smaller systems such as the caseins.

In spite of their current limitations, molecular mechanics calculations, coupled with computer graphics modeling and incorporating chemical intuition, provide a powerful means of examining the possible consequences of point mutations upon protein properties, such as heat stability, activity, or binding constants. While they cannot currently be applied to every system of interest, for those systems where they are appropriate theoretical modeling calculations can provide information about protein systems which is not available from any experimental technique. Theory can also be used to make practical predictions about properties of hypothetical modified proteins, helping make the actual experimental production of mutants more efficient. Due to the importance of molecular motions and aqueous solvation in biological systems, in many cases in protein engineering it will be necessary to include dynamics and solvent in molecular mechanics modeling studies used to make such practical predictions of protein behavior. The use of molecular dynamics simulations to model flexibility and solvation in food protein systems can thus be expected to grow as more powerful computers become available for the study of modified proteins.

Acknowledgments

The author thanks his research collaborator, M. Karplus, for helpful discussions. The stereo diagrams in Figure 4 were produced by the HYDRA molecular graphics program written by R. E. Hubbard of the University of York.

References

1. Perry, L., and R. Wetzel, *Science 226*:555 (1984).
2. Shih, H., J. Brady, and M. Karplus, *Proc. Nat. Acad. Sci. USA 82*:1697 (1985).
3. Privalov, P.L., *Adv. in Protein Chem. 33*:167 (1979).
4. Creighton, T.E., in "Progress in Biophysics and Molecular Biology", J.A.V. Butler and D. Nobel, (eds.) Pergamon, Oxford p. 231 (1978).
5. Chothia, C., *Ann. Rev. Biochem. 53*:537 (1984).
6. Karplus, M., in "Structure and Dynamics of Nucleic Acids, Proteins, and Membranes", E. Clementi and S. Chin, (eds.) Plenum, New York p. 113 (1986).
7. Case, D.A., and M. Karplus, *J. Mol. Biol. 132*:343 (1979).
8. Karplus, M., and J.A. McCammon, *Annu. Rev. Biochem. 53*:263 (1983).
9. McCammon, J.A., and S.C. Harvey, "Dynamics of Proteins and Nucleic Acids", Cambridge University Press, Cambridge, 1987.
10. Brooks, C.L., M. Karplus, and B.M. Pettitt, *Advances in Chem. Phys.*, Vol. LXXI, 1988.
11. Scheraga, H.A., *Biopolymers 22*:1 (1983).
12. Rao, V.S.R., P.R. Sundararajan, C. Ramakrishnan, and G.N. Ramachandran, in "Conformation in Biopolymers, Vol. 2", G.N. Ramachandran, (ed.), Academic Press, London, p. 727 (1967).
13. Brant, D.A., and P.J. Flory, *J. Mol. Biol. 23*:47 (1967).
14. Brant, D.A., *Ann. Rev. Biophys. Bioeng. 1*:369 (1972).
15. McCammon, J.A., B.R. Gelin, and M. Karplus, *Nature 267*:585 (1977).
16. Rossky, P.J., and M. Karplus, *J. Am. Chem. Soc. 101*:1913 (1979).
17. Brooks, B.R., R.E. Bruccoleri, B.D. Olafson, D.J. States, S. Swaminathan, and M. Karplus, *J. Comput. Chem. 4*:187 (1983).

18. van Gunsteren, W.F., H.J.C. Berendsen, J. Hermans, W.G.J. Hol, and J.P.M. Postma, *Proc. Natl. Acad. Sci.* USA *80*:4315 (1983).
19. Scheraga, H.A., *Biopolymers 20*:1877 (1981).
20. Brady, J. and M. Karplus, *J. Am. Chem. Soc. 107*:6103 (1985).
21. Rahman, A., *Phys. Rev. 136*:A405–A411 (1964).
22. Stillinger, F.H., and A. Rahman, *J. Chem. Phys. 60*:1545 (1974).
23. Palinkas, P., W.O. Riede, and K. Heinzinger, *Z. Naturforsch 32a*:1137 (1977).
24. Rossky, P.J., M. Karplus, and A. Rahman, *Biopolymers 18*:825 (1979).
25. Elber, R. and M. Karplus, *Science 235*:318 (1987).
26. Paterson, Y., G. Nemethy, and H.A. Scheraga, *Annals New York Acad. Sci. 367*:132 (1981).
27. Metropolis, N., A.W. Rosenbluth, M.N. Rosenbluth, A.H. Teller, and E. Teller, *J. Chem. Phys. 21*:1087 (1953).
28. Levitt, M., *J. Mol. Biol. 168*:595 (1983);*168*:621 (1983); *170*:723 (1983).
29. Harvey, S.C., H. Prabhakaran, and J.A. McCammon, *Biopolymers 24*:1169 (1985).
30. Singh, U.C., S.J. Weiner, and P. Kollman, *Proc. Natl. Acad. Sci.* USA *82*:755 (1985).
31. Brady, J.W., *J. Am. Chem. Soc. 108*:8153 (1986).
32. Post, C.B., B.R. Brooks, M. Karplus, C.M. Dobson, P.J. Artymiuk, J.C. Cheetham, and D.C. Phillips, *J. Mol. Biol. 190*:455 (1986).
33. Kuntz, I.D., and W. Kauzmann, *Adv. Protein Chem. 28*:239 (1974).
34. Goldammer, E.V., and H.G. Hertz, *J. Phys. Chem. 74*:3734 (1970).
35. Shirley, W.M., and R.G. Bryant, *J. Am. Chem. Soc. 104*:2910 (1982).
36. Wong, C.F., and J.A. McCammon, *J. Am. Chem. Soc. 108*:3830 (1986).
37. Bash, P.A., U.C. Singh, R. Langridge, and P.A. Kollman, *Science 236*:564 (1987).
38. Rao, S.N., U.C. Singh, P.A. Bash, and P.A. Kollman, *Nature 328*:551 (1987).
39. Papiz, M.Z., L. Sawyer, E.E. Eliopoulos, A.C.T. North, J.B.C. Findlay, R. Sivaprasadarao, T.A. Jones, M.E. Newcomer, and P.J. Kraulis, *Nature 324*: 283 (1986).
40. Sawyer, L., *Nature 327*:659 (1987).
41. de Vos, A.M., M. Hatada, H. van der Wel, H. Krabbendam, A.F. Peerdeman, and S.-H. Kim, *Proc. Natl. Acad. Sci.* USA *82*:1406 (1985).

Chapter Two
Protein Structure in Solution

I.D. Kuntz

Department of Pharmaceutical Chemistry
University of California, San Francisco
94143-0446

Proteins have quite complex structures when crystallized. Relatively simple features are put together in a bewildering array. The basic repetitive structural elements where recognized long ago by Pauling as alpha helices and beta sheets. More recently, turns of various types have been shown to be essential for the 3-dimensional folding. The most comprehensive organizing principle for globular proteins has been the "hydrophobic core" idea put forward by Kauzmann. The initial complexity was reduced by recognizing that the polar turns are located at the protein-solvent interface while the hydrophobic beta sheets are buried within the protein. A higher level of organization is also governed by thermodynamics: folding domains consisting of 150-250 amino acids are a general consequence of the number and distribution of hydrophobic residues. All of these ideas come directly from an impressive amount of crystallographic data, primarily on globular proteins. Over 500 structures have been determined.

In spite of such results, a question that continually is asked is, how important are the crystal forces in influencing the observed structures. How much will the structure of a protein change, if it could be measured directly in aqueous solution?

A method based on Nuclear Magnetic Resonance (NMR) is now able to provide structural information of reasonably high quality. The procedure requires 10-20 milligrams of protein of high solubility (1-3 millimolar). It is currently restricted to molecular weights below 20,000 daltons. The process takes 3-12 months. The steps involved are: acquisition of data from two basic types of NMR experiments—correlated scalar coupling (COSY) and nuclear Overhauser effect measurements (NOESY); assignment of spin systems to amino acid types; assignment of amino acids to specific sequence positions; use of NOE data to supply internal distances; use of computer programs such as Distance Geometry and molecular dynamics to build the actual structures. The results are structures roughly comparable to 3 angstrom resolution crystallography.

What has been learned so far? All possibilities have been seen from close agreement between the two techniques to striking discrepancies. At this early stage it appears that some proteins are quite rigidly constructed and only a few sidechains move in response to crystal packing. Others show considerable flexibility, especially in loop regions, which often become undefined in solution. In a few cases, the disagreement is so severe that the possibility is being considered that the crystallographic structure was incorrectly solved.

It is clearly a very exciting time for structural research. NMR provides a very powerful window into a new world: quantitative structural studies in noncrystalline environments.

Chapter Three

Interrelationship of Molecular and Functional Properties of Food Proteins

Srinivasan Damodaran

Department of Food Science
University of Wisconsin
Madison, WI 53706

The breakthroughs in the fundamental understanding of plant genetics and physiology and improved cultivation practices have enormously increased food production in the past four decades. However, despite the progress in food production, the lack of parallel developments in the postharvest handling and the insufficient basic understanding of the bio-material properties of food components (e.g., proteins) have limited efficient and total utilization of food ingredients.

For instance, it is estimated that the current world supply of food proteins may indeed be adequate enough to feed the world population (1); yet ironically, protein malnutrition is prevalent in many parts of the world. One of the underlying reasons for this situation is the fact that about 70% of the world protein supply is derived from plant sources, such as cereals and legumes (2). Although many vegetable proteins, especially legume proteins, are cheap and abundant, extensive utilization of these proteins in food systems has not been fully accomplished. Most of these plant proteins are being used in animal feed to produce animal proteins. Utilization of plant proteins directly in human foods is limited because these proteins lack certain requisite functional behavior in fabricated foods. In this respect, proteins of animal origin, e.g., meat, milk and egg proteins, which possess excellent functional properties, are often used as ingredients in formulated foods to perform various functional roles. A majority of formulated foods are largely of foam, emulsion and gel type products. Therefore, for enhanced utilization of plant proteins in fabricated foods, these proteins must possess important functional properties such as solubility, foaming, emulsifying and gelling properties.

The functional properties of a protein are fundamentally related to its physical, chemical and structural/conformational properties. These include: size, shape, amino acid composition and sequence, charge and their distribution, hydrophilicity/hydrophobicity ratio, secondary structure content and their distribution (e.g., α-helix, β-sheet and aperiodic structures), tertiary and quaternary arrangement of the polypeptide segments, inter- and intra-subunit cross-links (e.g., disulfide bonds), and the rigidity/flexibility of the protein in response to external conditions. The excellent fiber forming properties of actomyosin, the gelling behavior of gelatin, the viscoelastic properties of wheat gluten, the curd forming properties of caseins, the excellent whippability and emulsifying properties of egg albumin are attributable mainly to the unique physicochemical properties of these proteins. Factors such as processing conditions, the method of isolation, environmental factors (e.g., pH, temperature, ionic strength, etc.), and interaction with other food components do alter the functional properties of a protein. Such alterations, however, are simply manifestations of alterations in the conformational and structural features of the protein caused by these external factors.

Knowledge of the fundamental relationship between the conformational properties of food proteins and their functional behavior in food systems is either lacking or poorly understood. To increase the utilization of novel food proteins in conventional food products, the greatest challenge for the food scientists is to understand, at the fundamental level, the molecular bases for the expression of the functional properties of food proteins (3). Such a basic understanding would not only lead to development of better processing techniques so as to retain and/or improve protein functionality, but also to future developments in the genetic engineering strategies to alter the conformational characteristics of the under-utilized food proteins in order to improve its functionality.

Conformational Properties of Proteins

Proteins are polymers composed of twenty naturally occurring amino acids linked through peptide bonds. What makes these biopolymers different from other polymers and versatile in terms of their structure and biological function, is not only the unique properties of the individual amino acid components, but also the sequence and the pattern of distribution of amino acid residues in the polymer. Each protein has a unique sequence and distribution of amino acid residues which is defined as the primary structure of the protein. Given an amino acid composition, it is theoretically possible to construct hundreds of billions of proteins having a different amino acid sequence and distribution pattern in the primary structure; each of these proteins would have unique structural and functional properties.

The functional properties of either biologically important enzymes or commercially important food proteins is dependent on the final three-dimensional structure of the protein. The formation of a unique three-dimensional structure is the net effect of various attractive and repulsive interactions within the protein molecule and the interaction of the protein chain with the surrounding solvent medium. The various noncovalent interactions which contribute to the folding of the protein and maintenance of the folded structure are the hydrogen bonding, electrostatic, van der Waals and hydrophobic interactions. For a protein to assume a unique three-dimensional structure, which imparts the unique physicochemical properties to the protein, the favorable free energy change from the above noncovalent interactions should be greater than the unfavorable free energy resulting from the loss of conformational entropy of the polypeptide in the folded state, compared to the unfolded state. In other words, the net thermodynamic stability of the native structure of the protein may be expressed as:

$$\Delta G_N = \Delta G_{hb} + \Delta G_{ele} + \Delta G_{h\phi} + \Delta G_v - T\Delta S_{conf}$$

where ΔG_{hb}, ΔG_{ele}, $\Delta G_{h\phi}$, and ΔG_v are the free energy contribution from hydrogen bonding, electrostatic, hydrophobic and van der Waals interactions. The $T\Delta S_{conf}$ is the unfavorable change in the free energy resulting from loss of conformational freedom of the polypeptide.

The net thermodynamic stability of the native structure of many proteins is only about 10–20 Kcal/mole. Hence, any perturbation, causing even a small decrease in the free energy of stabilization, would profoundly affect the structural stability, and consequently, the functional property of the protein. In food systems such changes may be brought about by agents such as heat, salts (ionic strength and the type of ions), pH and the interaction of other components with the protein. To understand the processing induced changes in the structure and functional properties of food proteins, it is imperative to understand the thermodynamic stability of food proteins and the nature of the forces involved, and their relative magnitude of contribution to the stability of the protein.

Proteins contain numerous functional groups which have the potential to form hydrogen bonds. The α-helical and β-sheet structures in proteins are in fact stabilized by the intra- and inter-segment hydrogen bonds between the peptide groups. Since the energy required to break a hydrogen bond in proteins is about 4 Kcal/mole, it can be presumed that the numerous hydrogen bonds in proteins would contribute enormously to the stability of proteins.

In spite of the demonstrated evidence of the existence of hydrogen bonds in proteins, there has been considerable skepticism regarding the importance of these hydrogen bonds both as a driving force for protein folding and for maintaining the stability of the folded native structure. This is because in aqueous solutions (since water itself is a very good donor and acceptor of hydrogen bonds) there is no thermodynamic incentive for the polypeptide chain to form hydrogen bonds between its functional groups. Using N-methyl acetamide (NMA) as a model compound (which is analogous to peptide groups in proteins) it has been shown that in the aqueous medium the change in the free energy for the formation of hydrogen bond between the NMA molecules is about 0.75 Kcal/mole (4). One can assume, therefore, that in the absence of other favorable interactions, the actual contribution of hydrogen bonds to protein folding and to the stability of the folded state would be marginal. Since the hydrogen bond is primarily ionic in character, the stability of hydrogen bonds in the interior of proteins would be expected to be higher; this is attributable to the low polarity of these regions, created by association of the hydrophobic amino acid residues. In other words, it can be argued that the hydrogen bonds in proteins are mainly in a pseudo-stable state; any change in the internal low dielectric environment (e.g., caused by the destabilization of hydrophobic interactions) would profoundly affect the stability of hydrogen bonds (5).

At neutral pH both the acidic (glutamate and aspartate) and basic (lysine and arginine) side chains are in the ionized form. The interaction between these oppositely charged residues forms the basis of electrostatic interactions in proteins. Depending on the relative number of the acidic and basic amino acid residues, a protein has either a net negative or positive charge at a given pH other than the isoelectric pH. In most proteins, most of these charged groups are present on the surface and are exposed completely to the aqueous medium. Because of hydration and the partial dielectric screening effect of the solvent on these charges, the electrostatic repulsion or attraction between these charges is greatly reduced.

It is now well recognized that the major force responsible for the folding and maintenance of the folded structure is the hydrophobic interaction between the constituent nonpolar amino acid residues (6). The origin of hydrophobic interactions in proteins is the thermodynamically unfavorable hydration of the apolar residues in proteins; the hydrophobic hydration of the apolar residues increases the structural order of liquid water. In addition, the relative orientation of the water molecules in the hydration shell of the hydrocarbon is different and incompatible with that of the relative orientations of the water molecule in the bulk water structure (7). These changes in the structural state of water molecules greatly decrease the entropy of the system, which leads to a large increase in the free energy. In order to overcome these unfavorable decreases in the entropy and to reduce the free energy, the system tends to force the apolar side chains out of the aqueous environment. This is accomplished partially by the grouping of the apolar residues, which releases some of the low entropy water molecules to their high entropy state. Such entropy driven association of apolar

TABLE 3.1
Free energies of transfer of amino acid side chains from water to organic solvents at 37°C.

Amino acid	$-\Delta G_{tr}$ ($H_2O \rightarrow NMA$)	$-\Delta G_{tr}$ ($H_2O \rightarrow$ Ethanol)	$-\Delta G_{tr}$ ($H_2O \rightarrow$ Surface)	ΔG_{tr} ($H_2O \rightarrow$ Hexane)
Glycine	0	0	0	0
Alanine	0.666	0.801	0.208	-0.436
Leucine	2.550	2.150	2.560	2.800
Isoleucine	2.294	3.170	2.350	2.756
Valine	1.685	1.841	1.623	1.750
Methionine	2.324	1.633	1.529	—
Phenylalanine	3.016	2.881	2.423	2.590
Tyrosine	3.576	2.673	2.330	—
Tryptophan	4.270	3.817	2.090	—
Cysteine	1.285	1.477	0.468	—
Serine	0.533	-0.031	0.406	-0.800
Threonine	0.774	-0.031	0.540	-0.041
Histidine	0.658	0.801	0.125	0.600
Arginine	0.624	0.780	0.125	-1.500
Lysine	1.236	1.602	0.364	0.260
Aspartic acid	2.389	0.582	0.208	1.290
Glutamic acid	1.896	0.593	0.312	1.020
Aspargine	1.358	-0.010	-0.083	—
Glutamine	0.205	-.100	-0.166	—

ΔG_{tr} values are in Kcal/mole (9).

residues in aqueous solutions is known as hydrophobic interactions. In proteins, the hydrophobic interaction between the apolar side chains causes folding of the polypeptide side chain. In other words, the hydrophobic interactions act as the driving force for the folding of the polypeptide from the unfolded to a native folded state. During this folding process the protein molecule folds in such a way that the majority of the apolar side chains are buried in the interior and the polar groups are exposed to the energetically favorable aqueous environment.

Direct measurement of the hydrophobic free energy of amino acid residues in proteins is very difficult. However, assuming that the internal environment of a protein is analogous to that of ethanol, Nozaki and Tanford (8) calculated the hydrophobicity of amino acid residues in terms of their free energy of transfer from ethanol to water. However, since the interior of a protein is neither homogeneous in composition nor uniform in its polarity, there has been considerable skepticism regarding whether or not the hydrophobic free energies of amino acid residues obtained from the water → ethanol transfer indeed represented the hydrophobic free energies of transfer from the aqueous phase to the interior of proteins. Recently, it has been shown that the transfer free energies of apolar amino acid residues from the water → gas phase (dielectric constant 1) and from the water → N-methylacetamide (dielectric constant 190) were almost the same as those of the water → ethanol (dielectric constant 25) (9) (Table 3.1). This strongly indicates that the hydrophobicity of amino acid side chains is not highly dependent on the polarity of the organic phase to which they are transferred, but is solely a manifestation of the thermodynamics of interaction of liquid water with the apolar side chains. In other words, it does not matter what the internal dielectric conditions or the variations in the polarity of the protein interior are; the hydrophobic free energy of the amino acid side chains would not be affected.

Recently, a concept known as accessible surface area has been introduced to help understand the role of hydrophobic interactions in protein stability (10,11). The accessible surface area is defined as the extent to which the surface of a protein is accessible to the surrounding solvent. It has been shown that, despite the differences in the amino acid sequence and the secondary structure among globular proteins, a protein folds in such a way that the accessible surface area is proportional to the hydrophobic free energy and also to their molecular weight by the relation (Fig. 3.1) (12-14).

$$A_n = 11.1 \, M^{2/3}$$

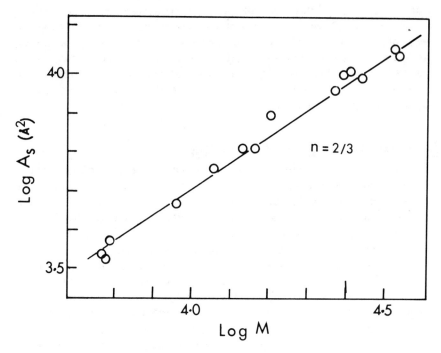

Fig. 3.1. Proportionality of loss of accessible surface area to the molecular weight of proteins. The proteins analyzed were: insulin, rubredoxin, pancreatic trypsin inhibitor, HIPIP, calcium binding protein, ribonuclease S, lysozyme, staphylococcal nuclease, papain, chymotrypsin, concanavalin A, subtilisin, thermolysin, carboxypeptidase A (12).

Since the relationship between the accessible surface area and the molecular weight of the unfolded polypeptide chain follows the relation, $A_u = 1.44\ M$ (13), it is possible to calculate the net hydrophobic surface area, i.e., A_u-A_n, buried during folding of monomeric globular proteins (15). Assuming that the hydrophobic free energy is about 22 cal/mole/Å^2 area of hydrophobic surface (14), the apparent contribution of hydrophobic free energy to the stability of the protein can be calculated.

Hydrophobic interactions in proteins are temperature dependent. Since these are entropy driven interactions, these interactions are stronger at higher temperatures and weaker at lower temperatures. Oligomeric food proteins, in which the subunits are held together predominantly by hydrophobic forces, undergo dissociation at low temperatures, mainly because of weakening of the hydrophobic interactions. For example, prechilling of milk at 4°C causes dissociation of β-casein from the micelles; this dissociation and other changes in the physical state of the casein micelles affect subsequent rennet coagulation of milk at 30°C (16). Several enzymes undergo low temperature denaturation owing to destabilization of hydrophobic interactions (17). The freeze toughening of fish muscles during storage might also be related to destabilization of actin and myosin.

Conformational Flexibility of Proteins.

The backbone of proteins is made up of covalent single bonds. Since the dihedral angle of a single bond has 360° rotational freedom, one would expect the polypeptide backbone to have a high degree of flexibility. However, because of delocalization of the electrons involved in the peptide bond, the peptide bonds linking the amino acid residues have a partial double bond character. Because of this, the rotational freedom of the peptide bond is reduced to about 6°. Since about one-third of the covalent bonds in the polypeptide backbone are made up of the peptide bond, the restriction on the rotational freedom of the peptide bond drastically reduces the flexibility of the protein backbone. Furthermore, the various bulky side chains of the amino acid residues provide steric hindrance to rotation of the other covalent single bonds in the backbone, i.e., C_α-C and C_α-N bonds, and thus further decrease the flexibility of the polypeptide chain. In other words, the selection of the peptide bond, i.e., CO-NH bond, as the linkage of the monomeric units, and the selection of the twenty different side chains can be regarded as a deliberate design by nature aimed to control the rigidity/flexibility of the protein macromolecule. This is essential not only for the functioning of enzymes as biocatalysts, but also for the functioning of the structural proteins such as actin, myosin and collagen.

The structure of a protein can be categorized into four levels. These are defined as the primary, secondary, tertiary and quaternary structures. While the primary structure refers to the amino acid sequence of the protein, the secondary and tertiary structures refer to the local conformations of polypeptide segments (e.g., α-helix, β-sheet and random structures) and the arrangement of the entire polypeptide in the three dimensional space, respectively.

The secondary structure of a polypeptide segment is predominantly determined by the amino acid sequence of the segment. The driving force for the formation of folded secondary structures in proteins is derived from the short range residue-residue interactions, segment-solvent interactions (hydrophobic forces) and the limitations on the rotational freedom of the single covalent bonds in the segment. The formation of α-helical and β-sheet structures at various segments of the protein is related to the local hydrophobicity (18). If a segment contains more hydrophilic residues, the formation of neither a α-helix nor β-sheet is favored (18). On the other hand, hydrophobic amino acids such as methionine, leucine and isoleucine favor helix formation when the neighbor residues are hydrophilic. When the local hydrophobicity is very high, the favored conformation is β-sheet (18). In other words, for the formation of secondary structures, the segments should contain an appreciable number of hydrophobic residues, and the hydrophobic interactions between these residues in the segment should act as the driving force for the local folding of the segment. However, once formed, the intra- and inter-segment hydrogen bonds add to the stabilities of α-helical and β-sheet structures, respectively.

Polypeptides containing very high content of proline residues tend to exist in the random state. For example, 35 out of 209 residues in β-casein, and 17 out of 199 residues in α_5-casein are proline residues (19). The uniform distribution of these proline residues in the primary structure of these caseins effectively precludes formation of either α-helical or β-sheet structures in these proteins, and makes them highly flexible.

Several methods to predict the secondary structures of polypeptides from their amino acid sequences have been developed (20-22). These approaches are based on the statistical analysis of the probability of the existence of various amino acid residues in α-helical, β-sheet and random coil structures in many globular proteins, for which the crystallographic structures are known. From such analysis, the statistical probability parameters for each amino acid have been developed; using these parameters it is possible to predict whether a given amino acid residue is a former, breaker or indifferent for α-helix, β-sheet structures in proteins. It has been shown that glutamate, methionine, alanine and leucine are strong α-helix formers, whereas valine, isoleucine and tyrosine are strong β-sheet formers. In a peptide segment, if a strong α-helix former is followed by at least four weak helix formers, then the probability of that segment being in the α-helical form will be high. Similarly, if a strong β-sheet former is followed by either strong or weak β-sheet formers, then the probability of that segment being in β-sheet structure will be high. For many proteins, the information on the secondary structure obtained from the predictive methods have been shown to be in good agreement with those obtained from crystallographic studies.

Although the three dimensional crystallographic structures for many food proteins are not available at the present time, the amino acid sequences of many important food proteins, such as caseins (19), B-lactoglobulin (19), α-lactalbumin (19), glycinin (23), B-conglycinin (24), phaseolin (25) and pea vicilin (26), are known. In order to understand the structure-functionality relationship of these proteins, it is possible to predict their secondary structure from their amino acid sequences.

The optical rotatory dispersion (ORD) and the circular dichroism (CD) measurements in the far uv region can be used to estimate the secondary structure of proteins. While the ORD measurements are very useful in predicting the α-helical content of predominantly α-helical proteins (such as myoglobin, hemoglobin and serum albumin), the circular dichroism measurements are more useful in predicting both the α-helical and β-sheet content of proteins. The circular dichroic spectra of poly-L-lysine in the α-helical, β-sheet and random coil conformations have been studied in detail (27). The CD spectrum of α-helix contains three distinct ellipticity bands at 190.5, 207 and 221 nm, whereas the β-sheet conformation exhibits bands at 195 and 217 nm; in the random coil state polypeptides have a negative ellipticity band at about 195 nm. When a protein contains a mixture of these conformations, the determination of the secondary structure from the CD spectrum becomes very difficult. However, several computational methods have been developed to deduce the secondary structure of proteins from their CD spectra (28-31). These methods are mainly based on the statistical modeling of the CD spectra of several proteins for which the secondary structures are known from the crystallographic data.

While the secondary structure of a protein refers to local ordered conformations of polypeptide segments, the tertiary structure refers to the conformation of the entire polypeptide in the three dimensional space. In terms of the functionality of food proteins, it is the tertiary structure rather than the secondary structure itself, which is very important. The attainment of this unique structure is determined by the sum total of all the noncovalent interactions in the protein.

Molecular and Functional Properties of Food Proteins

The various functional properties of food proteins can be regarded as the manifestations of two important molecular properties of protein macromolecules, viz., their hydrodynamic and surface related properties. While the functional properties such as viscosity, gelation, etc., are manifestations of the hydrodynamic properties of protein macromolecules, the functional properties such as solubility, foaming, emulsification, fat and flavor binding, etc., are the manifestations of their surface active properties. The hydrodynamic properties are affected largely by the shape and size of the macromolecule and are independent of the details of the amino acid composition and distribution. In contrast, the surface active properties of proteins are influenced more by the amino acid composition/distribution and molecular flexibility, and less by the actual shape and size of the macromolecule.

Surface Active Properties.

Hydrophobicity/hydrophilicity. Many of the molecular and functional properties of food proteins are related to the content of hydrophobic and hydrophilic amino acids, and their distribution in the primary structure. Many attempts have been made in the past to relate the hydrophobic and hydrophilic amino acid content of proteins to their physical properties (32-35). Bigelow (35) proposed that the average hydrophobicity ($H\phi_{ave}$) and the charge frequency are the most important molecular features that have the greatest influence on the physical properties, such as solubility, of proteins. The solubility of a protein under a given set of solution conditions can be expressed as the manifestation of the equilibrium between the protein-solvent (hydrophilic) and the protein-protein (hydrophobic) interactions. That is:

$$\text{Protein-Solvent} \rightleftharpoons \text{Protein-Protein} + \text{Solvent-Solvent}$$

According to Bigelow (35), proteins having lower average hydrophobicity and higher charge frequency would have a higher solubility. Although this empirical relationship seems to be true for most proteins, it does not explain the solubility characteristics of certain proteins. For example, the charge frequency of myoglobin and serum albumin (0.34 and 0.33, respectively) are almost the same; however, the average hydrophobicity of serum albumin (1120 cal/mole residue) is greater than that of myoglobin (1090 cal/mole residue). On the basis of these values, one would expect that the solubility of serum albumin should be lower than that of myoglobin. However, on the contrary, while myoglobin is insoluble at its isoelectric pH, serum albumin is extremely soluble at its isoelectric pH. This suggests that the physical and chemical characteristics of the protein surface and the thermodynamics of its interaction with the surrounding solvent, rather than the average hydrophobicity or charge frequency of the molecule as a whole, are critically important for its functional behavior. Given the same hydrophobicity and charge frequency, two proteins can exhibit distinctly different solubility characteristics depending on their amino acid sequence and consequent differences in the spatial arrangement of the residues in their tertiary structures. In this respect, it may be stated that, while the extent of hydrophobic surfaces buried at the interior of the protein contributes to its structural stability, the extent of the hydrophobic surfaces exposed at the exterior determines the solubility and other solution related physico-chemical properties of the protein.

The surface characteristics of a protein are affected by its folding pattern, which in turn is dictated by the amino acid sequence and the solvent conditions. The folding of a polypeptide is guided by the thermodynamic consideration that the majority of the nonpolar residues be inside and the majority of the polar residues be at the surface exposed to the solvent, so that the global free energy of the protein is minimum under the given solution conditions. In keeping with this general rule, in globular proteins there is a general propensity for hydrophobic residues to be buried in the interior and the hydrophilic residues to be located mostly at the surface. However, in many native globular proteins, while almost all the hydrophilic residues are located at the surface, not all the hydrophobic residues are totally buried. Consequently, in most globular proteins, e.g., myoglobin, lysozyme and ribonuclease S, about 40–50% of the surface is found to be made up of nonpolar patches, distributed uniformly on the surface (10). In other words, even in small monomeric, soluble proteins it is physically impossible to contain all the hydrophobic surfaces in the interior of the protein. However, in spite of this seemingly large exposure of hydrophobic surfaces, these proteins exist in the soluble monomeric state. This could be because of much greater hydration and intermolecular electrostatic repulsive forces at the protein surface. However, when either the fraction of the hydrophobic patches on the surface exceeds a critical level, or when the hydrophobic patches are distributed asymmetrically on the surface (as in the case of caseins), the greater hydrophobic interaction between these surfaces would lead to self-association and formation of oligomeric structures in proteins (36). This is the principal reason for the oligomeric structures observed in many legume proteins.

Hydrophobic fluorescent probes have been used successfully to determine the surface hydrophobicity of proteins (37–41). The basic principle involved in these methods is that certain fluorophores exhibit dramatic sensitivity in their fluorescence characteristics to changes in the polarity of their environment; the quantum yield of fluorescence increases several fold in a nonpolar environment compared to that in solvent water. For example, probes such as 1-anilino-naphthalene-8-sulfonate (ANS) and cis-parinaric acid (38,39), which exhibit a very low quantum yield of fluorescence in water, become highly fluorescent on binding to hydrophobic regions in proteins. Based on this principle, Kato and Nakai (40) used cis-parinaric acid as the fluorescent probe to determine the relative surface hydrophobicity of several proteins. According to this method, an increasing amount of protein is added to a solution containing a known amount of cis-parinaric acid; and the fluorescence of the protein-parinaric acid conjugate at 420 nm, with excitation at 325 nm, is measured. The initial slope of the fluorescence intensity vs % protein concentration plot is taken as a measure of the surface hydrophobicity (S_o) of the protein.

The surface hydrophobicity values (S_o) of various proteins, measured by the cis-parinaric acid method, are given in Table 3.2. Comparison of

TABLE 3.2
Average and Surface Hydrophobicity Values of Various Proteins

Protein	Average Hϕ[a]	Surface Hϕ[b]
Actin	1000	—
Bovine serum albumin	1000	325
Conalbumin	980	—
Collagen (calf skin)	880	5
Elastin	990	—
Gliadin	1300	—
Hemoglobin (Bovine)	960	—
α-Lactalbumin	1050	—
B-Lactoglobulin	1050	426
Lysozyme (chicken)	890	—
Myosin	880	14
Ovalbumin	980	6
Soy isolate	822[b]	95
Pea isolate	277[b]	66
Sunflower isolate	597[b]	47
Casein	725[b]	28
Whey protein	387[b]	182
Gluten	349[b]	17

[a]From Ref. 35 (in cal/mole residue)
[b]From Ref. 43 and 74.

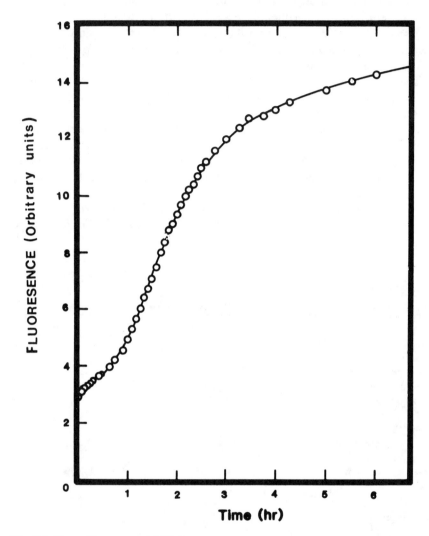

Fig. 3.2. Rate of increase of ANS (1-anilinonaphthalene-8-sulfonate) binding capacity of bovine serum albumin during refolding from the denatured state to the folded state (44).

Bigelow's average hydrophobicity values (35) with those of the surface hydrophobicity values of various proteins showed no correlation (Table 3.2). While several functional properties, such as foaming, emulsification, interfacial tension and solubility, of various proteins exhibited good correlation with the surface hydrophobicity values (40,42,43), no such correlations were found with the average hydrophobicity values.

Although the surface hydrophobicity as measured by the fluorescent probe binding method (40) shows reasonable correlation with many functional properties of food proteins, it is debatable whether the measured quantity is the hydrophobicity of the protein surface that is in contact with the solvent. Recently, it has been shown that in the reduced and denatured state bovine serum albumin exhibits very low affinity for ANS, albeit the exposed surface hydrophobicity is much greater than the native BSA; however, as the molecule refolded to the native state, the extent of binding of ANS, as measured from the increase of ANS fluorescence intensity, increased with the extent of refolding of BSA (44) (Fig. 3.2). This indicates that the binding of fluorescent probes to the protein requires a well defined hydrophobic cavity rather than groups of hydrophobic residues randomly distributed on the protein surface. These cavities are accessible to the fluorescent probes but not to the solvent. Therefore, the quantity that is measured by the parinaric acid probe may not be the true surface hydrophobicity of the protein, but the accessible hydrophobicity of the nonpolar cavities in the protein.

Foaming and Emulsification. A survey of the fabricated and processed foods would indicate that the majority of these foods are either foam, emulsion or gel type systems. An emulsion or a foam is a two phase system in which one of the phases is a dispersed phase and the other is the continuous phase. In the absence of a surfactant, because of the interfacial tension between the two phases, foams and emulsions collapse with time, resulting in separation of the phases. Surfactants, because of their amphiphilicity, adsorb and orient at fluid-fluid interfaces and significantly decrease the interfacial tension and thus stabilize foams and emulsions.

In formulated foods, especially in emulsion and foam based products, proteins are often used as the functional ingredient to perform the role of a surface active agent (3,45). Although all proteins are amphiphilic, they differ considerably in their surface active properties. These differences are attributable mainly to differences in their molecular properties such as amino acid composition and sequence, conformational flexibility and the hydrophobicity/hydrophilicity ratio.

The important initial step in the formation and stabilization of protein based foams and emulsions is the adsorption and spreading of protein at the surface or interface. Several studies on the adsorption of proteins at fluid-fluid interfaces have indicated that for a protein to be a good surfactant, it should possess the following molecular properties: 1. Higher rate of diffusion and adsorption at the interface. 2. Rapid unfolding and reorientation of the polypeptide segments at the interface. 3. Optimal intermolecular interactions at the interface to form a continuous cohesive, viscous film.

To understand the structure-function relationship of proteins with respect to their surface active properties, Graham and Phillips (46-50) systematically studied the adsorption and film forming properties of four structurally very different proteins — lysozyme, bovine serum albumin, k-casein and B-casein — at the air-water and oil-water interfaces. Among these proteins, B-casein adsorbed at the air-water interface and decreased the surface tension rapidly, and formed a dilute monolayer; whereas, both k-casein and lysozyme exhibited a lower rate of adsorption, but formed concentrated films. Furthermore, under identical surface concentrations, B-casein exhibited higher surface pressure (lower surface tension), whereas both lysozyme and bovine serum albumin exhibited lower surface pressures (Figure 3.3) (46). These results indicated that B-casein readily adsorbs, unfolds and reorients itself at the interface, occupies maximum surface area and thus exerts maximum surface pressure. In contrast, both lysozyme and bovine serum albumin adsorb slowly, unfold only partially, occupy lesser surface area per molecule and hence exert lower surface pressure.

The differences in the unfolding abilities of these three proteins were attributed to their unique structural properties (46,47). For instance, B-casein has a highly flexible random coil structure and contains no disulfide bonds. On the other hand, lysozyme and bovine serum albumin are highly structured and contain intra-molecular disulfide bonds. The highly flexible random coil state of B-casein enables the molecule to unfold completely under the prevailing thermodynamic conditions at the interface. The highly ordered, rigid conformational states of lysozyme and bovine serum albumin impair the flexibility and adaptability of these molecules at the interface.

Although the comparative studies on the adsorption characteristics of three different proteins have clearly demonstrated that the conformation of proteins influence their behavior at interfaces, the fundamental understanding of the influence of molecular properties on surface adsorption is far from complete. For instance, the rate of adsorption of a protein at a fresh interface is considered to be diffusion limited. Since the diffusion coefficient is inversely proportional to the cube root of molecular weight, a smaller protein would be expected to adsorb rapidly to an interface. However, it has been reported that the diffusion coeffi-

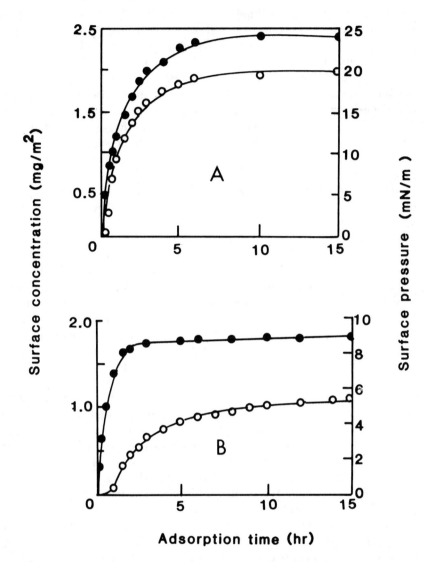

Fig. 3.3. Rate of change of surface pressure (O——O) and surface concentration (●——●) during adsorption of B-casein (A) and lysozyme (B) at the air-water interface (46).

cient for B-casein and k-casein, obtained from the initial rate of adsorption at a clean surface were 3.3 and 1.5×10^{10} m^2/sec, respectively (51).

Similarly, calculation of diffusion coefficients from the adsorption data for B-casein and lysozyme (45) showed that these proteins exhibit almost identical diffusion coefficients. Since both lysozyme (M_r=14,400) and k-casein (M_r=19,000) are smaller in size than B-casein (M_r=23,980), a higher rate of adsorption should be expected. The lack of correlation between the molecular size and the rate of adsorption of proteins at interfaces clearly indicates that although a higher rate of diffusion is desirable, it is not very critical for the surface activity of a protein. This is especially true in practical situations where application of mechanical force (either sparging or homogenization) is involved in the formation of a foam or an emulsion. Other molecular parameters, such as molecular flexibility and the hydrophobicity/hydrophilicity characteristics of the protein surface are of paramount importance in the surface adsorptivity of proteins.

It is generally assumed that a highly unfolded and random coil protein would possess higher molecular flexibility at the interface and would occupy greater surface area per molecule than a compact folded protein. However, recent studies on the adsorption of the structural intermediates of bovine serum albumin at the air-water interface do not support this view (52,53). For example, the structural intermediate, which was highly unfolded and devoid of disulfide bond, occupied less area per molecule at the air-water interface when compared to other structural intermediates. In fact, the study showed that, for serum albumin, neither the completely unfolded nor the native protein had the ability to occupy a large surface area at the interface; an optimum ratio of ordered to disordered structure seemed to be essential to occupy a large surface area and cause a greater decrease of surface tension per adsorbed molecule (52,53). This optimum ratio or order/disorder and rigidity/flexibility requirement for better surface activity might be different for different proteins, depending upon the amino acid sequence and the distribution pattern of hydrophilic and hydrophobic patches on the surface of the molecule.

The ability of a protein to adsorb, unfold and decrease the free energy of the surface/interface is fundamentally related to the topography and the physicochemical properties of the protein surface itself. Under ideal conditions, the rate of adsorption of proteins at interfaces is considered to be diffusion controlled and is described by the relation,

$$\Gamma = 2C_0 \, (Dt/3.14)^{1/2}$$

where 'T' is surface concentration, C_0 is the protein concentration in the

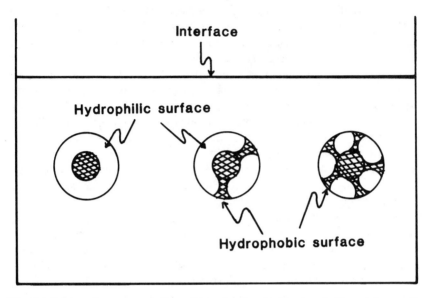

Fig. 3.4. Schematic representation of the role of surface hydrophobic patches on the probability of adsorption of proteins at liquid interfaces.

8 bulk phase, D is the diffusion coefficient and t is the time. The above model assumes that every collision of the protein molecule with the interface leads to adsorption. However, it can be argued that the success of every collision leading to adsorption should be related to the hydrophobicity/hydrophilicity of the protein surface. If the surface of a protein is extremely hydrophilic and devoid of hydrophobic patches, the tendency of the protein to adsorb at a surface or an interface, which is relatively nonpolar, will be very low. On the other hand, if the protein contains few hydrophobic patches on the surface, the protein would bind to the nonpolar surface/interface. The statistical probability of binding of the protein at the interface would increase with the extent of hydrophobic patches on its surface (Figure 3.4). In other words, the probability $'P_a'$, of adsorption of a protein at the interface will be directly proportional to the hydrophobicity $'H\phi'$ and inversely proportional to the hydrophilicity $'Hy'$ of the protein surface. The rate of change of surface concentration under these conditions can be expressed as

$$\Gamma = 2C_o \, (P_a) \, (Dt/3.14)^{1/2}$$

where $P_a \propto (H\phi/Hy)$.

The subsequent spreading and orientation of the absorbed molecule at the interface would depend on its thermodynamic stability at the interface. If the interfacial energy is lower than the activation energy of denaturation of the protein, the unfolding, spreading and reorientation of the protein molecule would proceed slowly, resulting in a lower rate of decrease in the interfacial tension. In general, monomeric proteins containing more surface hydrophobic patches will be thermodynamically unstable at the interface and would readily undergo unfolding and reorientation. The excellent correlation between the surface hydrophobicity and the foaming and emulsifying properties of proteins observed by Nakai et al. (40,42,43) clearly indicate that the presence of hydrophobic cavities or patches on the surface of proteins is essential for better surface adsorptivity and activity.

Although rapid adsorption, spreading and reorientation of the protein molecules at the interface is critical for the formation of foams and emulsions, these attributes may not be critical for the stability of foams and emulsions. For instance, even though B-casein has the ability to adsorb rapidly and decrease the surface tension, the stability of B-casein foams are weaker than that of bovine serum albumin or lysozyme (54). The half-lives of B-casein, serum albumin and lysozyme foams were 15, 32 and 200 minutes, respectively. It is ironic that the protein which has the ability to foam rapidly does not have the molecular properties which impart stability to the foam, whereas, the proteins which do not possess the desirable molecular flexibility to foam rapidly display the molecular characteristics which impart stability to the foam.

The thermodynamic stability of foams and emulsions is not only dependent upon the extent of reduction of the surface/interfacial free energy by the adsorbed protein molecules, but also on the rheological properties of the protein film (55,56). The rheological properties of protein films are, in turn, governed by the hydrodynamic properties, which are affected by the conformation of the protein in the adsorbed state. Since probing of the conformational state of the protein in the adsorbed state is difficult, it is not yet possible to predict the rheological properties of an adsorbed protein from its conformation in solution.

In qualitative terms, the surface rheology of adsorbed protein film is dependent on the extent of hydration and intermolecular interactions. Highly hydrophobic proteins, such as B-casein, form thin, less viscous films because of poor hydration. Furthermore, greater hydrophobic interaction between the protein molecules at the interface leads to aggregation and coagulation, resulting in poor mechanical and viscoelastic properties of the protein film. On the other hand, in the case of highly charged proteins where the hydration is not the limiting factor, the greater electrostatic repulsive interaction between the protein molecules causes destabilization of the protein films. In order for a protein film to have more mechanical stability and viscoelasticity, the protein in the adsorbed state should possess an optimum degree of hydration and intermolecular interactions.

Another parameter that would affect the viscoelasticity of the protein film is the physical entanglement of the adsorbed proteins. The degree of entanglement is dependent on the molecular weight (chain length) of the polypeptide. Thus, large, highly flexible proteins form extended gel-like network at the interface, which provides greater viscosity and mechanical strength to the protein film (57). Foams and emulsions formed with partially hydrolyzed proteins exhibit poor stability compared with unhydrolyzed proteins (58). The highly stable films formed by gelatin and the increase in the stability of gelatin based foams and emulsions with molecular weight of gelatin might be related to the superior rheological properties of gelatin films, which result from molecular entanglement and gel-forming characteristics (59).

Gelation.

Gelation refers to the transformation of the protein in the sol state into a gel-like structure by heat or other agents, in which the individual protein molecules interact with each other to form a three-dimensional network. Depending upon the type and extent of network formation, proteins either form irreversible opaque type gel (e.g., egg white) or reversible translucent type gel (e.g., gelatin and soy protein). The sequence of events that occur during this transformation can be depicted as in Figure 3.5 (60-62).

In the heat induced gelation of proteins, the protein in the sol state is heated above its denaturation temperature, which results in the formation of a progel state. The heat induced sol-to-progel state is usually an irreversible process which involves dissociation, denaturation and the unfolding of the protein molecule. Depending upon the molecular properties of the protein in the unfolded state, it undergoes two types of interactions: Proteins that contain high levels of apolar amino acid residues undergo hydrophobic aggregation; when the protein concentration is high enough, these aggregates coagulate to form an irreversible coagulum type gel. On the other hand, proteins that contain below a critical level of apolar amino acid residues form soluble aggregates,

Thermal Gelation of Proteins

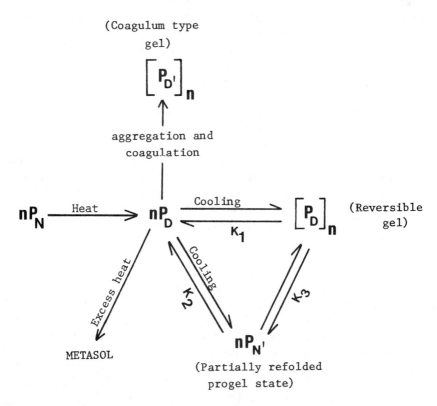

Fig. 3.5. Heat induced changes during thermal gelation of globular proteins. P_N is protein in the native state; P_D, the denatured state; $P_{N'}$, protein in partially refolded state (formed during the cooling phase; $[P_D]_n$, translucent type gel state; K_1, K_2, K_3 are the equilibrium constants; $[P_D']_n$, coagulum type gel state (61).

which set into a thermally reversible transparent type gel upon cooling. Excessive heating of the protein sol at temperatures far higher than the denaturation temperature, e.g., 125°C, leads to a metasol state which does not set into a gel upon cooling (62). This might be related to B-elimination of disulfide bonds and scission of peptide bonds involving aspartate residues at high temperatures (63,64).

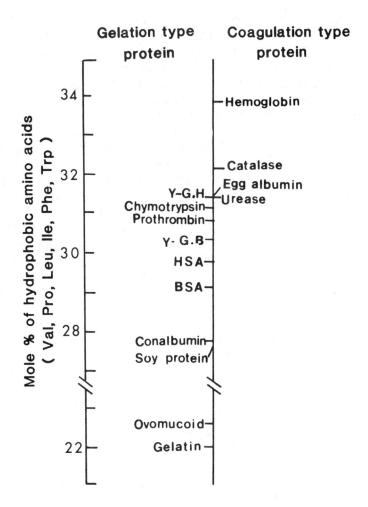

Fig. 3.6. Relationship between the mole percent of hydrophobic amino acid residues and the type of gel network (60).

Whether or not a protein would form a coagulum type gel or a translucent type gel is fundamentally related to Bigelow's (35) average hydrophobicity and the net charge of the protein. Shimada and Matsushita (60) showed that proteins which contain above 31.5 mole percent of hydrophobic residues (Val, Pro, Leu, Ile, Phe and Trp) form coagulum type gels, whereas those that contain less than 31.5 mole percent of the above apolar residues form translucent type gels (Figure 3.6). It is not evident why the above authors did not include other hydrophobic amino acid residues (i.e., Ala, Met and Tyr), in the hydrophobicity calculations.

Although this empirical rule seems to be valid in most cases, it should be emphasized that the ratio of net charge to hydrophobicity would be a more appropriate parameter than the hydrophobicity alone. For instance, although the mole percent of the apolar amino acid residues in B-lactoglobulin is about 32% (42% including Ala, Met and Tyr), the protein forms a translucent gel in the absence of salts (Damodaran, unpublished data). However, in the presence of NaCl, even at 0.05 M, B-lactoglobulin forms a coagulum type gel. This indicates that in spite of the higher content of hydrophobic residues, the higher electronegativity of B-lactoglobulin facilitates the formation of a transparent gel. Neutralization of the electrostatic forces in the presence of 0.05 M NaCl promotes hydrophobic aggregation leading to the formation of a coagulum type gel. Furthermore, the above empirical correlation assumes that thermal denaturation of proteins results in total exposure of all the hydrophobic residues. This need not be true for all the proteins. For instance, at zero ionic strength bovine serum albumin, which contains about 29 mole percent of Val, Pro, Leu, Ile, Phe and Trp, forms a translucent gel when heated above its denaturation temperature. However, in the presence of reducing agents such as cysteine or dithiothreitol, serum albumin forms a coagulum type gel. The reason being that in the former case, even though the protein has been heated above its denaturation temperature, the molecule apparently seems to regain its folded state during cooling (Figure 3.7). This results in partial internalization of the hydrophobic residues and thus precludes formation of a coagulum type gel. On the other hand, in the presence of a reducing agent the reduction of the disulfide bonds in serum albumin facilitates irreversible unfolding and exposure of hydrophobic residues, which leads to the formation of a coagulum type gel.

Gelation, either coagulum or translucent type, involves three dimensional network formation between the protein molecules or aggregates. However, unlike coagulum type gels, where the network formation is a random process, the translucent gels involve ordered association of the polypeptide chains (65), via noncovalent interactions, such as hydrogen bonding, ionic and hydrophobic interactions. Several studies have shown that the primary mode of cross-linking in the gel network, especially in translucent type gels, is the hydrogen bonding between the polypeptide chains (66,67). Although the involvement of electrostatic interactions in gel network formation is not known, it is conceivable that

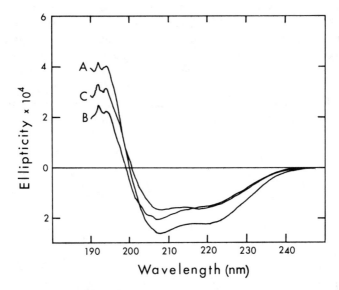

Fig. 3.7. Circular dichroism spectra of bovine serum albumin. A) CD spectrum of the unheated native protein. B) CD spectrum of dilute (0.02%) protein solution heated at 80°C for 30 min and cooled to room temperature. C) CD spectrum of the fluid obtained from 5 % BSA gel (heated at 80°C for 30 min, cooled at 4°C overnight) by centrifugation at 30,000 rpm in a Beckman T-55 swing rotor. The CD spectra indicate that the BSA samples heated at 80°C refolded and regained significant amount of helical structure during the cooling phase.

electrostatic repulsion between the molecules would promote destabilization, rather than stabilization of the network. Furthermore, the dielectric screening of the charges by their hydration shells would also decrease the role of ionic bridges in the gel network. The melting temperature of soy protein gels has been shown to increase in the presence of 0.5 M NaCl (66), indicating that neutralization of electrostatic repulsion between the protein molecules apparently increases the number of cross-links in the gel network. Hydrophobic interactions play an important role in coagulum type gel, however, its role in translucent type gels seems to be very limited (66).

The extent of the three dimensional network formation in protein gels depends on the intrinsic molecular properties of the protein *per se*, as well as other factors such as protein concentration, the rate of heating and cooling, environmental factors such as pH, ionic strength and solvent composition.

The formation of a self-supporting gel network that is stable against thermal and mechanical motions is dependent on the number of cross-links per monomer or per unit cell of the gel. This depends on both protein concentration and the number of loci available per molecule. In order to form a self supporting gel network, a minimum concentration of protein known as the least concentration endpoint (LCE) is required (68,69). Ferry (70) showed that the rigidity of gelatin gel is proportional to the square of the concentration. This relationship seems to be true

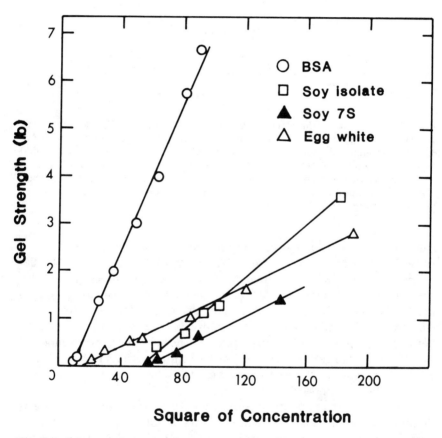

Fig. 3.8. Gel strength versus square of protein concentration plots for the heat induced gels of various proteins. All samples were heated at 90°C for 30 min, cooled overnight at 4°C.

even for globular proteins such as serum albumin and soy proteins (Figure 3.8). The least concentration endpoint of a protein can be conveniently determined from the x-intercept of the gel strength versus C^2 plot.

Under a given set of gelation conditions (like protein concentration, pH, temperature, ionic strength) the LCE of various proteins is dependent on their molecular properties. For gelatin, which is devoid of cysteine and cystine, the square root of the gel strength is proportional to the molecular weight (70). This might be related to the dependence of the extent of molecular entanglement and the number of cross-linking loci per unit cell of the gel on the chain length of the polypeptide. In the case of globular food proteins (which usually contain cysteine and cystine residues), it is difficult to establish the relationship between the gel strength and the molecular weight because of the formation of intermolecular disulfide bonds during heat induced gelation, which results in formation of a heterogeneous population of polypeptide species with various chain lengths. However, studies on the gelation of soy proteins and BSA in the presence of excess cysteine — which prevents formation of inter-and intra-molecular disulfide bonds during thermal gelation — did seem to show that the square root of gel strength is apparently proportional to the weight average molecular weight of the polypeptides in the sample (Figure 3.9). This seems to suggest that the rheological and physical properties of globular protein gels is more dependent on the molecular size, and less influenced by the chemical nature (i.e., amino acid composition and distribution) of the proteins *per se*. Further research, involving a wide range of globular proteins having a range of molecular size, is needed to establish the apparent relationship between the molecular size and the gel properties.

The ability of a protein to form intermolecular disulfide bonds during the heat treatment has often been considered a prerequisite for gelation of the protein (71-73). However, it should be pointed out that, of all the proteins, gelatin is known to be the best gelling protein; yet this protein is devoid of cysteine and cystine residues, which clearly implies that formation of disulfide bonds is not essential for gelation of proteins. In molecular terms, the apparent role of disulfide bonds in gelation may be related to their ability to increase the chain length of the polypeptide, rather than as a network former. The increase in the effective chain length of the polypeptide may increase the molecular entanglement in the gel structure, which might restrict the relative thermal motions of the polypeptides. In addition, the lower diffusion coefficient of the longer polypeptide species might prevent rupture of the weak noncoval-

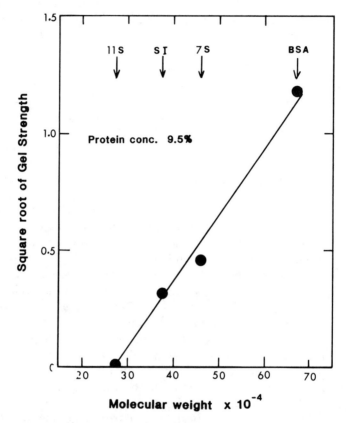

Fig. 3.9. Relationship between the square root of gel strength and the average molecular weight of various proteins. Protein concentration was the same, i.e., 9.5% in all cases. The gels were formed by heating at 90°C for 30 min in the presence of 400 mM cysteine, cooled overnight at 4°C. Cysteine was included to prevent inter- and intramolecular disulfide bonds. The weight average molecular weights were calculated based on the subunit composition and their molecular weight, and the weight fraction of various species in the protein sample.

ent interactions and thus stabilize the gel network. Both these molecular effects might increase the total number of noncovalent bonds between the polypeptides (70). In other words, it is the physical constraints imposed by the length of the polypeptide chain, rather than the disulfide bonds *per se*, that are responsible for the perceived influence of disulfide bonds on gel strength. Conversely, one can also improve the gelation of a cysteine or cystine lacking protein by increasing its effective chain length via other chemical cross-linking methods. The longer the chain of the cross-linked protein, the greater the gelling power will be at a given protein concentration, and the lower the least concentration endpoint for the formation of gel structure will be.

Conclusions

The expression of a functional property is a complex process, in which the physical, chemical and conformational properties of the protein *per se* and the nature and extent of its interactions within itself and with the surrounding environment play a crucial role. Although all the molecular properties of a given protein act in a concerted manner in expressing its various functional properties, in each of the functional properties certain molecular properties of the protein might play a more predominant role than the others in expressing that function. Identification of the major molecular parameters for the expression of a function is essential, and a fundamental understanding of their mechanistic role in expressing that function is equally essential. This understanding will facilitate the development of strategies to improve the functional properties of under-utilized proteins through chemical, physical and genetic modifications.

Acknowledgment

Support from the National Science Foundation Grant No. CBT-8616970 is gratefully acknowledged.

References

1. Waterloo, J.C., and P.R. Payne, *Nature* 258:113 (1975).
2. FAO., Production Yearbook, FAO, Rome (1979).
3. Kinsella, J.E., *CRC Crit. Rev. Food Sci. Nutr.* 7:219 (1976).
4. Klotz, I.M., and J.S. Franzen, *J. Amer. Chem. Soc.* 84:3461 (1962).
5. Damodaran, S., *Intl. J. Peptide Protein Res.* 26:598 (1985).
6. Tanford, C., The Hydrophobic Effect: Formation of Micelles and Biological Membranes, Wiley Interscience, New York, (1973).
7. Franks, F., in "Water Relations of Foods", Duckworth, R.B., (ed.), Academic Press, New York, p. 3, (1975).
8. Nozaki, Y., and C. Tanford, *J. Biol. Chem.* 276:2211 (1971).
9. Damodaran, S., and K.B. Song, *J. Biol. Chem.* 261:7220 (1986).
10. Lee, B., and F.M. Richards, *J. Mol. Biol.* 55:379 (1971).
11. Richards, F.M., *Ann. Rev. Biophys. Bioeng.* 6:151 (1977).

12. Janin, J., *J. Mol. Biol. 105*:13 (1976).
13. Janin, J., and C. Chothia, *Biochemistry 17*:2943 (1978).
14. Chothia, C., *Nature 248*:338 (1974).
15. Chothia, C., *Nature 254*:304 (1975).
16. Quist, K.B., *Milchwissenschaft 34*:467 (1979).
17. Brants, J.F., in "Thermobiology", A.H. Rose, (ed.), Academic Press, New York, p. 25 (1967).
18. Kanehisa, M.I., and T.Y. Tsong, *Biopolymers 19*:1617 (1980).
19. Swaisgood, H.E., in "Developments in Dairy Chemistry-1". P.F. Fox, (ed.), Elsevier Applied Sci. Publ., London & New York, p. 1 (1982).
20. Chou, P.Y., and G.D. Fasman, *Ann. Rev. Biochem. 47*:251 (1978).
21. Anfinsen, C.B., and H.A. Scheraga, *Advan. Protein Chem. 29*:205 (1975).
22. Nemethy, G., and H.A. Scheraga, *Quat. Rev. Biophys. 23*:239 (1977).
23. Marco, Y.A., V.H. Thanh, N.E. Tumer, B.J. Scallon, and N.C. Neilsen, *J. Biol. Chem. 259*:13436 (1984).
24. Doyle, J.J., M.A. Schuller, W.D. Godette, V. Zenger, R.N. Beachy, and J.L. Slightom, *J. Bio. Chem. 261*:9228 (1986).
25. Slightom, J.L., S.M. Sun, and T.C. Hall, *Proc. Natl. Acad. Sci.*, U.S.A. *80*:1897 (1983).
26. Lycette, G.W., A.J. Delauney, J.A. Gatehouse, J. Gilroy, R.R.D. Croy, and D. Boulter, *Nucleic Acids Res. 11*:2367 (1983).
27. Townend, R., T.F. Kumosinski, S.N. Timasheff, G.D. Fasman, and B. Davidson, *Biochem. Biophys. Res. Commun. 23*:163 (1966).
28. Provencher, S.W., and J. Glockner, *Biochemistry 20*:33 (1981).
29. Hennessey, J.P., and W.C. Johnson, *Biochemistry 20*:1085 (1981).
30. Manavalan, P., and W.C. Johnson, *Anal. Biochem. 16*:76 (1987).
31. Chang, C.T., C.S.C. Wu, and J.T. Yang, *Anal. Biochem. 91*:13 (1978).
32. Waugh, D.F., *Adv. Protein Chem. 9*:326 (1954).
33. Kauzmann, W., *Adv. Protein Chem. 14*:1 (1959).
34. Fisher, H.F., *Proc. Natl. Acad. Sci.*, U.S.A. *51*:1285 (1964).
35. Bigelow, C.C., *J. Theoret. Biol. 16*:187 (1967).
36. Van Holde, N., in "Food Proteins", J. Whitaker, and S. Tannenbaum, (eds.), Avi Publ. Co., Westport, CT., p. 1 (1971).
37. Hahn, T.R., and P.S. Song, *Biochemistry 20*:2602 (1981).
38. Sklar, L.A., B.S. Hudson, and R.D. Simoni, *Biochemistry 16*:5100 (1977).
39. Sklar, L.A., B.S. Hudson, and R.D. Simoni, *Proc. Natl. Acad. Sci.*, U.S.A. *72*:1649 (1975).
40. Kato, A., and S. Nakai, *Biochim. Biophys. Acta 624*:13 (1980).
41. Stryer, L., *Science 162*:526 (1968).
42. Nakai, S., L. Ho, N. Helbig, A. Kato, and M.A. Tung, *Can. Inst. Food Sci. Technol. 13*:23 (1980).
43. Voutsinas, L.P., E. Cheung, and S. Nakai, *J. Food Sci. 48*:26 (1983).
44. Damodaran, S., *Biochim. Biophys. Acta 914*:114 (1987).

45. Kinsella, J.E., and S. Damodaran, in "Criteria of Food Acceptance", J. Solms, and R.L. Hall, (eds.), Forster Publ., Zurich, p. 296 (1980).
46. Graham, D.E., and M.C. Phillips, *J. Colloid Interface Sci. 70*:403 (1979).
47. Graham, D.E., and M.C. Phillips, *J. Colloid Interface Sci. 70*:415 (1979).
48. Graham, D.E., and M.C. Phillips, *J. Colloid Interface Sci. 70*:427 (1979).
49. Graham, D.E., and M.C. Phillips, *J. Colloid Interface Sci. 76*:277 (1980).
50. Graham, D.E., and M.C. Phillips, *J. Colloid Interface Sci. 76*:240 (1980).
51. Benjamins, J., J.A. Feijter, M.T.A. Evans, D.E. Graham, and M.C. Phillips, *Faraday Disc. Chem. Soc. 59*:218 (1975).
52. Song, K.B., and S. Damodaran, *J. Agric. Food Chem. 35*:236 (1975).
53. Damodaran, S., and K.B. Song, *Biochim. Biophys. Acta 954*:253 (1988).
54. Graham, D.E., and M.C. Phillips, in "Foams", R.J. Ackers, (ed.), Academic Press, London, p. 237 (1975).
55. Rivas, H.J., and P. Sherman, *Colloids Surf. 11*:155 (1984).
56. Boyd, J.V., C. Parkinson, and P. Sherman, *J. Colloid Interface Sci. 41*:359 (1972).
57. Joly, M. in "Surface and Colloid Science", E. Matijevic, (ed.), Wiley, New York, p. 79 (1972).
58. Alder-Nissan, J., and H.S. Olsen, American Chemical Society Symposium Series 92, p. 125 (1979).
59. Dickinson, E., A. Murray, B.S. Murray, and G. Stainsby, in "Food Emulsions and Foams", E. Dickinson, (ed.), Royal Society of Chemistry, London, p. 86 (1987).
60. Shimada, K., and S. Matsushita, *J. Agric. Food Chem. 28*:413 (1980).
61. Damodaran, S., *J. Agric. Food Chem. 36*:262 (1988).
62. Catsimpoolas, N., and E.W. Meyer, *Cereal Chem. 47*:559 (1970).
63. Ahren, T.J., and A.M. Klibanov, *Science 228*:1280 (1985).
64. Volkin, D.B., and A.M. Klibanov, *J. Biol. Chem. 262*:2945 (1987).
65. Hermansson, A.M., *J. Texture Studies 9*:33 (1978).
66. Babajimopoulos, M., S. Damodaran, S.S.H. Rizvi, and J.E. Kinsella, *J. Agric. Food Chem. 31*:1270 (1983).
67. Eldridge, J.E., and J.D. Ferry, *J. Amer. Chem. Soc. 58*:992 (1954).
68. Trautman, J.C., *J. Food Sci. 31*:409 (1966).
69. Acton, J.C., M.A. Hanna, and L.D. Satterlee, *J. Food Biochem. 5*:101 (1981).
70. Ferry, J.D., *J. Amer. Chem. Soc. 70*:2244 (1948).
71. Huggins, C., D.F. Tapley, and E.V. Jensen, *Nature 167*:592 (1951).
72. Nakamura, T., S. Utsumi, and T. Mori, *J. Agric. Food Chem. 32*:349 (1984).
73. Utsumi, S., and J.E. Kinsella, *J. Food Sci. 50*:1278 (1985).
74. Voutsinas, L.P., S. Nakai, and V.R. Harwalker, *Can. Inst. Food Sci. Technol. 16*:185 (1983).

Chapter Four
Structure:Function Relationships in Food Proteins, Film and Foaming Behavior

J.E. Kinsella and L.G. Phillips

Institute of Food Science
Cornell University, Ithaca, NY

The widespread importance of proteins as structural and functional components of traditional and new foods has been amply described (1-6) and the use of proteins as versatile functional ingredients is now widespread in the food industry (7). However, food applications are generally based on empirical observations and compared to knowledge of biochemical functions of proteins, only limited information is available concerning relationships between the structure and functional properties of ingredient proteins. Elucidation of important structure/function relationships and the effects of stabilization forces on functional behavior is needed for the rational modification of proteins by genetic engineering, enzymatic, chemical and physical methods to improve and/or stabilize their behavior in specific applications (8). A concerted effort to elucidate relationships between protein conformation and function should result in useful practical information. Research is needed to systematically determine the primary, secondary and tertiary structures of food proteins and relate these to particular physical behavior (9).

Protein Structure

The unique physicochemical and functional properties of different food proteins (e.g., viscoelastic gluten, amphiphilic caseins, structural myofibrillar proteins and the network/gelling tendency of collagen/gelatin) reflect differences in the primary structure, their folding behavior (i.e., conformation and protein:protein interactions). Molecular shape and reactivity (charge, hydrophobicity, molecular flexibility etc.) and molecular size account for many properties of food proteins (9). The particular structure of a protein is dynamic reflecting the immediate thermodynamic conditions; however, for most food applications, proteins are

TABLE 4.1
The Major Secondary Forces Involved in Protein:Protein Interactions in Films and Foams

Interaction type	Energy of interaction kJ/mol	Interaction distance A°	Groups involved	Temperature effect	Occurrence in films
Hydrophobic interactions	3–10	3–5	Apolar residues aliphatic, aromatic in aqueous system	increase	yes
van der Waals	1–10	1–4	Induced dipoles closely aligned groups	?	yes
Hydrogen bonds	8–40	2–3	Amide, carbonyl hydroxyl	decrease	yes
Electrostatic	42–84	2–3	Oppositely charged groups NH_3:COO^-	decrease	yes
Covalent S-S	320–380	1–2	Cystine	—	—

usually viewed as being in either of two major states, viz., native or denatured.

Polypeptides, in their native state, fold in a characteristic fashion when in an aqueous environment to form localized secondary structures (α-helix, β-pleated sheet, β-turns or random coil) to minimize negative entropy effects in the system, i.e., folding is hydrophobically driven (10). Secondary structures, e.g., α-helixes, are stabilized by hydrogen bonds. Secondary structures progressively fold to form tertiary structures which minimize contact of apolar groups and hydrophobic segments with water. Thus polar groups occupy the surface while apolar groups are generally folded into the interior when sterically and thermodynamically possible (10). The tertiary structure of proteins is stabilized by van der Waals' interactions between symmetric closely apposed segments, by hydrogen bonding, electrostatic interactions and in some instances, by disulfide bonds. Quaternary structures are formed by the association of folded polypeptides usually via surface hydrophobic interactions as in caseins, electrostatic interactions as in β-lactoglobulin (β-Lg) dimers and/or via disulfide linkages as in glycinin (7,9,11).

Because functional properties reflect the unique structures of proteins, the non-covalent forces responsible for structures (Table 4.1) greatly affect functions and are particularly important in the rearrangements and the new intermolecular associations required for food applications as in films or gels. Thus, modification of the functional groups on proteins, alteration of pH or ionic strength, the presence of surfactants and the prevailing temperature etc. can markedly affect the magnitude of these forces and alter protein conformation and functional behavior (9). These forces play important roles in the behavior of proteins in films and foams (1,3-5).

Interfacial Behavior of Proteins

The interfacial behavior of proteins is important in determining quality attributes of many foods (e.g., milk, meat emulsions, mayonnaise, spreads, ice cream, frozen desserts, cakes, breads, whipped toppings etc.). The structure of many of these products depends upon the formation and stability of interfacial films which facilitate mixing, impart structure and contribute to sensory qualities. Although foods are heterogeneous in composition and the nature of interfacial films varies with the product in question, proteins are important film-forming components in many food products. Proteins being amphipathic, with flexible molecular structures, possess surface active properties which range from very effective to poor in terms of reduction of surface tension, film formation, film properties, foam and/or emulsion stabilization. This reflects the heterogeneous structure, variable conformation, varying molecular flexibility, environmental factors and the versatile range of dynamic and physical properties required in each phase of film formation and stabilization (16).

Relatively little is known about the dynamic behavior of proteins in food foams; however, based on studies of pure proteins, useful information is accumulating concerning desirable physicochemical criteria of surface active proteins (1-7 and Chapters (UNDS) and (UNDS)).

Film Formation

Proteins are amphipathic flexible macromolecules which can undergo dynamic conformational changes and adjust to new environments. Hence proteins are good surface active agents for foam (and emulsion) formation. The surface active properties are related to their ability to lower the interfacial tension between air/water interfaces. This reflects the ease with which proteins can diffuse to, adsorb, unfold and rearrange at an interface (6). Thus, size, native structure and solubility in the aqueous phase are closely correlated with surface activity of the proteins in model systems (1-5). Solubility is a primary prerequisite for rapid film formation under normal conditions though in food systems high shear mixers may facilitate film formation from hydrated protein aggregates and dispersions. In a food system the soluble protein molecules should rapidly diffuse to the new interface being continually formed by mechanical agitation, adsorb, reduce interfacial tension, penetrate the interface, unfold to some extent, spread and extensively interact with contiguous proteins to form a continuous cohesive film. The reduction of interfacial tension facilitates the creation of new surface area. This sequence of events occurs very rapidly (msec) and in fact all of these events are occurring during film formation in a foam (Table 4.2).

The spontaneous adsorption of a protein at an interface is initially diffusion dependent and is accompanied by an increase in surface pressure (decrease in surface tension), and for dilute solutions (10^{-3} g/100 ml) and low surface pressures (0.1 mNm^{-1}) up to 0.7-2 mg protein/m^2 is adsorbed (6). The diffusion of proteins from solution to the interface

TABLE 4.2

Molecular Characteristics Desirable in Proteins for Films and Foams

Soluble	— facilitates rapid diffusion to interface
Large	— allows more interactions in the interface, stronger films
Amphipathic	— provides unbalanced distribution of charged and apolar residues for improved interfacial interactions
Flexible domains	— facilitate phase behavior and unfolding at interface
Interactive regions	— the disposition of different functional segments facilitate secondary interactions in the air, the interface and aqueous phases
Disposition of charged groups	— affects protein:protein interactions in the film and charge repulsion between neighboring bubbles
Retention of structure	— enhances overlap and segmental interactions in films
Polar residues	— provide hydratable (glycosyl) or charged residues to keep bubbles apart, binding and retain water

is generally favorable thermodynamically because some of the conformational and hydration energy of the protein is lost at the interface (6,7,12-15). Entropic factors provide a driving force for successful collision with and adsorption to the interface by dehydration of hydrophobic segments and a gain in entropy of hydrophobic segments upon unfolding in the apolar air phase. Hydrophobic segments easily adsorb at the air-water interface with varying loss in tertiary structure (5,6). Initially, at low protein concentrations, there is a limited barrier to adsorption and for protein molecules that are readily adsorbed at the interface, the rate of adsorption is mostly diffusion controlled, but as the concentration of the adsorbed protein increases there is an activation energy barrier to adsorption which may involve electrostatic, steric and osmotic effects at the interface. Under these conditions, the ability of protein molecules to penetrate and create space in the existing film and rearrange at the surface becomes rate-determining (14-19).

Only a part of a protein adsorbs initially and then it may subsequently unfold further as hydrophilic and hydrophobic relationships are altered at the interface. This may reflect the size and structure of the particular protein, net charge (pH), protein stabilizing forces, the relative amount of hydrophobic domains and their flexibility. Thus, depending upon conditions, proteins vary in rate and extent of adsorption (e.g., for lysozyme and κ-casein at a surface concentration of 1.5 mg/m^2 the surface pressures were approximately 5 and 20 mNm^{-1}, respectively) (20). Differences may reflect an energy barrier to insertion of new protein into the film even at low-surface pressures because of electrostatic and steric factors (1,6). Thus, stable globular proteins, e.g., lysozyme, may have difficulty penetrating the interface (1). Denatured soluble globular proteins (e.g., bovine serum albumin (BSA) or α-lactalbumin) increased surface pressure more rapidly (3×) than the native proteins because a greater number of segments could penetrate the interface possibly because of increased flexibility (5).

Some unfolding of proteins occurs at an interface, the extent being governed by thermodynamic conditions, protein size, type and flexibility, the relative concentration (especially 4-9 mg/m^2), steric and physical constraints in a packed film and environmental factors (i.e. pH etc.). Because unfolding is a time-dependent process and may be slower than diffusion and successful adsorption, it is conceivable that only limited unfolding of most proteins occurs in an interfacial film. This has been confirmed by the retention of tertiary structure in certain films which apparently improves film strength (1,20-23). Few or several segments of a molecule may adsorb (trains) at the interface and intervening loops and tails may occupy the air or water phase. Conceivably if electrostatic repulsions are not excessive, the tails and loops interact to form a continuous film (12). Song and Damodaran (13) showed that an optimum structural balance (i.e., 50% each α-helix and random coil) was observed with unfolded BSA for surface pressure development and area covered. Thus, a totally disordered molecule was not optimum for adsorption and disordered molecules do not occupy maximum area; in fact, they may occupy less area than a partly structured molecule. Thus, retention of ordered structure is important in interfacial behavior and properties of protein film (1,4,6). Thus, only certain segments of a protein may unfold in the interface while the bulk of the molecule which remains in aqueous phase retains tertiary structure or the thermodynamically most stable structure enforced by water and secondary interactions.

Thus, β-lactoglobulin films show tertiary structure and bovine serum albumin apparently retaining tertiary structure in emulsion films (20-22,25,26).

The proportion of a protein adsorbed decreases as the protein concentration and the thickness of films increases (1,4-6). The thickness of films may range from 3-6 nm depending upon protein structure, conformation, and concentration (1,4-6).

The capacity of proteins to unfold at an interface depends very much on the conformational stability of the flexible segments of protein molecules (16,17). Where there is extensive intramolecular associations and disulfide bonding, unfolding at the interface tends to be limited, and formation of an interfacial membrane takes longer, as with soy proteins (17). Molecular flexibility (configurational energy gain) may facilitate multiple points of adsorption (1,6). Extensive disulfide-linked proteins which cannot readily unfold may be less surface active. Reduction of such proteins may improve their surface activity (18).

The extent of protein unfolding at an interface can be monitored by analyses of spread films which usually contain from 1-8 mg of protein·m^{-2}, depending upon the tertiary and quaternary stability of the protein (1,16,17). Proteins with a net negative charge in a film may further adsorb cationic proteins to form a thicker, stronger membrane (12). Generally larger molecules adsorb more slowly but in time, because of thermodynamic effects, they may displace smaller molecules which initially adsorb more rapidly (12). This, of course, depends on respective binding energies and configurational entropic factors (e.g., a molecule with extensive train and/or large apolar segments in the air phase would be difficult to displace from a film). The extent to which this occurs depends on surface pressure; e.g., at higher surface pressures desorption is favored (12) perhaps because there is less 'complete' unfolding and fewer trains adsorbed in the interface. In food systems, a heterogeneous population of proteins may exist in an interface with molecules in various states of unfolding. In mixed systems, proteins can effectively compete for sites and displace previously adsorbed molecules; i.e., casein displaces gelatin from interfaces (19).

Conceptually, at the interface, the more apolar hydrophobic segments of the protein occupy the apolar air phase whereas the polar-charged regions occupy the polar aqueous phase. However, with the exception of amphiphilic, random coil β-caseins, the structure and intramolecular linkages in most proteins prevent their facile unfolding and reorientation in the interface. Thus, there is a gradation in behavior depending upon concentration, molecular flexibility, rigidity of tertiary structure, location and disposition of hydrophobic, polar domains and charged residues (1-7). Many globular proteins whose structures are stabilized by hydrophobic association (and the concomitantly enhanced hydrogen bonding, electrostatic and van der Waals interactions) can rearrange in the interface (because of the entropy gain), with parts of the apolar domains relocating in the air phase. Bovine serum albumin and β-lactoglobulin unfold to a limited extent; however, extensive unfolding is restricted by structure stabilizing intramolecular disulfide bonds. This is in contrast to the limited surface activity of lysozyme and glycinin where unfolding is apparently very limited because of the number and location of the intramolecular disulfide linkages (1,11,17-21).

The kinetics of protein adsorption at an interface is measured by surface concentration and surface pressure, as a function of time (1,4). Proteins migrate and adsorb very rapidly (msec) and β-casein is more surface active than serum albumin, β-Lg and lysozyme (6). This reflects the relative rates of diffusion and molecular 'flexibility' of these proteins (1-4,20-24). Thus, β-casein at very low concentrations (0.01 mg/100 ml) spreads rapidly to form a film. At low concentrations and low surface pressures, there is ample area for each molecule (>38 A^2/residue) to spread in the interface with very few loops protruding into either phase. As surface concentration and surface pressure increase (to 7 mN/m), folding and looping of the β-casein occur. Monolayer coverage forming a film of 60°A thickness of tightly packed folded molecules (>7.7 A^2/residue) is complete when the concentration is 0.1mg/100ml. Further adsorption is limited because of charge repulsion and steric factors until the substrate concentration is 1 mg/100 ml when multilayer formation occurs (1). Lysozyme adsorbs at a lower rate than β-casein and surface pressure is much less than β-casein at any given surface concentration, reflecting limited unfolding of the lysozyme. At high surface concentration, lysozyme can form a film although it retains extensive native structure (20-24).

For effective formation of a cohesive film (with the requisite physical and mechanical properties for stabilizing foams), extensive protein:-protein interactions are critical in forming a continuous three-dimensional network. The nature, type and extent of interactions are determined by the protein(s) involved and environmental conditions. Obviously the nature of interactions differs in these three phases: air, interface and aqueous media. The extent of interaction in each phase and the relative contribution of each to stable film formation have not been established. The extent of unfolding, total exposed surface area, nature of these surfaces and their steric and physicochemical compatibilities influence the potential for interactions. Presumably the more extensive the interdigitation of segments and the greater the number of interactions between the polypeptides, the stronger will be the resultant film.

Conceivably the forces which stabilize protein structure are also involved in network formation and stabilization. Thus, in the aqueous phase, hydrophobic interactions and hydrogen bonding may be involved; in the apolar air phase, hydrogen bonding and electrostatic interactions if present would be enhanced. The active involvement of coulombic charge-charge interactions in films is indicated by the fact that film properties are affected by pH, tending toward maxima in the isoelectric pH range (3,4). During film formation all of these phenomena must occur very rapidly for effective entrapment of air during bubble formation in preparing foams.

Film Properties

The properties of the film (i.e., mechanical strength, viscoelastic and restorative properties which are critical in stabilizing foams) are determined by the concentration of protein (film thickness) and the number and nature of the secondary interactions. At approximately similar concentrations globular bovine serum albumin forms stronger films than the random structured β-casein (16) presumably because of the greater number and variety of interactions. In addition, the bulky residual structure may improve viscosity of BSA at the interface (1). However, the number of protein:protein interactions must not be excessive or coagulation might occur and result in destabilization of the film, especially in foams (4).

The viscoelastic properties of films determine the capacity of films in foams to absorb and accommodate shocks and deformation without rupture of the film. Viscosity, yield stress and elastic properties are associated with flexibility of films and enable them to adjust, expand, deform to a limited extent in response to localized stresses without physical disruption or breakage (1,6,19–30). The molecular properties determining viscoelastic properties have not been determined. Conceptually the reversible coiling of the polypeptides in the film interface and the facility with which new secondary interactions can form as new polypeptide segments which become exposed with stretching of the film may partly account for viscoelasticity. These properties are critical during foam formation when the film once formed should be able to withstand the shear distortion and shock effects of the mixer/whipper blades without collapsing. Furthermore, once formed they must stabilize the foam against mechanical shocks, vibration, drainage and temperature fluctuations during storage and use (1,3,4,6,30).

Films have very high surface viscosity reflecting the energy required to move molecules in the film from one point to another (12). The surface viscosity or resistance to shear stress of the surface film is an indication of its mechanical strength and is an important parameter related to the stability of films and foams (1,26). Protein films are very thin (1-10 nm) at 1-8 m^2/mg and upon compression the proteins in surface films coagulate easily reflecting extensive protein:protein interactions (5,6,30). Though films reach equilibrium with the subphase in a matter of hours, the development of viscosity of film takes much longer (31). Surface viscosity increases with thickness of film, type of protein, aging and pH. Above certain concentrations (>1 mg/100ml) multilayers of proteins can adsorb as gelled protein layers and these multilayer films possess marked viscosity, resistance to shear and lower compressibility (larger dilatational modulus) than β-casein films (20-24).

The manner in which surface tension forces respond to (i.e., oppose) changes in surface area is measured by dilatational modulus (ϵ), an important criterion of foam stability. This is generally higher for globular proteins than for casein (24). The capacity of proteins to respond to stress may also be estimated by tensiolaminometry (32). The response to surface expansion may reflect further unfolding of the protein in the interface and/or adsorption of new segments of the protein as the local surface tension is increased (i.e., the Gibbs-Marangoni effect) (1,5,6).

Benjamins et al. (23) studied the effect of aging on the elasticity of β-casein and κ-casein films. The dilatational modulus of κ-casein was larger than that of β-casein and increased by a factor of three with aging, whereas the dilatational modulus of β-casein films changed little with time because of limited interaction. The κ-casein presumably unfolds less than β-casein at the air/water interface because it has a more ordered tertiary structure but may rearrange with time. Significant protein:protein interactions, repulsions and attractions occur between segments of polypeptide chains which extend both above and below the plane of the air/water interface in surface protein films. Whereas caseins form weak films with low viscosity, globular β-lactoglobulin, α-lactalbumin and serum albumin formed stronger viscoelastic films (19) reflecting greater interactions and possible entanglements between the globular proteins which retained tertiary structure in the film and presented greater resistance to flow (24). Maxima in shear rheological properties of bovine serum albumin and lysozyme occurred at concentrations of 1-10 mg/100 ml corresponding to interface concentrations of 3.5 mg/m^2 (21-24).

Surface pressure development, protein:protein interactions, film thickness and viscoelastic properties are greatly influenced by net charge on the protein (i.e., pH of the medium). Generally stronger, more rapidly formed films are obtained at pH values close to the isoionic pH of most proteins (3-6,26-29). For example, β-lactoglobulin undergoes a progressive expansion and increasing flexibility between pH 3 and 9. This is associated with improved surface activity. The rate of surface pressure development is rapid at pH 7.0 reaching equilibrium at 20 minutes versus 24 hours at pH 3.0; i.e., the looser conformation enhanced interfacial activity (33). This has also been observed for films made from modified β-lactoglobulin (26,28).

Kim and Kinsella (27) observed that pH-affected surface pressure development, film elasticity and yield stress of BSA films with maxima were slightly above the isoelectric pH range 5-6 (Table 4.3). Significantly the maximum foam stability also coincided with this pH indicating the importance of minimum electrostatic repulsion governing protein:protein interactions (possibly via hydrophobic interactions) in the film (27-29). Stability is governed by the rheological properties of the interfacial film, the film strength, and the number and types of interactions. Generally globular proteins which retain structure form stronger films than disordered proteins. Viscous films (i.e., surface viscosities of 1,100 compared to 8 and 0.5 mN m^{-1} for caseinate and β-casein, respectively) are formed by β-lactoglobulin. Films formed near the isoelectric point of the major proteins are more condensed and stronger (1-5, 20-29) (Table 5.3). Films composed of mixtures of soluble proteins with differ-

TABLE 4.3
Effect of pH on Film and Foaming Properties of Bovine Serum Albumin

pH	Surface pressure	Surface yield stress (dyne/cm)	Film elasticity	Drainage half-life (min)
4.0	2.8	3.0	2.2	5.0
5.0	15.0	3.8	5.0	8.0
5.5	19.0	4.0	5.2	9.6
6.0	14.0	4.3	5.4	8.5
7.0	10.0	3.0	2.3	6.3
8.0	2.0	2.2	1.8	6.0

ent net charges and hydrophilic:hydrophobic residues are usually more stable (e.g., egg white) (34). The formation of disulfide bonds during film formation may enhance film stability (8,12). The strength of protein films tends to increase on aging, reflecting rearrangements and increased interactions between component molecules in the film (12).

Foam Formation and Stabilization

The behavior of proteins at interfaces influences the formation and the stabilization of foams which depend to a great extent on the properties of the interfacial film. The extent of molecular interactions and the properties of the film depend on the particular protein and the prevailing conditions. A number of different forces may be more or less desirable at different stages of film formation; electrostatic interactions (attractive forces) at strategic locations may be critical to film network formation and cohesion but if excessive, these may result in coagulation. Furthermore, net electronegativity on the aqueous surface of the film, by causing repulsion between contiguous bubbles, is desirable for foam stability. Factors which are desirable for optimum film properties in simple systems may retard film formation or cause destabilization in foams; for example, many rheological properties which improve film stability are maximum in the isoelectric pH range of specific proteins, where proteins tend to be poorly soluble.

Foaming

During foam formation, a number of sequential phases (involving many reactions) are involved. Initially, proteins diffuse to the air-water interface, adsorb, concentrate and reduce surface tension; structural rearrangement of component polypeptides occurs at the interface; i.e., polar moieties orient toward the water and apolar/hydrophobic segments unfold into air, and finally, interdigitation between the polypeptides occurs via various noncovalent (and perhaps covalent) interactions to form a continuous cohesive film. The ability to rapidly diffuse to the interface, reorient with limited aggregation and form a viscous film is critical for the formation of protein-based foams. The tails and loops of neighboring molecules should interact favorably to form a continuous cohesive film. Because protein-based foams depend upon the intrinsic molecular properties of the protein, the foaming properties of proteins vary widely (1-6, 14-26).

The physical stability of the film once formed is the major determinant of foam stability. This reflects the intrinsic properties of the protein: film thickness, extent and nature of protein:protein interactions, and extrinsic factors, i.e., temperature, pH, ionic strength, and viscosity of the continuous phase (1-6). More condensed, tightly packed, thicker films (3-5 mg/m^2) formed by using high concentrations (>1 mg/100 ml) of proteins, by adjusting pH close to pI, or by mixing acidic and basic proteins demonstrate superior mechanical strength (greater protein:protein interactions), possess better rheological and viscoelastic properties and more stability. Films composed of bulky proteins which retain more tertiary structure are usually more stable than proteins lacking structure; this is reflected in more stable foams (1-4). Such films have better viscoelastic properties (dilatational modulus) and can adapt to physical perturbations without rupture. This is illustrated by β-lactoglobulin which forms strong viscous films while casein films show limited viscosity due to weaker protein:protein interactions and lack of bulky structure (20-23).

Egg white may be superior to other proteins because of its heterogeneous composition and range of physical properties with each protein or protein:protein complex performing a distinct but complementary function. Thus, the globulins are surface active and readily envelope and incorporate air bubbles when whipped; the ovomucin:lysozyme complex imparts film strength (dynamic viscoelasticity) and with ovomucoid imparts thermal stability and water retention while the ovalbumin/conalbumin provide heat-setting structure (34). The hydrophilic glycoprotein moieties of ovomucin and ovomucoids may improve orientation in the film aqueous layer and enhance foam stability by increasing lamellar viscosity and imparting steric effects (prevention of droplet contact).

Several intrinsic and extrinsic factors affect the foaming properties of proteins: e.g., protein structure and conformational stability, secondary interactions, pH, salts, oxidizing/reducing agents, properties of the protein mixture, protein:protein interactions in the film, temperature and degree of modification (1-5,12,16,17,30,34). The formation of a good foam requires the rapid formation of cohesive viscoelastic film with high dilatational and shear moduli, i.e., a thick condensed film containing globular proteins with residual tertiary structure and extensive interactions to minimize leakage of air and coalescence of adjacent bubbles.

Foams are composed predominantly of air bubbles enveloped by a thin continuous film with each bubble separated by lamella, i.e., a thin layer of water held within the capillaries between adjacent bubbles.

When foams are initially formed, the air bubbles are spherical (high internal pressure) and lamella are thick containing large amounts of water (Kugelschaum foam). With time the liquid drains from the foams, the lamella thins, air bubbles pack closer and assume polyhedral shapes (30,35). Drainage of fluid from the lamella is the main destabilizing force as it allows the bubbles to become closer where, if the film is permeable, disproportionation occurs and large bubbles grow at the expense of small ones (Ostwald ripening) (35). To minimize this, retention of moisture in the lamella and strong impermeable films are required.

The stability of foams reflects a balance between forces within the film (which impart viscoelasticity) and the various forces (e.g., electrostatic and steric repulsion between adjacent bubbles, osmotic and volume restriction effects, viscous phenomena in the lamellar phase and capillary action), all of which tend to stabilize foams. Drainage, gravity effects, hydrostatic phenomena, and Van der Waals and coulombic attractive forces represent destabilizing forces. The disjoining pressure tends to be positive when it retards drainage (1,4,16,35). The negative pressure at the plateau borders may help retain moisture in the foam as drainage progresses though initially it may pull water from the film. The rate of drainage is retarded by bulky, polar substituent groups (which possess good water-binding capacities) penetrating the aqueous lamellar phase and by adding sugars and hydrocolloids which avidly bind water and decrease flow (4).

The capacity of the film to resist thinning is a function of its surface tension and rheological properties. The Gibbs-Marangoni effects, reflecting surface elasticity and surfactant mobility, are important criteria in adjusting to localized stresses and preventing rupture (1,4,12,14,35). The rheological properties of the interfacial film (i.e., high viscoelastic moduli) are critically important in stabilizing foams while the surface charge and bulky groups which provide electrostatic repulsion and steric hindrance contribute to their stability (16,20-24). The age-strengthening of protein films extends the lifetime of protein foams (30).

Foam stability (i.e., retention of volume and moisture) is a reflection of film integrity, permeability to gas and film mechanical strength. A thicker, denser film retards gas diffusion, slows disproportionation, coalescence, rupture and eventual collapse. More condensed thicker films are formed by certain proteins depending upon pH, (e.g., BSA versus casein). Usually thicker films are formed closer to the isoelectric pH when protein concentration is not limiting (rarely encountered in food systems). The adsorption of denatured proteins may improve stability (12).

Several factors affect foam stability: thickness of film, mechanical strength (extensive interactions), viscoelastic properties of film, steric repulsion between bubbles, osmotic phenomena, disjoining pressure, electrostatic repulsion, van der Waals' attraction, viscosity of aqueous phase, plateau border, capillary effects, permeability of film (disproportionation) and drainage rates. The pH of the aqueous phase affects foaming by determining the magnitude and nature of the net charge on proteins. The number and disposition of the charges which vary with pH and protein affect protein:protein interaction in films and repulsion between foam bubbles. It is not yet known how to balance these desirable but different phenomena in food protein foams.

The foaming properties of protein may be related to surface hydrophobicity. Thus, because hydrophobic regions or segments (that are normally restricted in the interior of globular proteins) gain considerable entropy by unfolding into the apolar air phase in a foam, they favor adsorption and intermolecualr hydrophobic interactions between contiguous polypeptides in the interface and may be important in stabilizing films. Kato et al. (36) observed positive relationships between the surface hydrophobicity of several proteins and foaming power (i.e., overrun), but no correlation with foam stability. Townsend and Nakai (37) reported a relationship between average hydrophobicity but not surface hydrophobicity. Thus, while a strong correlation exists between hydrophobicity and surface activity, there is only a weak correlation between hydrophobicity and foaming. However, great variability can be expected because of the variable disposition of apolar amino acids in different proteins, the proportion and disposition of hydrophilic and polar residues, the differences in molecular flexibility between molecules, and the presence of disulfide bonds (e.g., β-casein versus β-lactoglobulin). Partial unfolding of some proteins improves their foaming properties and limited reduction of β-lactoglobulin (38) improves its foaming properties perhaps by allowing greater hydrophobic interactions.

Both the intrinsic properties of proteins and extrinsic factors (e.g., pH, ionic strength) affect the formation, properties and behavior of protein films in foams. Few pure proteins fulfill all the putative properties and criteria required for the successful formation of films, and usually mixtures of proteins are superior in food foams (3,5,12).

Protein Modification

Because of the importance of molecular size, shape, conformation flexibility, and functional groups (charged, polar, hydrophobic, thiol) in film formation, numerous studies concerning the effects of modification on film and foaming properties of proteins have been conducted. Studies on modification of protein may provide useful information concerning the mechanisms of foaming and an approach for improving foaming properties.

Charge modification.

Manipulation of pH can be exploited to alter charges on proteins and enhance surface active properties. Thus, initial surface pressure development by proteins is markedly pH dependent showing a sharp maximum around pH values close to the respective isoelectric points (1,4,12,20,26,40). The rate of adsorption of protein at the interface is increased near the isoelectric point of a protein when the protein remains soluble. The adsorbed proteins exert less electrostatic repulsion at the interface; the compact protein molecules can pack more easily into the interfacial film, and extensive protein:protein interaction is facilitated. This is reflected in enhanced foaming properties of proteins (3,4,17,18,26,28,30,39).

Because the net charge on protein molecules affects the solubility and extent of protein:protein interactions, alteration of the charge by chemical modification may improve foaming properties. The progressive succinylation of glycinin (41), or β-lactoglobulin (38) increased the unordered structure, electronegativity, specific viscosity, hydration, solubility and altered the foaming properties of the modified proteins. Film strength as reflected in surface yield stress of glycinin was increased by succinylation up to <50% but decreased following extensive succinylation. The surface yield stress also decreased, especially with increasing pH, indicating that the increased net negative charge by causing excessive repulsion reduced the formation of a continuous cohesive viscoelastic film (41). Dynamic film elasticity measured by tensiolaminometry revealed that limited succinylation of glycinin initially enhanced elasticity, but at high levels of succinylation, elasticity decreased (Table 4.4). These data reflect the diminished cohesiveness of the film as net charge repulsion became excessive (41). The foaming behavior and foam stability significantly increased at low levels of succi-

TABLE 4.4
Effect of Succinylation on Surface Active Properties of Succinylated Glycinin

Property	Extent of succinylation, %			
	0	25	50	100
Net charge (pH 7.5)	-250	-290	-330	-400
Specific viscosity × 10^{-2}	0.5	0.8	1.5	8.0
Surface hydrophobicity	200	330	360	290
Surface pressure dyne/cm 5 min	3.0	16.0	14.5	5.5
Surface yield stress (dyne/cm)				
pH 5	—	5.3	5.0	4.5
pH 6	3.5	4.6	4.0	0.2
pH 8	3.2	4.1	3.5	0.2
Film elasticity (dyne/cm)	3.9	4.6	4.4	1.9
Foam stability (half life, min)	5.0	14.5	11.0	8.0

nylation, up to 50%, but at 100% succinylation, foam formation was poor, while foam stability was greater than with native glycinin. The enhanced stability of the bubbles may have been caused by the greater water binding and retention in the lamella and by charge repulsion which impeded coalescence of neighboring bubbles (41). Succinylation of 27% of the lysine groups of β-Lg improved the surface properties and rates of foam development but greatly depressed foaming properties (38).

The amidation or esterification of proteins, e.g., β-Lg, causes an increase in random structure while increasing the isoionic pH from 5.2 up to pH 10 depending upon extent of modification (42). This research showed that populations of proteins with varying charges could be prepared. Subsequently, Poole et al., (40,43,44) exploited this observation to improve the foaming properties of food proteins (Table 4.5). Amidation of the carboxylic acid groups of β-lactoglobulin using the carbodiimide method increased the isoelectric point of the protein and progressively improved its action in complementing and enhancing foaming performance of other proteins. The best results were obtained with amidated β-Lg which had a pI of 9.9 and was as effective as clupeine, at concentrations of 0.3% (40). The carboxylic groups of β-Lg were also modified via carbodiimide with glycine ethylester, glucosamine, ammonium chloride and arginine methyl ester to increase the isoelectric pH. These modifications yielded a heterogeneous population of β-Lg with up to 67% of available carboxylic acid groups being modified with glycine ester and up to 87% with arginine ester. When added to ovalbumin at concentrations of 0.5% in the presence of sucrose (10%) at pH 7.0, the derivatives with pI above 9.0 significantly improved foaming properties (i.e., from 40% expansion in the controls up to 620% for glycine and arginine ester derivatives of β-Lg) (12). These were as effective as clupeine and the foam stability was concomitantly improved (12). Amidation improved foaming but glucosamine impaired foaming properties (12). Significantly these modified basic proteins were very effective in stabilizing foams in the presence of oil though they were not as tolerant as clupeine (5,12,40,44).

Glycosylation of the ϵ amino group of lysine with maltosyl residues using the cyclic carbonate method and of carboxylic groups with various sugars, using the carbodiimide method modified the charges and enhanced the polar properties of β-Lg (45-47). The rates of adsorption and rearrangement of the protein molecules in the interface were estimated from surface pressure data. The area cleared during adsorption tended to increase with pH and with extent and type of modification of β-Lg, reflecting the expanded molecular structure and bulkiness of the added carbohydrate groups. The surface films appeared to be more condensed at lower pHs and more expanded at higher pH values. The carbohydrate moieties enhanced the hydrogen bonding between the proteins and solvent and altered a number of charges on the protein which caused changes in the nature and magnitude of the forces acting between molecules in the interfacial film.

In addition, changes in the hydrodynamic volume of proteins reduced the rate of diffusion of the modified protein, and the loss of conformational energy following modification (i.e., less secondary structure) may have resulted in a decreased gain in free energy of these proteins upon unfolding at the interface. Foams made from glycoslyated β-Lg showed greater stability than those made from native β-Lg perhaps because of greater water retention and steric stabilization of the bubble film (26,46).

Protein:protein interactions.

The prevailing pH markedly affects film formation and protein:protein interaction in films indicating the importance of electrostatic interactions in foaming. Because most food proteins possess a net negative charge at pH 6-7, cohesive film formation may be impaired. In fact, this represents a challenge in devising a good foaming protein, namely, how

TABLE 4.5

Effects of Basic Proteins and Modified β-Lactoglobulin on Foaming Properties of Bovine Serum Albumin

Protein	PI	Relative foaming properties	
		Expansion %	Stability %
BSA		480	20
BSA + Clupeine	12	700	80
BSA + lysozyme	10.7	140	10
BSA + amidated β-Lg	8.1	200	10
BSA + amidated β-Lg	9.5	180	15
BSA + amidated β-Lg	9.9	620	70
Ovalbumin	—	40	50
BSA + β-Lg GE[b]	6.7–9.5	300	80
BSA + β-Lg GE	>9.1	620	95
BSA + β-Lg AE[b]	>9.1	620	95
BSA + β-Lg NH$_2$	8–10	560	80
BSA + Clupeine	12	620	40

[a]Data from Poole and Fry (40).
[b]GE — glycine ester; AE — arginine ester of β-lactoglobulin; 0.5% with 10% sucrosee. 5% BSA plus 0.1% basic protein pH 7.

to balance molecular charges to ensure solubility, molecular flexibility, effective adsorption at the interface against an electrostatic barrier, extensive interaction to form a cohesive film and also to impart some net charge repulsion between the outer surface of adjacent droplets. The design of proteins with an appropriate disposition of charged and hydrophobic groups which orient properly in a film to ensure balanced interactions is a goal that may be achievable via recombinant DNA techniques. However, pending such a development, chemical modification or blending of proteins may provide an interim solution. Thus, Poole et al. (40,44) have successfully used small amounts of basic proteins — lysozyme (pI 10.7), clupeine (pI 12), histones and amidated β-lactoglobulin — to significantly improve the foaming properties of other food proteins around pH 7.0 (5,12,40,44). At appropriate concentrations, depending upon charge difference, these basic proteins can interact with ovalbumin, serum albumin or β-lactoglobulin via electroattraction and form complexes that result in the formation of more stable films as reflected in enhanced foam formation and stability (Table 4.5). In addition, these mixtures were effective in improving the tolerance to lipids of these

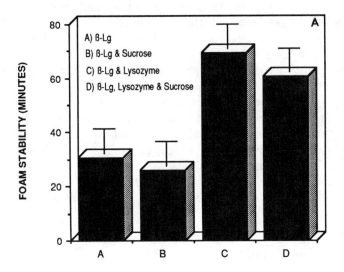

A

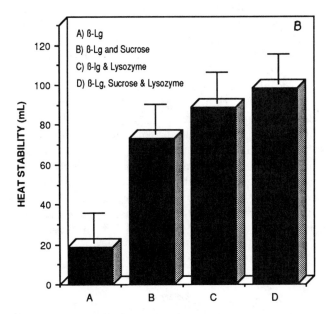

B

Fig. 4.1. The effects of lysozyme (0.5%) and sucrose (10%) on foam stability (A) and heat stability of foams (B) made from β-lactoglobulin (β-Lg).

foams (12). Furthermore, the addition of clupeine overcomes the suppressive effects of sucrose on the foaming of egg albumen (43,44).

As the pH was decreased from pH 7 to pH 5, the effectiveness of clupeine decreased reflecting the low net negative charge on the food proteins and hence decreased protein:protein interaction. This reflects the importance of net charge difference for optimum effect; thus, clupeine (pI 12) was better than lysozyme with a pI of 10.7. However, there is some evidence that the molecular size of the basic protein is also important because proteolysis eliminated the effect of clupeine (40). The addition of lysozyme (0.5%) markedly reduced the drainage rate (i.e., increased the stability of foams formed by whipping β-lactoglobulin) (Fig. 5.1a). Most significantly these foams were remarkably stable to heating up to 90°C (Fig. 5.1b) (Phillips and Kinsella, unpublished data).

Disulfide bond reduction.

Disulfide bonds stabilize the tertiary structure of proteins and may impede unfolding rearrangement and the interaction of proteins at the interface by conferring certain structural constraints (e.g., lysozyme, glycinin). The reduction of the disulfide linkages in proteins can increase molecular flexibility and enhance surface active properties. Thus, Song and Damodaran (13) reported that partial reduction of BSA improved its interfacial properties with best results being obtained for molecular species with equal proportions of random coil and α-helix structures. Similar observations were made by Kim and Kinsella (18) who showed that reduction of approximately 50% of the disulfide bonds of glycinin with dithiothreitol and blocking with iodoacetamide significantly improved (threefold) interfacial properties while further reduction caused no further improvement. Incidentally, reduction caused a fourfold increase in surface hydrophobicity of the glycinin reflecting greater exposure of tryosine and tryptophan. Partial reduction was associated with slight improvements in film viscoelastic properties and improved foaming properties especially at lower pH (i.e., pH 6 versus pH 7.0) (18) (Table 4.6).

Proteolysis.

Proteolysis improves solubility, provides small molecules that rapidly adsorb at the interface and facilitates rapid foaming. However, such foams tend to be unstable, very sensitive to antifoaming agents (lipids)

TABLE 4.6
Effects of Reduction of Disulfide Bonds on Surfactant Properties of Glycinin

Properties	Glycinin		
	Native 20SS bonds	Reduced 13SS bonds	Fully reduced 0SS bonds
Relative hydrophobicity	220	900	870
Specific viscosity × 10^{-3}	4.2	8.0	7.3
Relative fluorescence at 336 nm	46	52	57
Surface pressure (dyne/cm, 5min)			
pH 6	2	16	133
pH 7	5	14	15
pH 8	4	13	15
Surface film yield stress (dyne/cm)			
pH 6	3.5	4.6	5.0
pH 7	3.7	3.9	4.7
pH 8	3.2	3.5	4.0
Foam stability (half-life, min)			
pH 6	3	16	60
pH 7	6	13	25
pH 8	7	10	12

and lack heat-setting properties. Good foaming preparations should possess a wide range of molecular sizes with the large molecules providing better stability (2,4). Foam stability is generally decreased by proteolysis because of the changes in molecular structure and size during hydrolysis (47,48). Numerous studies have empirically explored various enzymes and proteins (soy, caseins, whey, gluten) with variable results (40,47-52). Limited enzyme hydrolysis improves the foaming properties of most proteins, especially large denatured proteins that have limited solubility. Soy protein isolate treated with neutral protease had slightly increased water absorption and foaming properties, but foam and emulsion stabilities were decreased (52).

The specific foam volume of whey protein concentrate treated with pepsin, pronase or prolase was increased by very limited hydrolysis, but decreased by more extensive hydrolysis. Foam stability was greatly decreased by limited hydrolysis (48,49). Horiuchi et al. (51) concluded that protein hydrolysates with large surface hydrophobic regions adsorb more readily at interfaces, and rates of surface desorption are lower. However, secondary structures, as measured by optical rotatory dispersion and infrared spectra, and the content of the total hydrophobic amino acids in protein hydrolysates showed positive correlations with foam stability. Systematic studies which elucidate the molecular characteristics of the products and include determination of foam formation and stabilization by standardized methods are needed.

Sucrose.

Sucrose may improve or impair foaming though the mechanisms are not clear. Sucrose enhances adsorption of certain proteins at an air/water interface (53) and it may minimize surface denaturation thereby enhancing film strength and viscoelasticity. Sucrose enhances foaming of certain proteins (e.g., ovalbumin near its isoelectric point) and of mixtures of proteins with clupeine (5,40). Thus, ovalbumin, a poor foaming protein at all pH values, had markedly improved foaming properties at pH 5.25 in the presence of sucrose (40% by weight) where stable foams were formed. Sucrose greatly enhances the heat stability of β-Lg foams (38) (Fig. 4.1b).

Conclusion

Proteins because of their dynamic structures and amphiphilic nature possess varying interfacial properties. Though the ideal molecular structure for a foaming protein (or perhaps more accurately a mixture of proteins) is not clearly known, the foregoing overview provides some clues of molecular structure to function relationships. More precise data are needed to aid the design of the ideal foaming protein which could then be produced by recombinant DNA techniques.

In this context, the food scientist needs to understand not only the chemical but also the nature of the physical interactions occurring in foods, the kinetics and time dependence of such reactions. This is particularly true in describing colloidal behavior of food macromolecules, proteins, starches etc. and controlling events involved in formation of foams, gels and emulsions. While much descriptive information concerning the behavior of food mixtures has been gained from empirical observations, the more recent use of well-controlled model systems is beginning to provide more fundamental information concerning the functional behavior of food macromolecules. This approach can be further developed to study more complicated mixtures. To facilitate and validate this approach it is desirable for scientists to agree on appropriate standardized methodology (7). The food industry should become more active in the support of such efforts which could greatly improve the value of published data.

There is a need for much additional and systematic research on the film and foaming properties of proteins. This research should include: elucidation of molecular basis of surface activity; modification of protein structure and/or addition of functional groups to enhance interfacial activity; studies of the structure of proteins at interfaces and in films; elucidation of protein:protein interactions in films and quantification of the forces involved; differentiation of steric, electrostatic and hydrophobic phenomena in foam stabilization; development of methodologies for studying interfacial behavior of food proteins and relating these to foaming properties and ultimately the design of surface active proteins.

Acknowledgment

Support of National Dairy Board and Wisconsin Milk Marketing Board is gratefully acknowledged.

References

1. Graham, D.E., and M.C. Phillips, in *Foams*, edited by R.J. Akers, Academic Press, New York, NY, 1976, p. 237.
2. Kinsella, J.E., *CRC Crit. Rev. Food Sci. Nutr. 7*:219 (1976).
3. Kinsella, J., *Food Chem 7*:273 (1981).
4. Halling, P.J., *CRC Crit. Rev. Food Sci. Nutr. 15*:155 (1981).
5. Mitchell, J.R., in *Developments in Food Proteins*, Vol. 4, edited by B. Hudson, Applied Science/Elsevier, London, England, 1986, p. 291.
6. MacRitchie, F., *Adv. Protein Chem. 32*:283 (1978).
7. Kinsella, J.E., *CRC Crit. Rev. Food Sci. Nutr. 21*(3):197 (1984).
8. Kinsella, J.E., and D. Whitehead, *Proc. Int. Dairy Congr.*, 1987.
9. Kinsella, J.E., in *Food Proteins*, edited by P. Fox, and S. Codon, Appl. Sci. Press, London, England, 1981.
10. Tanford, C., *Adv. Protein Chem. 23*:121 (1968).
11. Kinsella, J.E., S. Damodaran, and B. German, in *New Protein Foods*, Vol. 5, edited by A. Altschul and H. Wilkie, Academic Press, New York, NY, 1985, p. 116.
12. Dickenson, E., *Food Hydrocolloids 1*:3 (1986).
13. Song, K.B., and S. Damodaran, *J. Agric. Food Chem. 35*:236 (1987).
14. Joly, M., in *Surface and Colloid Science*, edited by E. Matijeic, Wiley-Interscience, New York, NY, 1972, pp. 1–77.

15. Padday, J.F., in *Surface and Colloid Science*, Vol. 1, edited by E. Matijeic, Wiley-Interscience, New York, NY, 1969, pp. 39-99.
16. Phillips, M.C., *Food Technol. 35*:50 (1981).
17. Tornberg, E., *ACS Symp. Ser. 92*:105 (1979).
18. Kim, S.H., and J.E. Kinsella, *J. Food Sci. 52*:128 (1987).
19. Boyd, J.V., J.R. Mitchell, L. Irons, P.R. Mussellwhite, and P. Sherman, *J. Colloid Interface Sci. 34*:76 (1973).
20. Graham, D.E., and M.C. Phillips, *Ibid. 70*:403 (1979).
21. Graham, D.E., and M.C. Phillips, *Ibid. 70*:427 (1979).
22. Phillips, M.C., M.T. Evans, D.E. Graham, and D. Oldani, *Colloid Polym. Sci. 253*:424 (1975).
23. Benjamins, J., J.A. De Feijter, M.T.A. Evans, D.E. Graham, and M.C. Phillips, *Discuss. Faraday Soc. 59*:218 (1975).
24. Graham, D.E., and M.C. Phillips, *J. Colloid Interface Sci. 76*:240 (1980).
25. Briggs, M.S., D.G. Cornell, R. Dluhy, and L. Gierasch, *Science 233*:206 (1986).
26. Waniska, R.D., Ph.D. Thesis, Cornell University, Ithaca, NY, 1982.
27. Kim, S.H., and J.E. Kinsella, *J. Food Sci. 50*:1526 (1985).
28. Waniska, R., and J.E. Kinsella, *J. Agric. Food Chem. 33*:1143 (1985).
29. Barbeau, W.E., and J.E. Kinsella, *Colloids Surfaces 17*:167 (1986).
30. Cumper, C.W.N., *Trans. Faraday Soc. 49*:1360 (1953).
31. Castle, J., E. Dickenson, B.S. Murray, and G. Stainsby, in *Proteins at Interfaces*, edited by J. Brash and T. Horbette, ACS Symp. Series 343, Am. Chem. Soc., Washington, DC, 1987, p. 118.
32. German, J.B., T.E. O'Neill, and J.E. Kinsella, *J. Am. Oil Chem. Soc. 62*:1358 (1985).
33. Shimizu, M., M. Saito, and K. Yamauchi, *Agric. Biol. Chem. 49*:189 (1985).
34. Johnson, T., and M.E. Zabik, *J. Food Sci. 46*:1237 (1981).
35. Adamson, A.W., *Physical Chemistry of Surfaces*, 3rd edn., Wiley, New York, NY, 1975.
36. Kato, A., K. Komatsu, K. Fujimoto, and K. Kobajashi, *J. Agric. Food Chem. 33*:931 (1985).
37. Townsend, A.M., and S. Nakai, *J. Food Sci. 48*:588 (1983).
38. Phillips, L., M.S. Thesis, Cornell University, Ithaca, NY, 1988.
39. Kinsella, J.E., and D. Whitehead, in *Proteins at Interfaces*, edited by J. Brash and T. Horbett, ACS Symp. Series 343, Am. Chem. Soc., Washington, DC, 1987, p. 629.
40. Poole, S., and J.C. Fry, in *Developments in Food Proteins*, Vol. 5, edited by B. Hudson, Applied Science/Elsevier, London, England, 1987, p. 257.
41. Kim, S.H., and J.E. Kinsella, *J. Food Sci. 52*:1341 (1987).
42. Mattarella, N.L., L. Creamer, and T. Richardson, *J. Agric. Food Chem. 31*:968 (1983).
43. Poole, S., S.I. West, and C.L. Walters, *J. Sci. Food Agric. 35*:701 (1984).
44. Poole, S., S.I. West, and C.T. Fry, *Food Hydrocolloids 1*:227 (1987).

45. Waniska, R.D., and J.E. Kinsella, *Int. J. Prot. Peptide Res.* *23*:467 (1984).
46. Waniska, R.D., and J.E. Kinsella, *J. Agric. Food Chem.* *32*:1042 (1984).
47. Gunther, R.C., *J. Am. Oil Chem. Soc.* *56*:345 (1979).
48. Kuehler, C.A., and C.M. Stine, *J. Food Sci.* *39*:379 (1974).
49. Adler-Nissen, J., *Proc. Biochem.* *12*:18 (1977).
50. Adler-Nissen, J., *J. Agric. Food Chem.* *24*:1233 (1976).
51. Horiuchi, T., D. Fukushima, H. Sugimoto, and T. Hattori, *Food Chem.* *3*:35 (1978).
52. Puski, G., *Cereal Chem.* *52*:655 (1975).
53. MacRitchie, F., and A.E. Alexander *J. Colloid Sci.* *16*:61 (1961).

Chapter Five
Film Properties of Modified Proteins

Srinivasan Damodaran

Department of Food Science
University of Wisconsin-Madison
Madison, WI 53706

Adsorption of proteins at fluid interfaces and their behavior in the adsorbed state play an important role in many biological and technological processes. In biological systems, where many cellular processes occur mainly at the liquid-liquid interfaces, adsorption and subsequent conformational changes in enzymes and proteins play a crucial role in the regulation of metabolic pathways. In technological processes, protein adsorption at interfaces has both desirable and undesirable effects. For example, in food and pharmaceutical industries, proteins are often used as desirable surface active agents in foam and emulsion type products. On the other hand, adsorption of proteins onto surfaces of heat exchangers (fouling) progressively decreases the efficiency of heat transfer in food processing operations. Novel and highly surface active proteins may also find their use as emulsifiers in tertiary oil extraction in the near future.

The important initial step in the formation of protein based foams and emulsions is the adsorption and spreading of the protein at the surface or interface (1). Although all proteins are amphiphilic, they differ very much in their surface or interfacial adsorptivity. These differences are apparently due to differences in the ionic, hydrophobic and conformational characteristics of proteins. To better understand the surface active properties of proteins, it is fundamentally important to define how each of these molecular parameters, individually and in combination, effect the dynamics of the adsorption and spreading process. These relationships are not well understood.

The dynamics of protein adsorption at fluid interfaces is very different from that of simple, low molecular weight amphiphiles. While in the case of small amphiphiles, such as aliphatic alcohols and alkyl sulfates, the entire molecule adsorbs and orients itself between the polar and nonpolar phases, the adsorption of protein macromolecules proceeds through attachment of several polypeptide segments; in most cases, a greater portion of the molecule remains suspended into the aqueous phase in the form of 'loops' and 'tails'. It can be argued that the retention of the molecule in the adsorbed state during the initial stages of adsorption against thermal motion would depend upon the number of segments involved in the attachment and the sum total of the free energy of adsorption of all the segments. If this energy is far greater than the thermal energy, kT (where k is the Boltzman constant and T is the temperature), the molecule would adsorb irreversibly to the surface/interface.

The number of segments that are involved in the attachment of the protein to the surface and the free energy of adsorption of each segment depend on the various molecular factors such as surface hydrophobicity of the molecule, the amino acid composition of the segments and the conformational state of the macromolecule. Once adsorbed, the subsequent behavior of the protein molecule in the adsorbed state depends on its conformational stability under the thermodynamic conditions that prevail at the surface/interface.

In phenomenalogical terms, proteins that are highly flexible will readily unfold, reorient and occupy a greater surface area at the interface and thus reduce the interfacial free energy. On the other hand, in the case of proteins containing relatively rigid, ordered structure, if the excess free energy surface/interface is smaller than the activation free energy for denaturation of the protein, the denaturation and reorientation of the protein, the denaturation and reorientation of the protein at the interface would occur at a slower rate. However, if the excess free energy of the surface/interface is greater than the activation free energy of denaturation of the protein, the denaturation and reorientation of the protein would occur at a faster rate. The ultimate ability of the protein to rapidly decrease the interfacial tension would be related to the rate of denaturation and reorientation as well as the area occupied by the molecule at the interface. These, in turn, are fundamentally related to the structural properties of the protein. However, the fundamental relationship between the structure and function of protein macromolecules with respect to their adsorption is not well understood.

Literature Review

Because of the fundamental importance of the behavior of proteins at fluid interfaces in chemistry, engineering, biology and medicine, there has been considerable interest both in the theoretical and experimental aspects of research in this area. However, most of the earlier investigations on the surface behavior of proteins were limited to studies on the effect of spread protein monolayers on the surface tension of water (2-4). Since the spread monolayer studies do not represent the dynamic process of transport and adsorption of proteins from the subphase, the knowledge obtained from those studies has limited significance in terms of understanding the mechanism(s) of surface/interfacial adsorption of proteins in biological as well as in food systems. Until recently there have been only a few studies on the kinetics of adsorption of proteins from the aqueous phase to the air-water or oil-water interfaces (5-15). However, the emphasis of most of those studies were on the role of solution conditions on the kinetics of adsorption; neither the influence of protein conformation on adsorption kinetics nor the state of protein conformation in the adsorbed phase has been elucidated

In order to understand the influence of protein conformation on adsorption at liquid interfaces, Graham and Phillips (1, 16, 17) recently studied the kinetics of adsorption of four structurally different proteins — lysozyme, bovine serum albumin, β-casein and κ-casein—at the air-water interface. Among these proteins, β-casein adsorbed rapidly to the air-water interface and formed a dilute monolayer, whereas both lysozyme and serum albumin exhibited lower rates of adsorption, but formed concentrated films. The differences in the adsorption behaviors of these proteins were attributed to their unique structural properties. For example, β-casein has a highly flexible random coil structure and contains no disulfide bonds, whereas lysozyme and serum albumin are highly structured and contain intramolecular disulfide linkages. These differences in the molecular conformations of these proteins are believed to directly affect their adsorption kinetics and spreading at interfaces. However, it should be emphasized that although these excellent studies did demonstrate conclusively that the adsorption characteristics of proteins depend, in general terms, on the molecular nature of the proteins, the fundamental understanding of the relationship between the protein conformation and its adsorptivity at interfaces is still very elusive.

Several investigators have studied the influence of several factors, such as pH, temperature, ionic strength, hydrophobicity/hydrophilicity ratio etc., on the film forming properties of several proteins (18-26). The film forming tendency and the emulsifying activity of proteins have been shown to be related to the polar-apolar amino acid residue ratio and the surface hydrophobicity of proteins (19,23). The proteins having low polar/apolar ratios unfold completely at liquid interfaces and exhibit higher emulsifying activity. However, hydrophobicity alone cannot ensure higher surface activity, because highly hydrophobic proteins would tend to coagulate at an interface. Thus, it is conceivable that an optimum ratio of polar-apolar ratio and optimum conformational characteristics are required to exert maximum surface adsorptivity and activity. However, these relationships are not well understood.

Several chemical modification methods have been studied to improve the film forming properties of food proteins (27-30). Chemical modifications involving succinylation and glycosylation have been shown to decrease the rate of adsorption and the film forming properties of proteins at fluid interfaces (29,30). However, the emulsifying capacity and activity of these modified proteins were substantially improved. These observations indicate that the greater hydration potential of these modified hydrophilic proteins decreases the thermodynamic driving force for adsorption at the relatively nonpolar interface. However, under the conditions used for emulsification, the mechanical energy input overcomes the energy barrier for adsorption at the interface; once adsorbed, the reorientation of the molecule, with the hydrophobic groups toward the oil phase and the hydrophilic groups toward the aqueous phase, stabilizes the interface. Furthermore, the greater electrostatic repulsion between the emulsion particles and probably the steric hindrance from the hydrated modifying groups stabilize the emulsion particles against coalescence. Although chemical modification of food proteins significantly improves several functional properties, such as solubility, foaming, emulsification, viscosity etc., the nutritional quality and safety of such modified proteins are questionable. Hence, alternate approaches should be developed to understand the molecular bases for adsorption of proteins at interfaces and to improve the foaming and emulsifying properties of novel food proteins.

One can study the influence of solution conformation of proteins on the kinetics and thermodynamics of adsorption at liquid interfaces using two different approaches: (a) Study the adsorption behavior of various proteins and relate differences in their behavior to their conformational differences, (b) alter the structure of a single protein to various extents by physical means and study the influence of each conformational state on the adsorption and behavior at interfaces. Of these two approaches, the second approach is novel, because this involves modification of the protein conformation only; since neither the amino acid composition nor the sequence is altered, the observed differences in the surface/interfacial adsorption behavior can solely be attributed to conformational differences.

In order to understand the role of protein conformation on its adsorption at interfaces, we prepared seven structural intermediates of bovine serum albumin and studied the kinetics of adsorption of these intermediates at the air-water interface.

Experimental

The basic scheme for the preparation of the structural intermediates of bovine serum albumin (BSA) is shown in Figure 5.1. In essence, BSA was first reduced and denatured in urea solution containing an excess

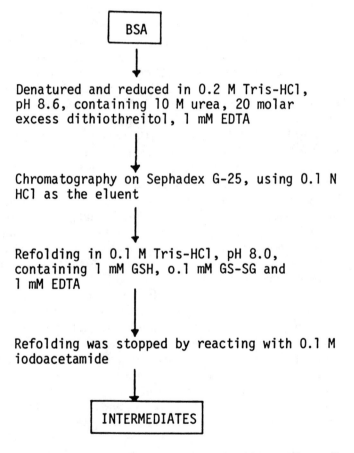

Fig. 5.1. General scheme for the preparation of strucutral intermediates of bovine serum albumin.

amount of dithiothreitol. The reduced and denatured BSA was allowed to refold in a redox regeneration buffer made up of 0.1 M Tris-HCl containing 1.0 mM reduced glutathione and 0.1 mM oxidized gluthathione. The final protein concentration in the regeneration solution was about 1.0 µM. Under these conditions, the denatured and reduced BSA has been shown to undergo refolding the reformation of disulfide bonds (31-33). In order to trap the structural intermediates of BSA during the refolding process, the refolding was stopped at the predetermined intervals of time by blocking the free sulfhydryl groups with 0.1 M iodoacetamide. The blocking of free sulfhydryl groups at various time intervals during the refolding process has been shown to trap the protein in the intermediate structural state (34). Seven structural intermediates of BSA were prepared by blocking the refolding process at 0, 0.5, 1, 2, 3, 6 and 24 hr time intervals. The samples were dialyzed exhaustively against water (pH 7.0), lyophilized, and chromatographed on a Sephadex G-25 and lyophilized again.

Adsorption of these BSA structural intermediates at the air-water interface was studied by the Wilhelmy plate method (15,35), using a Cahn electrobalance equipped with a dynamic surface tension accessory, as previously reported (36). In all these experiments, 20 mM sodium phosphate buffer, pH 7.0, was used as the aqueous phase.

Structural features of the BSA intermediates.

The structural features of the BSA intermediates were analyzed in terms of the number of disulfide bonds, secondary structure content using circular dichroism spectroscopy and their ANS (1-anilinonaphthalene-8-sulfonic acid) binding capacity. These molecular properties for the seven BSA intermediates are listed in Table 5.1. Since the CD spectra of all the intermediates were typically the spectrum for α-helical protein, the percent recovery of ellipticity at 221 nm (Table 5.1) could be used as a measure of the α-helical content of these intermediates. The percentage regained of native structure was calculated from the ratio F_t/F_n, where F_t is the binding capacity of ANS to the intermediate and F_n is the binding capacity of ANS to the native BSA under identical conditions. Since the binding capacity of ANS to the intermediates is related to the extent of regeneration of hydrophobic regions (and/or the three dimensional topography of the protein), the above ratio was taken as a measure of the regeneration of the tertiary structure of the protein.

TABLE 5.1
Structural Properties of Bovine Serum Albumin Intermediates

Intermediate #	Refolding time (hr)	$(\theta)_{221}$ (deg. cm/decimol)	% $(\theta)_{221}$ recovery	% Regain of native structure	Number of S-S bonds regained
1	0	- 7,693	36.0	6.5	0
2	0.5	-10,496	49.5	9.3	6.5
3	1.0	- 6,100	28.7	13.1	8.5
4	2.0	-15,300	72.0	20.7	12.7
5	3.0	-19,160	90.0	24.4	13.3
6	6.0	-20,090	94.5	25.0	13.7
7	24.0	-20,410	96.2	30.0	15.2
Native	—	-21,220	100.0	100.0	17.0

[a]Data from Damodaran and Song (36).

Adsorption.

The rate of change of surface pressure at the air-water interface of dilute solutions of BSA intermediates is shown in Figure 5.2. The subphase protein concentrations for all the intermediates, except the intermediates 1 and 2, were about 0.001%. The subphase concentrations for the intermediates 1 and 2 were 0.0003% and 0.00055%, respectively. The rate of increase of surface pressure (π), as well as the steady state values of the intermediates, were significantly greater than those of the native BSA. However, among the intermediates the steady state values and the rate of change apparently decreased with the extent of the folded state of the intermediate. In other words, the higher the compact folded state of the intermediate, the lower its rate of adsorption at comparable subphase concentrations. It should be pointed out that if the experiments with the intermediates 1 and 2 were carried out at 0.001% subphase concentration, the rate and extent of adsorption of these intermediates would have been higher than those of the other intermediates.

The data in Figure 5.2 clearly indicate that under identical experimental conditions, the differences in the conformational states of BSA intermediates profoundly affect the kinetics of their adsorption at the air-water interface. To obtain more insight into the role of protein conformation in the mechanism of its adsorption and the subsequent molecular behavior at the interface, the adsorption data were analyzed as follows.

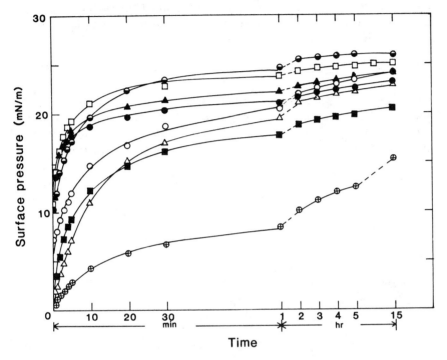

Fig. 5.2. Rate of increase of surface pressure during adsorption of albumin intermediates at the air-water interface. The substrate was 20 mM sodium phosphate buffer, pH 7.0. The subphase concentrations of various intermediates were: ○——○, intermediate 1, 0.308×10^{-3}%; △——△, intermediate 2, 0.559×10^{-3}%; ◕——◕, intermediate 3, 0.95×10^{-3}%; □——□, intermediate 4, 0.99×10^{-3}%; ●——●, intermediate 5, 0.996×10^{-3}%; ▲——▲, intermediate 6, 0.922×10^{-3}%; ■——■, intermediate 7, 0.973×10^{-3}%; ⊕——⊕, native BSA, 0.834×10^{-3}% (36).

The rate of arrival of protein molecules at the interface is considered to be diffusion controlled (1,3,5).

$$d\Gamma/dt = C_o (D_s/3.14\, t)^{1/2} \tag{a}$$

where γ is the surface concentration, C_o is the bulk concentration and D is the diffusion coefficient. However, in many adsorption studies it has been shown that the calculated diffusion coefficient, D_s, was always lower than the conventional diffusion coefficient, D_o, measured in solution (5-7). To account for this deviation, it was suggested that there might be an energy barrier, $\pi \Delta A$, at the interface, which might be related to the energy required to clear an area ΔA against the surface pressure in order for the molecule to adsorb (5,37). In other words, the rate of arrival of the protein at the interface is:

$$d\Gamma/dt = K\, C_o\, \exp.\, (-\pi \Delta A/kT) \qquad (b)$$

where $K=(D_o/3.14\, t)^{1/2}$, k is the Boltzman constant and T is the temperature. The basic assumption here is that for a protein molecule to adsorb at the interface, it should possess an energy equal to or greater than A. Assuming that $d\Gamma/dt = (d\pi/dt)(d\Gamma/d\pi)$, equation (b) can be expressed as

$$\ln\, (d\pi/dt) = \ln\, \{KC_o/d\Gamma/d\pi)\} - \pi \Delta A/kT \qquad (c)$$

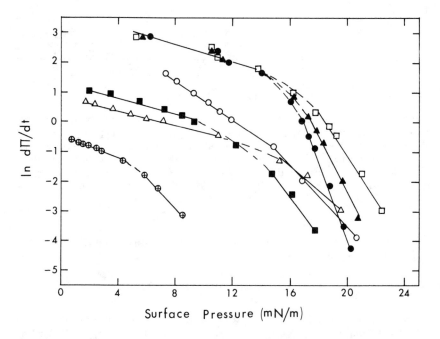

Fig. 5.3. The ln $d\pi/dt$ versus π plots for albumin intermediates. The $d\pi/dt$ values at each π value were obtained from the data in Figure 6.2 using a nonlinear curve fitting procedure. ○——○, intermediate 1; △——△, intermediate 2; □——□, intermediate 4; ●——●, intermediate 5; ▲——▲, intermediate 6; ■——■, intermediate 7; ⊕——⊕, native BSA (36).

According to equation (c), a plot of $\ln(d\pi/dt)$ vs π should be a straight line, the slope of which is $\Delta A/kT$. Analysis of the adsorption data of the BSA intermediates according to equation (c) is shown in Figure 5.3. The plots were non-linear; however, these curvi-linear plots could be divided into two linear regions. This apparently suggests that at least two molecular processes are involved during the formation of protein films at the interface. In phenomenalogical terms, these two molecular processes could be attributed to initial penetration and anchoring of the molecule at the interface (corresponding to the first slope) and the subsequent reorientation of the molecule at the interface (second slope).

The areas occupied by the BSA intermediates, calculated from the slopes of the first and second linear regions, are given in Table 5.2. It is interesting to note that the ΔA_1, i.e., the area cleared by the BSA intermediates to anchor themselves at the interface, is almost the same (i.e., about 60 $Å^2$). This strongly suggests that the initial penetration and anchoring of the protein molecule at the air-water interface is independent of the details of its conformation in the solution phase. If the surface area occupied by an amino acid residue is about 15 $Å^2$ (1,38), it follows that a peptide segment of about four residues only is needed for a protein to initially anchor itself at the interface.

In contrast to the behavior of ΔA_1, the ΔA_2, i.e., the area cleared during the rearrangement and reorientation of the adsorbed molecule, was very much affected by the state of the protein conformation in the solution phase. The relationship between ΔA_2 and the percentage recov-

TABLE 5.2
Surface Area Cleared per Protein Molecule during Initial Adsorption (ΔA_1) and Subsequent Rearrangement (ΔA_2) at the Air-Water Interface

Intermediate #	ΔA_1 $(Å)^2$	ΔA_2 $(Å)^2$
1	135.4	212.7
2	52.6	161.0
4	48.9	297.3
5	64.7	534.0
6	56.8	401.6
7	60.5	259.2
Native	77.5	204.5

[a]Data from Damodaran and Song (36).

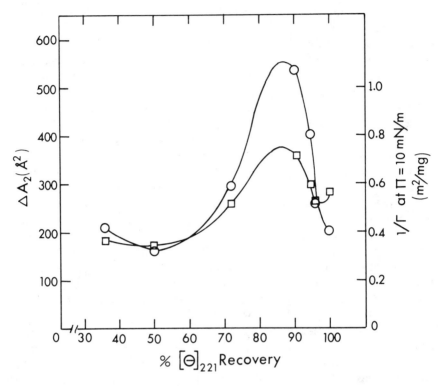

Fig. 5.4. Relationship between the percentage recovery of mean residue ellipticity at 221 nm and ΔA_2 (O—O), and $1/\Gamma$(□—□). The $1/\Gamma$ at 10 mN/m values were from the studies on spread monolayers of BSA intermediates.

ery of ellipiticity at 221 nm of the intermediates is shown in Figure 5.4. It should be noted that neither the completely unfolded (intermediate 1) nor the folded native BSA occupied a larger surface area at the interface during the rearrangement phase. In other words, the data apparently suggests that in order for a protein to occupy greater surface area at the interface, it should possess an optimum degree of folded state. For BSA this optimum structure seems to be the conformational state of the intermediate 5. However, for other proteins, depending upon their other physicochemical properties, the optimum conformation that would confer greater surfactivity would vary.

To confirm independently, whether the ΔA_2 calculated from the adsorption data indeed represents the area occupied by the molecule after the rearrangement and reorientation at the interface, and whether the ΔA_2 vs $[\theta]_{221}$ profile (Figure 5.4) indeed has any relevance, the properties of spread protein monolayers of the intermediates at the air-water interface were analyzed. The relationship between the reciprocal of surface concentration $(1/\Gamma)$ at a surface pressure of 10 mN/m vs $[\theta]_{221}$ is shown in Figure 5.4. Since $1/\Gamma$ is related to the surface area occupied by the protein molecule, the striking similarities between the two curves in Figure 5.4 indeed indicate that the ΔA_2 values calculated from the adsorption experiments represent the areas occupied by the intermediates during the rearrangement and reorientation phase of the adsorption process.

Effect of Protein Conformation on Diffusion Coefficient.

Since protein adsorption is considered to be diffusion controlled, to understand the role of protein conformation on its diffusion to the interface, diffusion coefficients of the BSA intermediates were calculated from the adsorption data using the equation,

$$\Gamma_t = 2C_o (D_s t/3.14)^{1/2} \tag{d}$$

which is the integrated form of equation (b). Plots of Γ vs square root of t were linear for all the intermediates in the time interval between 1 to 10 min (Fig. 5.5). Using this short time approximation, the diffusion coefficients of the BSA intermediates were calculated from the slopes. The relationship between the percent regain of native structure and the diffusion coefficient of the intermediates is shown in Figure 5.6. The diffusion coefficient, calculated from the adsorption process, increased dramatically with the degree of unfolded state of the intermediate. For example, the difference between the diffusion coefficients of the native BSA and the completely unfolded BSA (intermediate #1) was about two orders of magnitude.

Critical analysis of these results indicate a certain discrepancy in the simple theory of diffusion controlled adsorption of proteins. It should be pointed out that in the case of native BSA, the diffusion coefficient, D_s, calculated from the adsorption experiment is about 0.18×10^{-7} cm^2/sec; this is significantly lower than 5.94×10^{-7} cm^2/sec, which is the diffusion coefficient, D_o, of native BSA in solution (39). Moreover, the calculated diffusion coefficient, D_s, increases systematically with progressive unfolding of the protein molecule. This is contrary to what one would expect on the basis of the classical diffusion theory, $D = kT/f$, where f is the frictional coefficient. Since the frictional coefficient of a protein

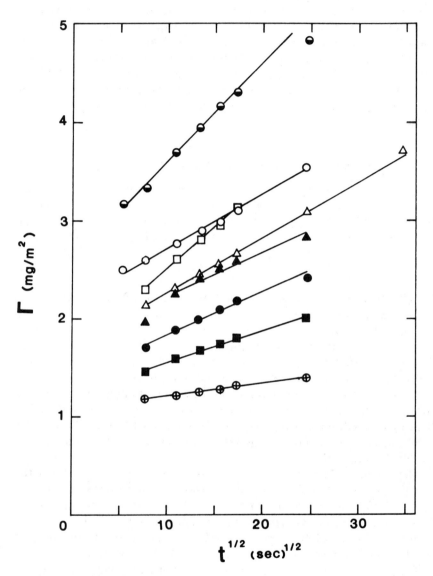

Fig. 5.5. Plot of Γ versus $t^{1/2}$ for BSA intermediates. The bulk phase concentrations of BSA intermediates are given in the legend for Figure 6.2: ○——○, intermediate 1; △——△, intermediate 2; ◓——◓, intermediate 3; □——□, intermediate 4; ●——●, intermediate 5; ▲——▲, intermediate 6; ■——■, intermediate 7; ⊕——⊕, native BSA.

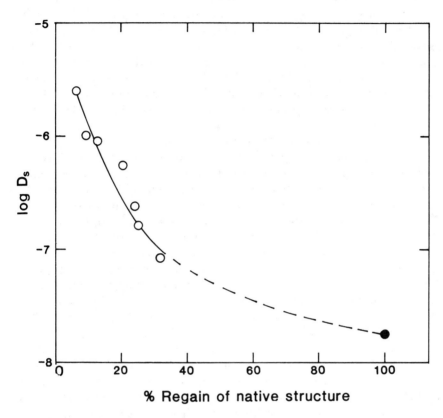

Fig. 5.6. Relationship between logarithm of diffusion coefficient and percentage regain of native structure (measured as percentage regain of ANS binding capacity) of BSA intermediates (36).

would be greater for the unfolded protein than the native state, one should expect a decrease in diffusion coefficient with progressive unfolding of the protein. The opposite trend observed here might strongly suggest that the simple diffusion theory cannot adequately describe the process of adsorption of proteins at liquid interfaces.

To explain the deviation of D_s, (calculated from adsorption) from D_o in solution, Ward and Tordai (37) proposed the idea that an energy barrier might exist at the surface for adsorption. According to this proposal, when the D_s is smaller than the D_o, it is not the diffusion but the energy barrier that plays the controlling role in the adsorption process.

However, although the energy barrier theory adequately explains the reasons for cases where $D_s < D_o$, it fails to account for the progressive increase in the D_s with the extent of unfolded state of the BSA intermediate. Furthermore, it should be pointed out that the D_s values for BSA intermediates 1, 2 and 3 were greater than either the D_s or the D_o of the native BSA. This, on the basis of the energy barrier theory, would mean the existence of a negative, instead of positive, energy barrier at the interface for the adsorption of the intermediates 1, 2 and 3. This is not reasonable.

In a purely diffusion controlled process, the rate of adsorption of a solute at the interface depends on the concentration gradient, the thermal energy of the system and the frictional coefficient of the solute (which is related to its hydrodynamic size and shape). It does not take into account specific solute-solvent interactions. In the case of D_o, since the solute-solvent interaction is constant at all points in the solution phase, it does not affect the diffusion coefficient measured in solution. However, in the vicinity of the interface, i.e., at molecular distances from the relatively nonpolar interface, the differences in the potential energy of the molecule between the solution phase and at the interface may act as an additional driving force for directed migration toward the interface. In other words, the apparent dependency of the calculated diffusion coefficient, D_s, on the structural state of BSA might be due to differences in the potential energy of the BSA intermediates in the solution phase, arising from differences in the energetics of solute-solvent interactions. This excess potential energy might be the sum total of the favorable and unfavorable interactions of the solute with the surrounding solvent. The favorable interactions arise from the surface hydrophilic groups and the unfavorable interactions arise from the hydrophobic hydration of the nonpolar patches on the surface of the molecule. The sign and magnitude of this subphase potential energy will depend on the sum of these interactions, i.e., $U_s = U_{h\phi} + U_{hy}$. In the native state, the surface of BSA is predominantly hydrophilic and hence the potential energy of the molecule in the subphase would be considerably low. Under these conditions, because of the marginal differences in the potential energies of the molecule between the interface and the subphase, (i.e., $U_i - U_s$, where U_i is the potential energy of the molecule at the interface), the protein might exhibit less tendency to adsorb to the interface. Furthermore, each collision of the molecule at the interface may not lead to adsorption, because of inappropriate orientation of the nonpolar patches towards the interface (Fig. 5.7). In addition, the existence of an energy barrier equal to $\pi \Delta A$, and the electrical potential

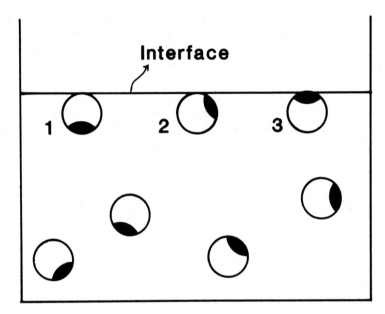

Fig. 5.7. Schematic representation of orientation effects on adsorption at the interface. The dark shades represent the nonpolar patches on the surface of protein molecule. (1) Orientation is unfavorable for adsorption; unsuccessful collision; (2) unfavorable orientation, but the probability of adsorption is better than the first case; (3) favorable orientation for adsorption; successful collision.

energy barrier at the interface might also operate against adsorption. These effects would decrease the rate of adsorption below that of the rate of diffusion. However, when the protein molecule is unfolded, exposure of nonpolar surfaces, which were initially buried in the interior of the protein, to the surrounding solvent water will alter the balance of favorable and unfavorable interactions of the protein with the solvent. The hydrophobic hydration of these nonpolar surfaces will greatly decrease the entropy of the water molecules in these hydration shells (40–42), and thus would increase the hydrophobic potential energy of the protein. The larger the exposure of nonpolar surfaces, the greater would be the hydrophobic potential energy of the unfolded molecule. We suggest that this subphase potential energy of the molecule, depending upon its sign and magnitude, (in addition to the thermal energy of the system) may act as the driving or retarding force for the migration of the protein from the subphase toward the interface. Furthermore, because of the preponderance of the nonpolar patches on the surface of the molecule (in the cases of the unfolded intermediates), the probability of each collision leading to adsorption would also increase. These effects may account for the higher rate of adsorption of the unfolded proteins.

There is evidence in the literature which tentatively supports the above arguments (37,43,44). Studies on adsorption of alkyl sulfates at the air-water interface showed that the calculated diffusion coefficient, D_s, increased with chain length (44). Addison (43) reported that the migrational velocity of n-alkanols, calculated from surface adsorption studies, increased with chain length. This agreed well with the observation that the diffusion coefficient, D_s, calculated from adsorption studies increased with chain length (Fig. 5.8). However, this behavior apparently contradicted the behavior of the diffusion coefficient in solution; the diffusion coefficient, D_o, in solution decreased with the chain length of alkanols. Furthermore, although the D_s values for amyl, hexyl and heptyl alcohols were lower than the corresponding conventional D_o values, the D_s for octanol was apparently higher than the conventional D_o (Fig. 5.8). The cross-over of the lines 1 and 2 in Figure 5.8 apparently indicates that the increase of D_s with chain length is related to the hydrophobic/hydrophilic potential energy balance of the molecule. In higher alcohols, the hydrophobic potential energy more than overcomes the hydrophilic potential energy and thus strongly drives the molecule towards the surface. This is very similar to the behavior of the BSA intermediates 1 to 3, in which the greater exposure of nonpolar surfaces results in rates of adsorption greater than the diffusion rate.

The D_s for octanol and the BSA intermediates 1 to 3 were apparently greater than the corresponding diffusion coefficients in solution. This is interesting because, according to classical chemical kinetics, the upper limit of chemical reactions/interactions is the diffusion rates of the reactants. However, several interacting systems have been shown to apparently exhibit abnormal kinetic behavior. For example, the association of *E. coli* lac repressor with its operator on DNA is considered to be a diffusion controlled process (47). However, it has been shown that the rate constant of association for this bimolecular interaction, in which the DNA is the immobile phase and the repressor is the mobile reactant, is far greater than the rate of diffusion of the repressor. This abnormal rate of association has been attributed to the electrostatic potential energy of interaction between the repressor and the DNA.

Based on the above arguments, we propose that the phenomenalogical rate of adsorption of amphiphiles can be empirically expressed as:

$$d\Gamma/dt = C_o \, (D_o/3.14 \, t)^{1/2} \exp{[-E + U_s)/kT]} \tag{e}$$

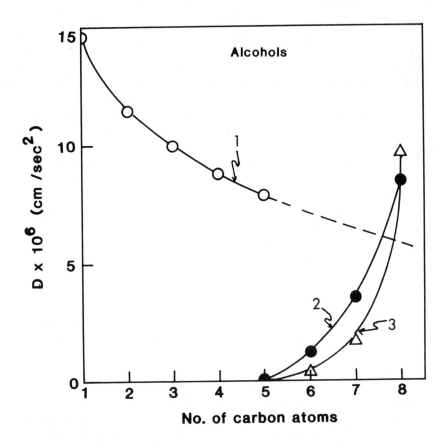

Fig. 5.8. Relationship between diffusion coefficients and chain length for n-alkanols. Curve 1 represents data for diffusion in solution (D_o) and Curve 2 represents diffusion coefficients calculated from surface adsorption (D_s) (From Figure 2 of ref. 37). Curve 3 represents the surface diffusion coefficients for alkanols predicted by equation (f). The following values, from the literature, were assumed for the various parameters in equation (f): The hydrophobic potential of a methylene group was assumed to be 540 cal/mol (45). The U_{hy} was assumed to be -4000 cal/mol, arising from the formation of one hydrogen bond between water and the hydroxyl group of the alkanol. For alkanols, since the electrostatic potential energy barrier is zero, the only energy barrier that might exist at the interface is $\pi \Delta A$. In this calculation, a value of $\pi = 5$ mN/m and $\Delta A = 24$ Å^2 were assumed (46). Temperature was 20°C.

or
$$(D_s)^{1/2} = (D_o)^{1/2} \exp. \, [(-E + U_s)/kT] \tag{f}$$

where E is the total energy barrier at the surface, which includes the pressure barrier and the electrical potential barrier; U_s is the potential energy of the molecule in the subphase arising from solute-solvent interactions, i.e. $U_s = U_{h\phi} + U_{hy}$. The hydrophobic component, $U_{h\phi}$, of U_s will be always positive and the hydrophilic component, U_{hy}, will be always negative. It can readily be seen that at a given surface energy barrier E, the magnitude of D_s directly depends on the sign and magnitude of U_s. If U_s is negative, D_s will be smaller than the value that would be predicted by the energy barrier alone. If U_s is positive and equal to E, then D_s will be equal to D_o. However, if U_s is positive and greater than E, equation (f) would predict that D_s would be greater than D_o. The sign magnitude of U_s is fundamentally related to the energetics of interaction of the nonpolar and polar components of the molecule with the subphase water molecules.

In order to examine the relevance of equations (e) and (f), the experimental data for alkanols (Fig. 5.8) were analyzed according to equation (f). For simple aliphatic molecules, the potential energy of the molecule in the subphase can be simplified as,

$$U_s = nU_{CH_2} + U_{hy} \tag{g}$$

where U_{CH_2} is the potential energy of a methylene group in the subphase and n is the number of methylene groups in the molecule. For a homologous series of alcohols, the U_s will approach positive values for longer chain lengths; hence equation (f) would predict an increase of D_s with chain length, which is in accord with the experimental results (Fig. 5.8). Using appropriate values for U_{CH_2}, U_{hy} and other variables, the D_s versus chain length profile predicted by the equation (f) is shown in Figure 5.8. The close agreement between the predicted and the experimental curves for the D_s versus chain length for alcohols (Fig. 5.8) strongly demonstrates that the subphase potential energy of the alkanol indeed plays an important role as a controlling factor in the rate of adsorption. The higher rate of adsorption (greater than the diffusion) observed for the BSA intermediates might also be related to an increase in the potential energy of the unfolded intermediates, arising from greater exposure of the nonpolar surfaces to the surrounding water molecules in the subphase.

The data presented here strongly suggest that the conformational state of a protein in the solution phase plays a fundamental role in its adsorption and behavior at an interface. However, it should be pointed out that although the rate of adsorption is apparently affected by the conformation, it is the conformation induced alterations in the physicochemical properties (hydrophobicity/hydrophilicity) of the protein surface and not the conformation *per se* that seems to be responsible for the greater surface adsorptivity of the unfolded proteins. Through judicial optimization of the properties of the protein surface via structural modification, it should be possible to improve the surfactivity of a protein.

Conclusion

The mechanism of adsorption and retention of proteins at liquid interfaces is very complex. The complexity of the process is further confounded by the multitude of molecular factors that affect the adsorption process. However, in spite of the seemingly insurmountable obstacles, continued efforts to understand the fundamental role of the solution conformation of a protein on its transport and adsorption at fluid interfaces is essential. In food systems, the majority of the fabricated foods are either foam or emulsion based products, and proteins play an important role in these products (48). Basic understanding of the structure-function relationship of proteins with respect to their surfactant properties might lead to development of novel strategies to improve the functionality of under-utilized proteins, such as legume proteins. For example, with the current and future advances in biotechnology, techniques will soon be available to engineer seed storage proteins in plants. Through appropriate site directed mutagenesis in the amino acid sequences of these proteins, it possible to induce conformational changes in the protein, accompanied by changes in the hydrophilic/hydrophobic properties of the protein surface, and thus improve their functional properties. However, the major impasse to developing such protein engineering strategies is the lack of fundamental information on the structure-function relationship of food proteins. In other words, it is not known how to ascribe a particular functional property to a specific conformation of the protein. Increased efforts in this area of research are needed.

Acknowledgment

Financial support from the National Science Foundation Grant #CBT-8616970 is gratefully acknowledged.

References

1. Graham, D.E., and M.C. Phillips, *J. Colloid Interface Sci. 70*:403 (1979).
2. James, L.K., and L.G. Augenstein, *Adv. Enzymol. 28*:1 (1966).
3. Miller, I.R., and D. Bach, *Surface Colloid Sci. 6*:185 (1973).
4. MacRitchie, F., *Adv. Protein Chem. 32*:283 (1978).
5. MacRitchie, F., and A.E. Alexander, *J. Colloid Interface Sci. 18*:453 (1963).
6. MacRitchie, F., and A.E. Alexander, *Ibid. 18*:458. (1963).
7. MacRitchie, F., and A.E. Alexander, *Ibid. 18*:464 (1963).
8. Yamashita, T., and H.B. Bull, *Ibid. 27*:19 (1968).
9. Khaiat, A., and I.R. Miller, *Biochim. Biophys. Acta 183*:309 (1969).
10. Gonzalez, G.., and F. MacRitchie, *J. Colloid Interface Sci. 32*:55 (1970).
11. Tornberg, E., *J. Sci. Food Agric. 29*:762 (1978).
12. Tornberg, E., *J. Colloid Interface Sci. 64*:391 (1978).
13. Benjamins, J., J.A. Feijter, M.T.A. Evans, D.E. Graham and M.C. Phillips, *Faraday Disc. Chem. Soc. 59*:218 (1975).
14. Bull, H., *J. Colloid Interface Sci. 41*:305 (1972).
15. Ward, A.J.I., and L.H. Regan, *Ibid. 98*:395 (1980).
16. Graham, D.E., and M.C. Phillips *Ibid. 70*:415 (1979).
17. Graham, D.E., and M.C. Phillips, *Ibid. 70*:427 (1979).
18. Phillips, M.C., M.T. Evans, D.E. Graham and D. Oldani, *Colloid Polymer Sci. 253*:424 (1975).
19. Birdi, K.S., *J. Colloid Interface Sci. 43*:545 (1973).
20. Mita, T., F. Ishido and H. Matsumoto, *Ibid. 59*:172 (1977).
21. Mita, T., F. Ishido and H. Matsumoto, *Ibid. 64*:143 (1978).
22. Horiuchi, T., D. Fukushima, M. Sugimato and T. Hattori, *Food Chem. 3*:35 (1973).
23. Kato, A., and S. Nakai, *Biochim. Biophys. Acta 624*:13 (1980).
24. Waniska, R., and J.E. Kinsella, *J. Agric. Food Chem. 33*:1143 (1985).
25. Nakai, S., L. Ho, N. Helbig, A. Kata and M.M. Tung, *Can. Inst. Food Sci. Technol. J. 13*:23 (1980).
26. Barbeau, W.E., and J.E. Kinsella, *Colloids Surfaces 17*:169 (1986).
27. Waniska, R., J. Shetty and J.E. Kinsella, *J. Agric. Food Chem. 29*:826 (1981).
28. Pearce, K., and J.E. Kinsella, *Ibid. 26*:716 (1976).
29. Kim, S.H., and J.E. Kinsella, *J. Food Sci. 52*:1341 (1987).
30. Waniska, R., Ph.D. Thesis, Cornell University, Ithaca, NY, 1981.
31. Johanson, K.O., D.B. Wetlaufer, R.G. Reed and T. Peters, Jr., *J. Biol. Chem. 256*:445 (1981).
32. Damodaran, S., *Int. J. Peptide Protein Res. 27*:589 (1986).
33. Damodaran, S., *Biochim. Biophys. Acta 914*:114 (1987).
34. Creighton, T.E., *J. Mol. Biol. 113*:329 (1977).
35. Gaines, G.L., Jr., *Insoluble Monolayers at Liquid-Gas Interfaces*, Interscience, New York, NY, 1966, p. 45.
36. Damodaran, S., and K.B. Song, *Biochim. Biophys. Acta 954*:253 (1988).
37. Ward, A.F.H., and L. Tordai, *J. Chem. Phys. 14*:453 (1946).

38. Ter-Minassian-Saraga, L., *J. Colloid Interface Sci. 80*:393 (1981).
39. Tanford, C., *Physical Chemistry of Macromolecules*, Wiley & Sons, Inc., New York, NY, 1961, p. 358.
40. Kauzmann, W., *Adv. Protein Chem. 14*:1 (1959).
41. Tanford, C., *The Hydrophobic Effect: Formation of Micelles and Biological Membranes*, Wiley Interscience, New York, NY, 1973.
42. Stillinger, F.A., *Science 209*:451 (1980).
43. Addison, C.C., *J. Chem. Soc.* p. 98 (1945).
44. Matura, R., H. Kimizuka, S. Miyamoto, R. Shimozawa, and K. Yatsunami, *Bull. Chem. Soc. Japan 32*:404 (1959).
45. Abraham, M.H., *J. Am. Chem. Soc. 101*:5477 (1979).
46. Aveyard, R., and D.A. Haydon, in *An Introduction to the Principles of Surface Chemistry*, Cambridge University Press, England, 1973, p. 104.
47. Riggs, A.D., S. Bourgeois, and M. Cohn, *J. Mol. Biol. 53*:401 (1970).
48. Kinsella, J.E., and S. Damodaran, in *Criteria of Food Acceptance* edited by J. Solms and R.L. Hall, Forster Publishing Ltd., Switzerland, 1981, p. 296.

Chapter Six

Glycosylation of β-Lactoglobulin and Surface Active Properties

Ralph D. Waniska[a] and John E. Kinsella[b]

[a]Department Soil & Crop Science
Texas A&M University
College Station, TX 77843

[b]Department Food Science
Cornell University
Ithaca, NY 14853

The structure/function relationships of glycoproteins are becoming increasingly important (1,2). Glycoproteins function in many food and biological systems to provide excellent emulsion and foam stability (3-5) and cell membrane specificity (1,6). These important functional properties of glycoproteins reflect the chemical and structural characteristics of the carbohydrate moieties and of the protein's content and disposition of amino acid residues, molecular size, shape and flexibility, secondary conformations, net charge, etc. The functionality of glycoprotein is also affected by the particular conditions prevailing in the system and interactions of proteins with other components.

Several hypotheses have been proposed to relate characteristics of proteins with their functional properties (6-15). Solubility of protein is required for surface activity, i.e., foaming and emulsifying properties. Then a balance of hydrophobic, charge density and flexibility of proteins is needed for improved surface activity (6,8,10,11,13,16-21). Chemically glycosylated proteins have increased solubility and heat stability as compared to native proteins (1,2,22-28). Basic information on the effect of hydrophilic, carbohydrate residues on protein functionality in food systems is needed.

Review of Literature

Chemical forces of proteins.

Various chemical forces affect the structure and functionality of protein (4,11,29,30). Polar groups of protein form ionic and hydrogen bonds and dipole-dipole interactions with other polar molecules, especially water,

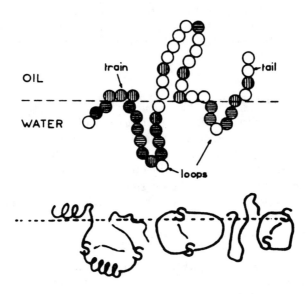

Fig. 6.1. Orientation of proteins at an interface. Schematic representation of nonpolar (⊖), polar (Φ) and neutral (○) residues of protein Ref. (38).

salts, and other polymers (Fig. 6.1). Nonpolar groups of protein are active in hydrophobic and van der Waal interactions with other molecules. Most polar groups of protein are exposed to the aqueous phase, while most nonpolar groups are clustered in the interior of the protein.

The conformational stability of protein results from a balance of their chemical forces along with intra- and intermolecular crosslinks (30-32). For example, the tertiary structure of protein expands at pHs greater than 2 units from its isoelectric point (IEP), which increases protein hydration and decreases the size of the hydrophobic area (33). Chemical modification of protein by acylation and alkylation increases the polar, (e.g., carboxyl, carbonyl and hydroxyl groups) and/or nonpolar groups (e.g., methyl, ethyl and propyl groups), respectively.

Interactions of proteins with other components in the food product ultimately determines the functional properties and applications of the protein (4,11,34). Interactions with water (solubility), hydrogen ion concentration (pH), salts, reducing/oxidizing compounds, hydrophobic compounds, interfaces, mechanical forces and temperature are primary factors affecting protein functionality. Protein solubility is critical for surface activity (11,13,35). Soluble proteins precipitate, flocculate, aggregate and/or polymerize when hydrophobic interactions increase and hydrophilic interactions decrease. This results because less energy is needed to form hydrophobic interactions than to hydrate nonpolar groups (29,31,36).

Surface activity of protein.

Hydrophobic and hydrophilic characteristics of protein enable it to orient at the interface and to stabilize the interfacial films in foams and emulsions (30,34,37). Nonpolar residues of protein preferentially interact with the hydrophobic side of the interface while polar residues interact preferentially with the aqueous side of the interface (Fig. 6.1) (38). Adsorption of proteins at the interface is driven by forces that normally stabilize its conformation, especially the energy of hydration of nonpolar groups. Initially, there is no barrier to adsorption of protein at the interface, and diffusion of protein to the interface limits adsorption. Soon, an interfacial tension and an electrostatic barrier develop which limit protein adsorption. Proteins arriving at the interface must then have sufficient energy to overcome the electrostatic barrier and to compress proteins already at the interface before absorption (7,35,37).

Proteins at the interface reorient or rearrange until a dynamic equilibrium is attained (37,39-42). Proteins tend to cover as much of the interface as possible since the free energy of the protein decreases with unfolding. Consequently, adsorbed proteins have high activation energies for desorption. Protein unfolds more at the oil than the air interface because more hydrophobic interactions occur with oil than air. However, because foaming properties correlated to the hydrophobicity of the heat denatured proteins, Townsend and Nakai (12) suggest that proteins unfold more at the air-water interface. Some α-helical and β-structures are retained by adsorbed proteins as indicated by hydrogen ion exchange, infrared spectroscopy, ellipsometry, fluorescence and circular dichroism (43-46).

Adsorption of protein.

The interfacial tension (g) of water decreases with the adsorption of protein at the interface (37,47). This is normally recorded as interfacial pressure (π, mN/M), $\pi = g_o - g_t$, where g is measured over time (t).

The first order rate constant of adsorption of protein at the interface (k) is determined by plotting $\ln(dn/dt)$ vs π where $dn/dt = (K)(c)(e)$ $-(\pi dA/kT)$, c is the bulk concentration of protein, k is Boltzmann's constant, T is the absolute temperature and πdA is the work of compression of the interfacial film required to create an area (dA) for protein to penetrate and adsorb (48,49). If no protein desorbs from the interface, values of dA provide an indication of the area of protein that penetrates the interface at different surface pressures. Small dA values observed for protein at > 15 mN/M π indicate only partial absorption of protein arriving at the interface (37,50).

Foams and emulsions are composed of dispersed air or lipid separated by water. However, the dispersed particles become less spherical and less stable as the water drains and the distances between the dispersed particles decrease (51-53). Forces acting on the interfacial films in foams and emulsions include hydrostatic, London-van der Waal, electrostatic and steric forces (10,34). The hydrostatic pressure is equal to the sum of the other forces, i.e., opposite of the disjoining pressure. London-van der Waals attractions decrease the distance between dispersed particles. Electrostatic and steric repulsions increase the amount of water held between the dispersed particles.

Rheology of interfacial film.

Monolayers of protein at interfaces exhibit considerable viscosity and elasticity despite their extreme thinness (54,55). Surface rheology of proteins is affected by intrinisic protein characteristics and extrinsic factors such as concentration, pH, agitation, etc. The compressibility and elasticity of interfacial films in the absence of shearing forces is measured by the dilatational modulus. The surface shear viscosity measures the friction between proteins in the film and between the interfacial film and adjacent layers of liquid.

Rheological properties of protein films in foams and emulsions contribute to their stability by resisting changes in interfacial area. In 1871 Marangoni postulated that interfacial pressure gradients caused the movement of molecules in the film. In 1878 Gibbs proposed that the surface elasticity was the driving force of the interfacial film during recovery from an applied force. A viscous interfacial film dissipates the disruptive effects of mechanical or thermal shocks better than a less viscous or elastic film (34,37). The retention of liquid in the lamella, which increases stability, is increased by bulky hydrophilic residues covalently linked or hydrogen bonded to the interfacial film (7,37).

Proteins with more hydrophobic residues usually exhibit better surface activity, foaming and emulsifying properties (8-11,13,16,18, 21,39,40,56). Ionic charge distribution of proteins vary depending upon pH and chemical modification, and more ionic repulsive forces generally correspond to decreased surface activity (11,12,19,20,57). Proteins with more hydrophilic, nonionic residues, e.g., glycoproteins, usually have more thermal stability (enzymes) and interfacial film stability (1,2, 28,58-60).

Rationale.

To better understand the structure-function relationship of proteins, hydrophilic residues were coupled to protein and their chemical, physical and functional properties were determined in this study. Since net ionic charge affects functionality of proteins, both the amino (+) and carboxyl (-) groups of protein were glycosylated.

Materials and Methods
Materials.

Bovine β-lactoglobulin (βLG) (crystallized and lyophilized, Sigma Chem. Corp.) was dialyzed against doubly distilled, deionized water containing 0.01% sodium azide, and lyophilized. Maltosyl- and β-cyclodextrinyl-β-lactoglobulin derivatives (MβLG and CβLG) were prepared by reacting cyclic carbonate derivatives of the carbohydrates with amino groups of βLG (Fig. 7.2) (61,62). Glucosaminyl- and glucosaminyloctaosyl-β-lactoglobulin derivatives (GβLG and G8βLG) were prepared using the carbodiimide method (Fig. 6.3) (62). Different levels of glycosylation were achieved by varying the amounts of activated carbohydrate or carbodiimide reagent.

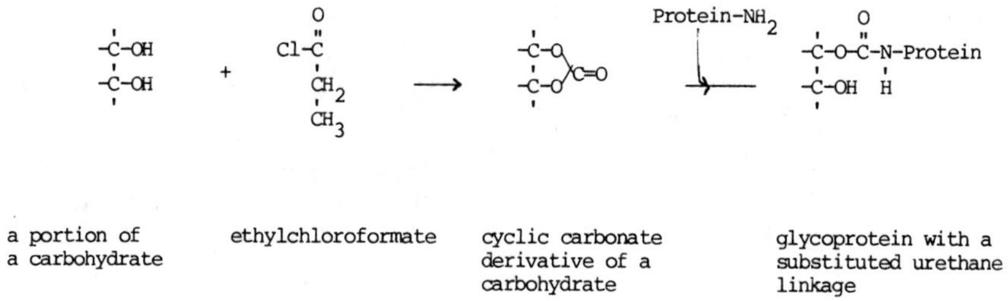

Fig. 6.2. Simplified sequence of reactions in the cylic carbonate method required to activate carbohydrates and to couple the cyclic carbonate derivatives to amino groups of protein.

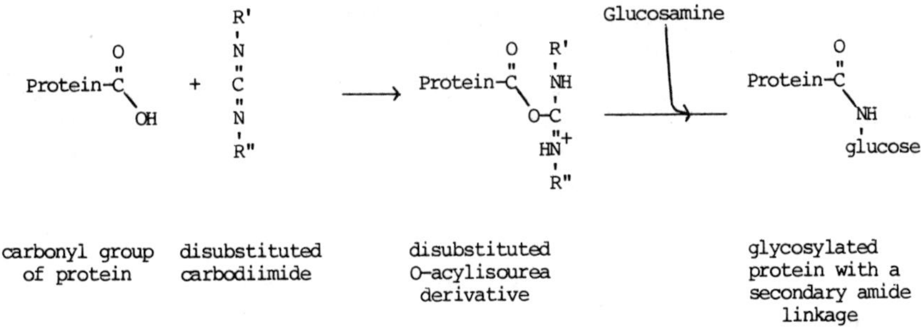

```
         R'                    Glucosamine
   O     N                                      O
   ‖     ‖            O   R'                    ‖
Protein-C  +  C   →  Protein-C   NH          Protein-C
       OH    ‖              O-C                     NH
         N                   ‖ +                     |
         R"                  HN                    glucose
                             |
                             R"
```

carbonyl group	disubstituted	disubstituted	glycosylated
of protein	carbodiimide	O-acylisourea	protein with a
		derivative	secondary amide
			linkage

Fig. 6.3. Simplified sequence of reactions in the carbodiimide method required to activate carboxyl groups of protein and to couple glucosamine (2-amino-2-deoxy-D-glucose) to the activated carboxyl derivative.

Analytical methods.

Protein was determined by the micro-Kjeldahl (63), biuret (64), and ultraviolet absorption methods (65) using βLG as the standard. Amino groups of hydrolyzed proteins were determined by trinitrobenzenesulfonic acid method (66). Carbohydrate residues of derivatized proteins were measured by the phenol-sulfuric acid method (67). Glucosamine residues of derivatized proteins were quantitated by difference of electrometric titration of carboxyl residues of βLG (68). Carbohydrate residues of G8βLG were determined by the phenol-sulfuric acid method after nitrous acid deamination (69). Glycoproteins were separated using sodium dodecylsulfate-polyacrylamide gel electrophoresis (SDS-PAGE) using a 5-20% linear gradient (70).

Glycosylated proteins were hydrolyzed using trypsin (EC 3.4.21.4; No. 9002-07-7; Type I, 10,000 units/mg protein) and α-chymotrypsin (EC 3.4.21.1; No. 9004-07-3; Type II, 40-50 units/mg protein) (Sigma Chem. Co., St. Louis, MO) (23,71). A 1:750 molar ratio of enzyme:protein was used to determine the rate of hydrolysis while a 1:250 ratio was used to determine the extent of hydrolysis after 18 hr.

Viscosities of protein solutions (0.1%, pH 6.3, 0.10 M sodium phosphate) were determined using an Ostwald-type viscometer at 25°C (72). Ultraviolet difference spectra of protein solutions (0.120 mM, pH 6.3, 0.10 M sodium phosphate, 0.01% sodium azide) were recorded using a Varian Spectrometer, Cary model 219 from 270 to 300 nm (72,73). Intrinsic fluorescence of protein solutions (0.00302 mM, pH 6.3, 0.10 M sodium phosphate, 0.01% sodium azide) were recorded with a Perkin-Elmer Spectrometer model 650-40 using 280 nm excitation and 300-400 nm emission wavelengths (72,74). Hydrophobicity of the surface of protein was determined using 0.092-0.459 molar ratios of protein to cis-parinaric acid (PNA) (9,72,75). Circular dichroic (CD) properties of proteins (0.01207 mM, pH 6.3, 0.010 M sodium phosphate, 0.01% sodium azide) were determined using a Varian Spectropolarimeter, Cary model 60 from 200-240 nm (72,76,77).

Determination of surface activity and surface viscosity.

The water was distilled twice, the second time from a glass apparatus. The protein solutions (0.10%) contained 0.02 N sodium citrate (pH 3-6) or 0.02 N sodium phosphate (pH 6-8), 0.01% sodium azide and 0.2 N sodium chloride. Surface tension (g) was determined using the Whilhelmy plate method (14,78). Surface pressure (II), rates of adsorption (K_1) and rearrangement (K_2) and the area cleared during adsorption (dA) were calculated (37,39,48,49,79-81).

The surface viscosity (n_s) of protein films was measured using a rotating ring on a viscometer (Model LVF, Brookfield Engineering Lab., Stroughton, MA) as described by Waniska and Kinsella (15).

Solubility, foaming, and emulsification properties.

Solubility of protein (0.010 g in 10.0 ml, 0.10M sodium chloride) was determined at pH 3-8 by ultraviolet absorption after 30 min (82). Foaming activity and stability were determined with a sparging apparatus that held 15.0 ml of 0.10% protein solution (83). Emulsions were prepared in a Janke-Kunkel model 10A blender (84). Emulsifying activity was determined by optical density at 550 nm while emulsion stability was determined after centrifugation (84,85).

Results

Chemical and physicochemical properties of glycosylated β-lactoglobulin.

Chemical analyses of glycosylated derivatives of βLG indicated that both methods of chemical modification (i.e., cyclic carbonate and carbodiimide methods) effectively glycosylated βLG (Tables 6.1 and 6.2) (62). The primary site of modification of the cyclic carbonate reaction was the amino group of βLG, while the primary site of modification of the carbodiimide reaction was the carboxyl group of βLG. However, some sulfhydryl and hydroxyl groups of βLG were also modified, especially at higher

TABLE 6.1
Extent of Glycosylation of β-Lactoglobulin with Maltose and β-Cyclodextrin Cyclic Carbonates

Treatment (cyclic carbonate: (protein-NH_2))	Number of amino groups modified	Number of sulfhydryl groups modified	Number of hydroxyl groups modified	Total number of sites modified	Molecular weight of carbohydrates	Number of maltose groups per protein
Maltose						
4.6:1	1.0	0.16	0.11	1.3	445	1.3
10:1	1.6	0.24	0.38	2.2	889	2.6
19:1	2.6	0.30	0.57	3.5	1,949	5.7
29:1	4.2	0.49	1.04	5.7	3,386	9.9
40:1	5.6	0.30	1.04	7.0	4,352	12.8
49:1	6.4	0.59	2.09	9.1	6,430	18.8
60:1	8.0	0.87	2.17	11.0	11,220	33.0
80:1	10.4	0.86	3.22	14.5	10,994	32.0
β-cyclodextrin						
5:1	2.7	N.D.	1.07	3.8	5,025	15.3[a]
(Total number of groups available for reaction)	(16)	(1)	(19)	(36)		

[a] There were 4.4 β-cyclodextrin groups per protein.

TABLE 6.2
Extent of Glycosylation of β-Lactoglobulin with Carbodiimide and Glucosamine or Glucosamineoctaose

Treatment (carbodiimide: Protein-COOH: glucosamine)	Number of carboxyl groups modified	Number of sulfhydryl groups modified	Number of hydroxyl groups modified	Total number of sites modified	Molecular weight of carbohydrates
Glycosamine					
0.33:1:20	3.2	—	0.39	3.6	515
0.5:1:20	4.2	0.07	2.07	6.3	1,014
1:1:20	8.8	0.06	2.99	11.8	1,017
2:1:20	12.4	0.16	3.48	16.0	2,303
10:1:100	16.6	0.52	4.78	21.9	2,673
Glycosamineoctaose					
0.5:1:20	12[a]	—	0.20	12.2	13,950
(Total number of groups available for reaction)	(25)	(1)	(19)	(45)	

[a] Determined by dividing the glucosamine concentration by eight, the average number of glucosamine residues in the oligosaccharide.

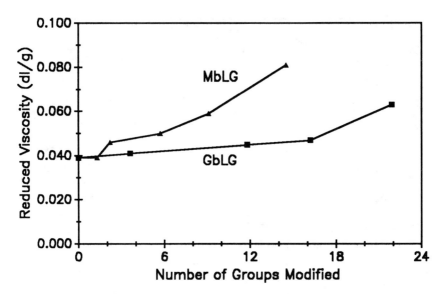

Fig. 6.4. Reduced viscosity of maltosyl-(▲) and glucoseaminyl-(■) β-lactoglobulin at 1.0% protein concentration.

reagent levels (62). The carbodiimide method more efficiently coupled carbohydrates to βLG as compared to the cyclic carbonate method. Glycosylation yielded 5.2 (cyclic carbonate method) and 16.6 glucose residues (carbodiimide method) using a 10:1 molar ratio of reagent to βLG residue. Both methods yielded glycosylated βLG that contained different charge distributions and amounts of carbohydrate residues.

Electrophoretic analyses of the glycosylated βLG indicated heterogeneity of glycosylation and in molecular size distribution (62). MβLG and GβLG had lighter and more diffuse protein bands than native βLG. The MW of MβLG was not affected by the extent of modification; however, faint bands were observed that corresponded to dimeric MβLG. The mobility of GβLG decreased as more carbohydrates were coupled to βLG. The change in mobility of GβLG probably resulted from decreased electronegativity as well as its larger MW. Thus, glycosylation of βLG was variable and it altered the mobility of βLG.

Several physicochemical properties of glycosylated βLG were determined, i.e., reduced viscosity, ultraviolet difference spectra, intrinsic fluorescence, hydrophobicity and circular dichroism (72). Reduced viscosity of glycosylated βLG increased as the extent of modification

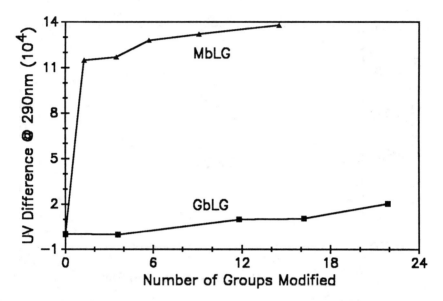

Fig. 6.5. Ultraviolet difference at 290 nm of maltosyl-(▲) and glucoseaminyl-(■) β-lactoglobulin.

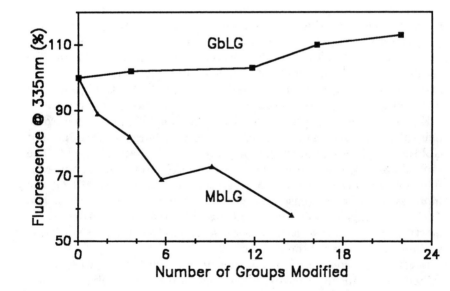

Fig. 6.6. Intrinsic fluorescence at 335 nm of maltosyl-(▲) and glucoseaminyl-(■) β-lactoglobulin.

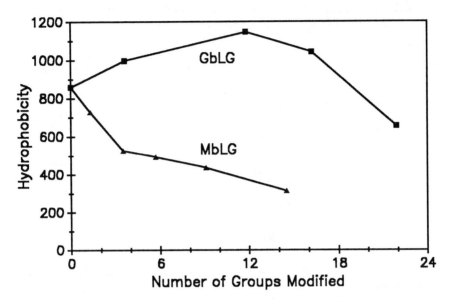

Fig. 6.7. Hydrophobicity of maltosyl-(▲) and glucoseaminyl-(■) β-lactoglobulin using *cis*-parinaric acid.

increased (Fig. 6.4). MβLG had higher reduced viscosities than GβLG because each residue of βLG that was modified had more than 1 maltosyl residue (Table 6.1). Extensively modified MβLG and GβLG derivatives had higher viscosities than expected because the conformation of βLG was partially unfolded.

The environments of aromatic amino acid residues of glycosylated βLG were indicated by ultraviolet difference spectra and intrinsic fluorescence (72,86). Difference spectra of MβLG were smooth curves with no characteristic peaks while those of GβLG had adsorption maxima at 282 and 290 nm. Aromatic amino acid residues of MβLG were increasingly exposed to polar environments as the extent of modification increased (Fig. 6.5). However, aromatic amino residues of some GβLG derivatives were exposed to only slightly more polar environments than those of native βLG. Glycosylated βLG did not change the wavelength of maximum fluorescence; however, intrinsic fluorescence of MβLG decreased as the extent of glycosylation increased (Fig. 6.6). Thus, tryptophyl, tyrosyl and phenylalanyl residues of MβLG were in more polar environments compared to these residues in βLG and GβLG.

TABLE 6.3

Content of Secondary Structure of Glycosylated β-Lactoglobulin as Estimated from Circular Dichroic Measurements

Protein[a]	Secondary conformations		
	α-helical (%)	β Structure (%)	Remainder (%)
Native β-lactoglobulin	10	40	50
Maltosyl-β-lactoglobulin			
(2.2)	6	43	52
(5.7)	6	43	50
(9.1)	4	42	54
(14.5)	2	42	56
Glucosaminyl-β-lactoglobulin			
(11.8)	8	48	43
(16.0)	8	43	49
(21.9)	2	40	59

[a]Number in parentheses indicates the number of β-lactoglobulin that were glycosylated.

Hydrophobicity of the surface of βLG was affected by glycosylation (Fig. 6.7) (72). Hydrophobicity of MβLG derivatives decreased as the extent of modification increased; however, the hydrophobicity of some GβLG derivatives was higher than that of native βLG. The hydrophobicity of GβLG derivative with 21.9 groups modified was lower than that of βLG. Thus, carbohydrates increased the hydrophilic and decreased the hydrophobic characteristics of βLG.

The proportion of secondary conformations of glycosylated βLG was determined using circular dichroism (Table 6.3) (72). The amount of α-helical structure of MβLG decreased with the extent of modification. The amount of α-helical structure of GβLG was similar to βLG except at the highest level of glycosylation. The β-structure of βLG was negligibly affected by glycosylation.

Enzyme hydrolysis of glycosylated βLG using trypsin and α-chymotrypsin indicated their digestibility (71). Rates of hydrolysis of extensively glycosylated MβLG and GβLG derivatives were higher than that of native βLG (Fig. 6.8). However, the extent of hydrolysis decreased as glycosylation increased (Fig. 6.9). These data indicate glycosylated βLG was partially unfolded which aided initial hydrolysis, but steric hindrance by carbohydrate residues limited the complete hydrolysis of modified βLG.

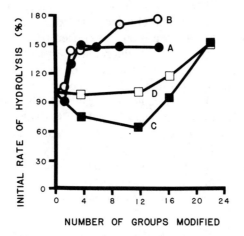

Fig. 6.8. Relative rate of hydrolysis of glycosylated β-lactoglobulin (βLG). Symbols correspond to maltosyl-βLG (● and ○) hydrolyzed by trypsin (A) and α-chymotrypsin (B) and to glucosaminyl-βLG (■ and □) hydrolyzed by trypsin (C) and α-chymotrypsin (D).

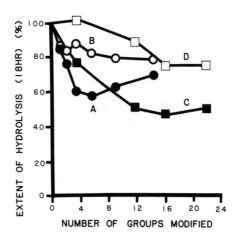

Fig. 6.9. Relative extent of hydrolysis of glycosylated β-lactoglobulin after 18 hr (Symbols — same as Fig. 6.8).

Thus, glycosylation of βLG markedly increased the hydrophilic character and increased the molecular weight of the protein. Physicochemical analyses revealed that the viscosity of βLG increased as the extent of modification increased; that the aromatic amino acids of MβLG and extensively modified GβLG were more exposed to polar environments; that the hydrophobicity of MβLG decreased but the hydrophobicity of the GβLG was unchanged; that the content of α-helical structure of MβLG and extensively modified GβLG decreased while the proportion of unordered structure increased. Enzymatic hydrolysis studies indicated that glycosylation increased the rate of hydrolysis but decreased the extent of hydrolysis. The hydropholic carbohydrate residues caused some alterations in the secondary and the tertiary structure of βLG. The balance of forces stabilizing the conformation of βLG was altered since some ionic groups were glycosylated and some hydrophobic amino acid residues were exposed to polar environments.

Surface chemical properties of glycosylated β-lactoglobulin.

The extent of adsorption of βLG at the air-water interface was affected by time and pH (78). The π of βLG increased over time (Fig. 6.10); but, the rate of change of π decreased with time. Values of π were higher near

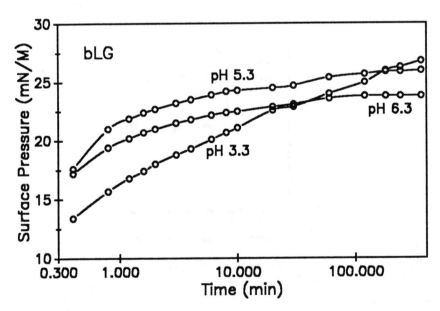

Fig. 6.10. Development of surface pressure of β-lactoglobulin films at the air-water interface at pH 3.3, 5.3 and 6.3.

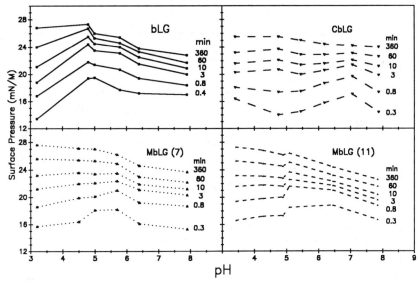

Fig. 6.11. Effect of pH on surface pressure of 0.10% solutions of β-lactoglobulin βLG (0), maltosyl-βLG with 7 (△) and 11 groups modified (▲) and β-cyclodextrinyl-βLG with 4 (CβLG, ▽).

the isoelectric point (IEP) of βLG than on either side of the IEP during the first 10 min (Fig. 6.11). This indicated that the net charge distribution of protein significantly affected its surface activity. Equilibrium values of π decreased as the pH increased from pH 3–8 (Fig. 6.11). The first order rate constants for adsorption and rearrangement (K_1) and for rearrangement (K_2) of βLG at the interface were also higher near the IEP than on either side of the IEP (Fig. 6.12 and 6.13). Values of dA and πdA were maximum near pH 8 (Fig. 6.14). Thus, the surface activity of βLG was optimum near its IEP, especially during the first few minutes of adsorption from solution.

Surface chemical properties of glycosylated βLG were affected by the age of the film and amount of carbohydrate residues (14). Lower π values were observed during the first 10 min of adsorption for glycosylated βLG compared to βLG (Fig. 6.11 and 6.15). Also, pH of the solution had less effect on π of glycosylated βLG compared to βLG. Equilibrium π of glycosylated βLG were either similar to or lower than those of βLG. The K_1 and K_2 were lower for glycosylated βLG than native βLG (Fig. 6.12 and 6.13). The dA for glycosylated derivatives were generally similar or slightly greater than for βLG (Fig. 6.14). Thus, glycosylated βLG exhibited decreased surface activity.

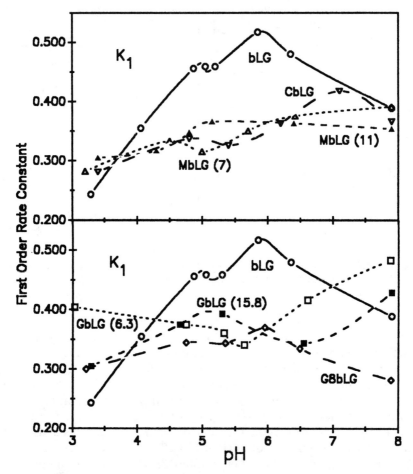

Fig. 6.12. Effect of pH on first-order rate constants of adsorption and rearrangement (K_1) of glycosylated β-lactoglobulin (0), maltosyl-βLG with 7 ($\triangle$) and 11 groups modified ($\blacktriangle$) and β-cyclodextrinyl-βLG with 4 (CβLG, $\triangledown$), glucosaminyl-βLG with 6.3 ($\square$) and 15.8 groups modified ($\blacksquare$) and glucosamineoctaose-βLG with 12.2 groups modified ($\lozenge$).

The n_s of βLG was also affected by the age of pH of the film (15). The n_s of βLG increased over time during the first few min (Fig. 6.16); then, the n_s decreased between 10-50% before stabilizing. The maximum n_s and rate of development of n_s of βLG was optimum near the IEP of βLG (Figs. 6.17 and 6.18).

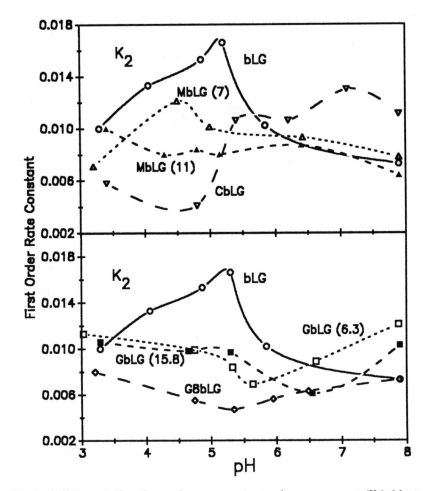

Fig. 6.13. Effect of pH on first-order rate constants of rearrangement (K_2) β-lactoglobulin (Symbols — same as Fig. 6.12).

The n_s of glycosylated βLG was affected by the age and pH of the film and extent of modification (18). The n_s of most derivatives were optimum near their respective IEPs (Fig. 6.19, Table 6.4). The n_s of GβLG with 15.8 groups modified was highest above pH 6.5, i.e., above its IEP. The n_s of MβLG derivatives were much lower than those of βLG while the n_s of CβLG, G8βLG and GβLG with 6.3 groups modified were only slightly lower than those of βLG. The rates of development of n_s of glycosylated

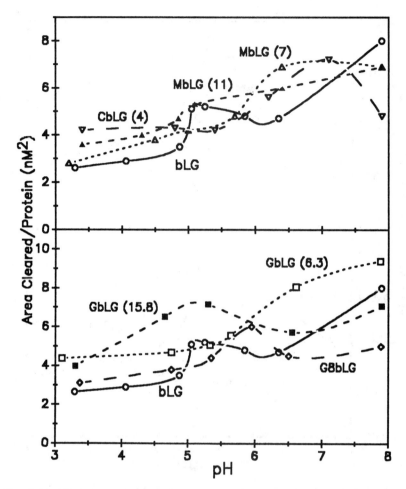

Fig. 6.14. Effect of pH on the average area cleared per protein during adsorption (0.4–8 min) of glycosylated β-lactoglobulin (Symbols — same as Fig. 6.12).

βLG were lower than for βLG (Fig. 6.18); however, the times to reach maximum n_s increased when compared to βLG (15).

Thus, surface activity of glycosylated βLG decreased after modification with carbohydrates. The rates of protein film formation and n_s development of glycosylated βLG decreased compared to βLG. Equilibrium π of the films were similar; but, the n_s of glycosylated βLG films were lower than those of βLG.

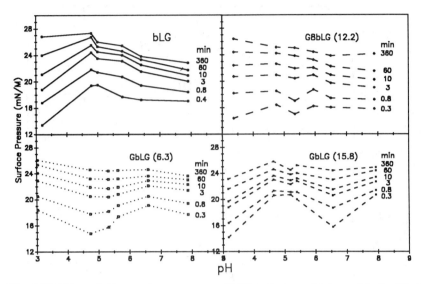

Fig. 6.15. Effect of pH on surface pressure of 0.10% solutions of β-lactoglobulin (βLG) (○), glucosaminyl-βLG with 6.3 (□) and 15.8 groups modified (■) and glucosamineoctaose-βLG with 12.2 groups modified (◊).

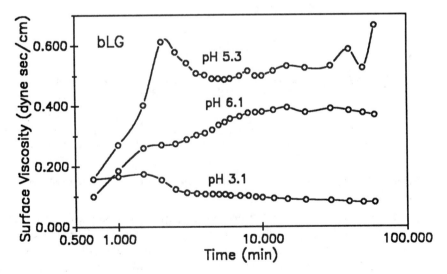

Fig. 6.16. Development of surface viscosity of β-lactoglobulin films at the air-water interface at pH 3.1, 5.3, and 6.1.

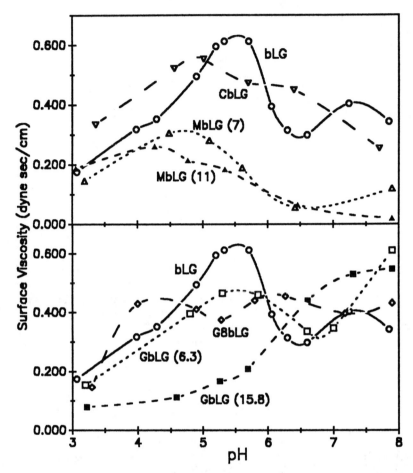

Fig. 6.17. Effect of pH on the maximum surface viscosity of films of glycosylated β-lactoglobulin (βLG) at the air-water interface (Symbols — same as Fig. 6.12).

Solubility, foaming and emulsification properties of glycosylated β-lactoglobulin.

The solubility, foaming and emulsifying properties of βLG were affected by pH (82). Solubility of βLG decreased slightly near its IEP (Fig. 6.19). The initial volume of liquid in the foam prepared with βLG was lower near its IEP; however, foam strength was poor at all pH values studied (Fig. 6.20). Emulsifying activity of βLG decreased near its IEP (Fig. 6.21); but, stability of emulsions improved near its IEP.

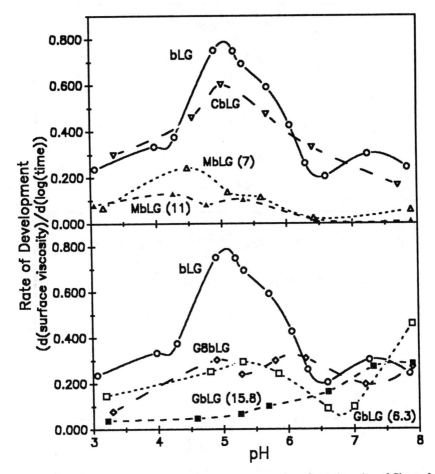

Fig. 6.18. Effect of pH on the rate of development of surface viscosity of films of glycosylated β-lactoglobulin (Symbols — same as Fig. 6.12).

Solubility of MβLG and CβLG decreased between pH 4-5 (82) (Fig. 6.19). Solubility of GβLG was above 98% while the solubility of G8βLG decreased slightly between pH 5-7. The pH of minimum solubility corresponded to the IEPs of the glycosylated proteins (Table 6.4).

Foaming properties of glycosylated βLG were affected by pH and extent of modification (82). Foaming capacity of glycosylated βLG was near 100% from pH 3-8. Initial volumes of liquid in the foam prepared with glycosylated βLG were higher than those of βLG. Drainage from the

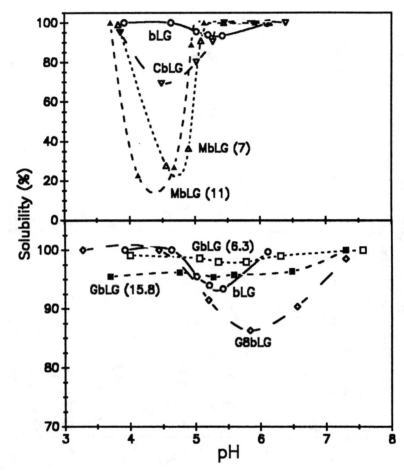

Fig. 6.19. Effect of pH on solubility of glycosylated β-lactoglobulin (Symbols — same as Fig. 6.12).

foams prepared with glycosylated βLG was lower than those of βLG. The strength of foams prepared with glycosylated βLG markedly improved (Fig. 6.20). However, strength of foams prepared with G8βLG was only slightly better than those of βLG.

Emulsifying properties of glycosylated βLG were affected by pH and extent of modification (82). Emulsifying activities of MβLG and CβLG were improved compared to native βLG (Fig. 6.21). This was especially noticeable near the IEP of MβLG. Stability of emulsions prepared with

TABLE 6.4
Chemical Properties of Glycosylated β-Lactoglobulin[a]

Residue	Amino/carboxyl groups modified[b]	Total sites modified	Weight of carbohydr. (g/mole)[c]	Molecular weight increase (%)[d]	Isoelectric point	Hydrophobicity[e]
Maltose	7.0	7.0	4352	24	4.6	480
Maltose	8.0	11.0	11220	61	4.4	380
β-Cyclodextrin	3.0	4.0	8150	45	4.9	—
Glucosamine	4.2	6.3	1014	6	5.9	1060
Glucosamine	12.0	15.8	2544	14	9.2	1060
Glucosamineoctaose	12.0	12.2	13950	76	11.1	—

[a]Data from Waniska and Kinsella (18, 19).
[b]Amino groups of β-lactoglobulin (βLG) were modified with maltose and β-cyclodextrin via the cyclic carbonate method. Carboxyl groups of βLG were modified with glucosamine and glucosamineoctaose via the carbodiimide method.
[c]The cyclic carbonate method yielded crosslinked maltosyl residues which increased the mass of the carbohydrate that was coupled to βLG.
[d]Molecular mass of βLG is 18300, IEP = 5.25 (17).
[e]Hydrophobicity = (Fluorescence Intensity/% protein) at pH 6.2. Native βLG has S_o = 860 FI/% protein.

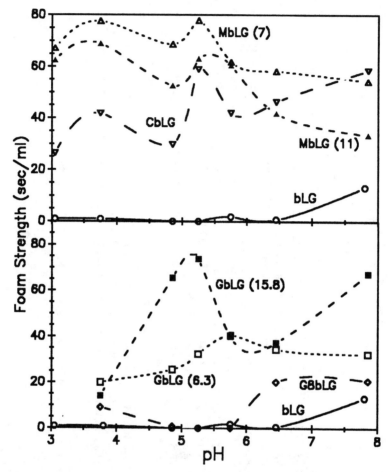

Fig. 6.20. Effect of pH on foam strength of glycosylated β-lactoglobulin (Symbols — same as Fig. 6.12).

the glycosylated βLG was also slightly improved compared to that of βLG.

Thus, functional properties of βLG were improved by glycosylation. Solubility of MβLG was reduced while the solubility of GβLG was near 100% from pH 3–8. Foaming and emulsifying properties of glycosylated βLG were improved as compared to those of βLG.

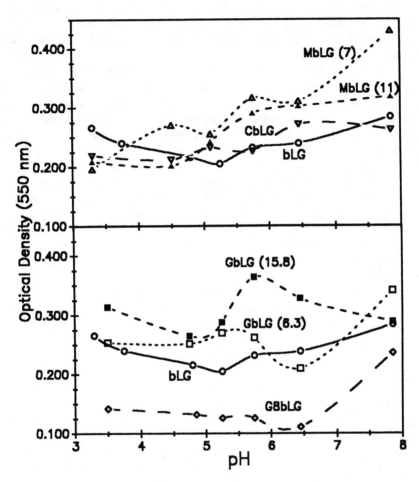

Fig. 6.21. Effect of pH on emulsifying activity of glycosylated β-lactoglobulin measured by optical density (Symbols — same as Fig. 6.12).

Discussion

Relationships between chemical, structural, interfacial and functional properties of glycosylated βLG-lactoglobulin.

Chemical, structural, interfacial and functional properties of βLG were affected by the covelant attachment of carbohydrate residues. Glycosylation of βLG altered the chemical forces stablizing βLG. Amino and carboxyl groups of the protein were coupled with neutral and amino sugars, respectively. Hence, the charge density of βLG was modified, i.e., MβLG and CβLG had fewer cations while GβLG and G8βLG had fewer anions; and the hydrophilic, nonionic character of βLG was substantially increased.

The conformation of βLG was affected by the additional hydrophilic residues since the forces stabilizing the polypeptide backbone were altered. The conformation of βLG with low levels of modification with glucosamine was not significantly altered; however, the conformation of βLG modified with maltose was altered even at low extents of modifications as indicated by several physicochemical criteria. The conformation of extensively modified MβLG and GβLG was partially unfolded as indicated by the decreased amounts of α-helical structures, increased exposure of aromatic amino acids and large relative viscosities. Hydrophobicity of MβLG decreased even at low levels of glycosylation. Binding sites for PNA were apparently occluded by carbohydrates that were linked to amino residues but not to carboxyl groups of βLG. Hence, glycosylation of amino groups of βLG affected its conformation more than glycosylation of carboxyl groups.

The conformations of many proteins were unfolded after chemical, physical or mechanical treatment (4,11,87-91). This results because the proteins contained increased electrostatic repulsions, decreased hydrogen bonds, decreased covalent crosslinks (e.g., disulfide bridges), and/or were exposed to hydrophobic surfaces, shear or high temperatures. Many proteins exhibited improved solubility, surface activity and functional properties after partial or complete modification or denaturation. This indicated that the balance of forces stabilizing the conformations of native proteins was less suitable for utilization as functional ingredients.

Enzymatic hydrolysis of MβLG and GβLG corresponded to the hypothesis that glycosylation with maltose altered the conformation of βLG more than with glucosamine. Kato et al. (13) reported increased protease susceptibility of proteins with a more flexible tertiary conformation. Glycosylation probably increased the flexibility of βLG, i.e., a partially unfolded tertiary conformation.

Rates of enzymatic hydrolysis of most chemically modified proteins decreased (23,24,92,93). Initial rates of hydrolysis decreased because the added residues reduced the binding coefficient of the enzyme for the substrate possibly by the formation of non-productive enzyme complexes and/or by product inhibition (24). Trypsin or α-chymotrypsin were not initially inhibited by glycosylated βLG, but during the course of hydrolysis these enzymes were inhibited by carbohydrate residues on βLG.

Surface chemical properties of βLG decreased after glycosylation. Glycosylated βLG had an increased bulk viscosity which decreased the rate of protein diffusing to, and adsorbing at, the air-water interface. Rheological properties of glycosylated βLG at the interface decreased because the carbohydrate moieties interacted preferentially with water (the bulk phase) and not with proteins at the interface.

Surface activity of partially heat denatured proteins increased as a function of hydrophobicity, solubility and protein flexibility (8,9,11-13,18,20,21,39-42,56,89,94,95). Solubility of proteins in water is desirable for emulsification, foaming and other physical characteristics which are normally evaluated on an equal soluble protein basis. The surface hydrophobicities of several proteins correlated with emulsification (11,21,56,95) while hydrophobicity of unfolded protein correlated with foaming properties (12). Some conformational flexibility of the protein correlates to improved surface activity since rearrangement of the protein conformation is necessary to expose hydrophobic and hydrophilic residues to the appropriate phase (13,56). Proteins with rigid conformations, i.e., proteins with many disulfide bonds, form condensed interfacial films with good rheological properties (11,20,94). However, foaming properties of rigid proteins improves after removing intramolecular crosslinks (20,95).

Foaming properties of βLG were improved by glycosylation. Interaction of glycosylated proteins in the interfacial film with water increased the amount of water in the foam initially and after draining. These hydrophilic interactions increased the strength and resiliency of the foam bubbles. Solubility of MβLG decreased near its IEP; however, foaming properties of MβLG were not appreciably effected by their decreased solubility. Classical surface chemical properties of glycosylated βLG did not correspond to foaming properties. Hydrophobicity of glycosylated βLG did not correlate with foaming properties; whereas, hydrophilicity of βLG corresponded to improved foaming properties.

Emulsifying properties of βLG were improved by glycosylation. Apparently, the hydrophilic residues on modified βLG increased the steric hinderance between the dispersed oil droplets which increased the stability of the emulsion even near the IEP of glycosylated proteins.

Conclusions

Chemical, structural, surface and functional properties of βLG were affected by the covalent attachment of carbohydrates. The conformation of βLG was partially unfolded. Glycosylated βLG had increased viscosity, exposure of aromatic amino acid residues and rates of enzymatic hydrolysis and decreased hydrophobicity and amount of ordered secondary structure.

The surface chemical properties of glycosylated βLG were altered. Equilibrium surface pressures were similar to that of βLG while their rates of adsorption and rearrangement at the interface decreased. The surface viscosity of glycosylated βLG decreased.

The foaming properties of glycosylated βLG improved significantly while emulsifying properties were slightly improved. Solubility of MβLG decreased between pH 4–5. Solubility of GβLG was not affected by pH. Functional properties of glycosylated βLG corresponded to their hydrophilicities. Carbohydrate residues of glycosylated βLG increased the amount of water bound to the interfacial film which stabilized the foam and emulsion particles. Decreased charge density and increased conformational flexibility probably contributed to their functionality. In summary, a protein that contains many hydrophilic and hydrophobic residues, that readily undergoes conformational change at the interface and that has a lower charge density should possess good foaming and emulsifying properties.

References

1. Schmidt, R.D., *Adv. Biochem. Eng. 12*:41 (1979).
2. Sharon, N., and H. Lis, *Chem. Eng. News*, March 30, p. 21 (1981).
3. Basch, J.J., Farrell, Jr. and R. Greenberg, *Biochim. Biophys. Acta 448*:589 (1976).
4. Kinsella, J.E., *CRC Crit. Rev. Food Sci. Nutr. 7*:219 (1976).
5. Lehnhardt, W.F., and F.T. Orthoefer, *U.S. Patent* No. 4,409,248 (1983).
6. Baier, R.E., in *Proc. Third Int'l. Congress Marine Corrosion and Fouling*, Gaithersberg, MD, 1978, p. 633.
7. Cumper, C.W.N., *Trans. Faraday Soc. 49*:1360 (1953).
8. Birdi, K.S., *J. Colloid Interface Sci. 43*:545 (1973).
9. Kato, A., and S. Nakai, *Biochim. Biophys. Acta 624*:13 (1980).
10. Phillips, M.C., *Food Technol. 35*:50 (1981).
11. Nakai, S., *J. Agric. Food Chem. 31*:676 (1983).
12. Townsend, A.M., and S. Nakai, *J. Food Sci. 48*:588 (1983).
13. Kato A., K. Kornatsu, K. Fujimoto and K. Kobayashi, *J. Agric. Food Chem. 33*:931 (1985).
14. Waniska, R.D., and J.E. Kinsella, *J. Colloid Interface Sci. 117*:251 (1987).
15. Waniska, R.D., and J.E. Kinsella, *Food Hydrocolloids 2*:59 (1988).
16. Horiuchi, T., D. Fukushima, M. Sugimata and T. Hattori, *Food Chem. 3*:35

(1978).
17. Brunngrabber, E.G., in *Neurochemistry and Neuropathology of the Complex Carbohydrates*, Thomas Publ. Co., Springfield, IL, 1979, p. 419.
18. Graham, D.E., and M.C. Phillips, *J. Colloid Interface Sci. 75*:427 (1979).
19. Mattarella, N.L., and T. Richardson, *J. Agric. Food Chem. 31*:972 (1983).
20. Hayakawa, S., and S. Nakai, *J. Food Sci. 50*:486 (1985).
21. Nakai, S., and E. Li-Chan, *J. Dairy Sci. 68*:2763 (1985).
22. Marshall, J.J., and M.L. Rabinowitz, *J. Biol. Chem. 251*:1081 (1976).
23. Lee, H.S., L.C. Sen, A.J. Clifford, J.R. Whitaker and R.E. Feeney, *J. Nutr. 108*:687 (1978).
24. Lee, H.S., L.C. Sen, A.J. Clifford, J.R. Whitaker and R.E. Feeney, *J. Agric. Food Chem. 27*:1094 (1979).
25. Puigserver, A.J., L.C. Sen, E. Gonzales-Flores, R.E. Feeney and J.R. Whitaker, *Ibid. 27*:1098 (1979).
26. Puigserver, A.J., L.C. Sen, A.J. Clifford, R.E. Feeney and J.R. Whitaker, *Ibid. 27*:1286 (1979).
27. Sakharov, I.Y., N.I. Larionova, N.F. Kazanskaya and I.V. Berezin, *Eng. Microb. Technol. 6*:27 (1984).
28. Kato, Y., T. Matsuda, N. Kato, K. Watanake and R. Nakamura, *J. Agric. Food Chem. 34*:351 (1986).
29. Kuntz, I.D., Jr., and W. Kauzmann, *Adv. Protein Chem. 28*:239 (1974).
30. Privalov, P.T., *Ibid. 33*:583 1979.
31. Bigelow, C.C., *J. Theoret. Biol. 16*:187 (1967).
32. Scheraga, H.A., in *Versatility of Proteins*, edited by C.H. Li, Academic Press, New York, NY, 1978, p. 119.
33. Bull, H.B., *Arch. Biochem. Biophys. 208*:229 (1981).
34. Kinsella, J.E., *Food Chem. 7*:273 (1981).
35. Halling, P.J., *CRC Crit. Rev. Food Sci. Nutr. 15*:155 (1981).
36. Ter-Minassian-Saraga, L., *J. Colloid Interface Sci. 80*:393 (1981).
37. MacRitchie, R., *Adv. Proteins Chem. 32*:283 (1978).
38. Phillips, M.C., *Chem. Ind. March 5*, p. 170 (1977).
39. Graham, D.E., and M.C. Phillips, *J. Colloid Interface Sci. 75*:403 (1979).
40. Graham, D.E., and M.C. Phillips, *Ibid. 75*:415 (1979).
41. Graham, D.E., and M.C. Phillips, *Ibid. 76*:227 (1980).
42. Graham, D.E., and M.C. Phillips, *Ibid. 76*:240 (1980).
43. Malcolm, in *Applied Chemistry at Protein Interfaces*, edited by R.E. Baier, Adv. Chem. Ser. No. 145, 1975 p. 338.
44. Morrissey, B.W., L.E. Smith, R.R. Stromberg and C.A. Fenstermaker, *J. Colloid Interface Sci. 56*:557 (1976).
45. Cornell, D.G., *Ibid. 88*:536 (1982).
46. Walton, A.G., and F.C. Maenpa, *Ibid. 72*:265 (1979).
47. Paddy, J.F., in *Surface and Colloid Science*, Vol. 1, edited by E. Matijeic, Wiley-Interscience, New York, NY, 1969, p. 39.
48. Tornberg, E., *J. Colloid Interface Sci. 64*:391 (1978).

49. Tornberg. E., *J. Sci. Food Agric. 29*:762 (1978).
50. Kitchener, J.A., and C.F. Cooper, *Chem. Soc. Q. Rev. 13*:71 (1959).
51. Davies, J.T., and E.K. Rideal, *Interfacial Phenomena*, Academic Press, New York, NY, 1963.
52. Bickerman, J.J., *Foams*, Springer Verlag, Berlin, W. Germany, 1973.
53. Phillips, M.C., *Water:A Comprehensive Treatise, Water in Disperse Systems 5*:133 (1975).
54. Joly, M., in *Surface and Colloid Science*, edited by E. Matijevic, Wiley-Interscience, New York, NY, 1972, p. 1.
55. Joly, M., *Ibid*, p. 78.
56. Nakai, S., L. Ho, N. Helbig, A. Kato and M.A. Tung, *J. Inst. Can. Sci. Technol. Ailment. 13*:23 (1980).
57. Halpin, M.I., and T. Richardson, *J. Dairy Sci. 68*:3189 (1985).
58. Moore, W.E., and J.L. Carter, *J. Texture Studies 5*:77 (1974).
59. Wasserman, B.P., and H.O. Hultin, *J. Food Biochem. 6*:87 (1982).
60. Canton, M.C., and D.M. Mulvihill, in *Proc. Int. Dairy Fed. Symp.* May 17-19, 1983, p. 339.
61. Doane, W.M., B.S. Shasha, E.I. Stout, C.R. Russell and C.E. Rist, *Carboyhdr. Res. 4*:445 (1967).
62. Waniska, R.D., and J.E. Kinsella, *Int. J. Peptide Protein Res. 23*:573 (1984).
63. AOAC, *Official Methods of Analysis*, 11th edn., Assoc, Official Anal. Chem., Washington, D.C., 1970.
64. Means, G.E., and R.E. Feeney, *Chemical Modification of Proteins*, Holden-Day Publ., San Francisco, CA, 1971.
65. McKenzie, H.A., in *Milk Proteins, Chemistry and Technology*, Vol. II, Academic Press, New York, NY, 1971, p. 257.
66. Fields, R., *Methods Enzymol. 25*:464 (1972).
67. Doubois, M., K.A. Gilles, J.K. Halmilton, P.A. Rebers and F. Smith, *Anal. Chem. 28*:350 (1956).
68. Tanford, C., *Adv. Protein Chem. 17*:70 (1962).
69. Lee, L.C., and R. Montgomery, *Arch. Biochem. Biophys. 82*:70 (1959).
70. Laemmli, U.K., *Nature 227*:680 (1970).
71. Waniska, R.D., and J.E. Kinsella, *J. Agric. Food Chem. 32*:1042 (1984).
72. Waniska, R.D., and J.E. Kinsella, *Int. J. Peptide. Protein Res. 23*:467 (1984).
73. Leach, S.J., and H.A. Scheraga, *J. Biol. Chem. 235*:2827 (1960).
74. Becker, R.S., *Theory and Interpretation of Fluorescence and Phosphorescence*, Wiley-Interscience, New York, NY, 1969.
75. Sklar, L.A., B.S. Hudson and R.D. Simon, *Biochemistry 16*:5100 (1977).
76. Greenfield, N., and G.D. Fasman, *Ibid. 8*:4108 (1969).
77. Blout, E.R., in *Fundamental Aspects of Recent Developments in Optical Rotary Dispersion and Circular Dichroism*, edited by F. Ciardilli and R. Salvadori, Heydon & Son, London, England 1973, p. 352.
78. Waniska, R.D., and J.E. Kinsella, *J. Agric. Food Chem. 33*:1143 (1985).

79. Goodrich, F.C., in *Progress in Surface and Membrane Science*, Vol. 7, edited by J.F. Danielli, M.D. Rosenberg and D.A. Canenhead, Academic Press, New York, NY, 1973, p. 151.
80. Buckingham, J.H., and C.S.W. Reid, *Rev. Rural Sci. 1*:65 (1974).
81. Turner, S.R., M. Litt and W.S. Lynn, *J. Colloid Interface Sci. 48*:100 (1974).
82. Waniska, R.D., and J.E. Kinsella, *Food Hydrocolloids 2*:59 (1988).
83. Waniska, R.D., and J.E. Kinsella, *J. Food Sci. 44*:1398 (1979).
84. Waniska, R.D., K.J. Shetty and J.E. Kinsella, *J. Agric. Food Chem. 29*:826 (1981).
85. Pierce, K.N., and J.E. Kinsella, *Ibid. 26*:716 (1978).
86. Donovan, J.W., in *Physical Principles and Techniques of Protein Chemistry*, Part A, edited by S.J. Lead, Academic Press, New York, NY 1969, p. 101.
87. Reese, E.T., and F.M. Robbins, *J. Colloid Interface Sci. 83*:393 (1981).
88. Voutsinas, L.P., S. Nakai and V.R. Harwalker, *Can. Inst. Food. Sci. Technol. J. 16*:185 (1983).
89. Voutsinas, P. Leandros, E. Cheung and S. Nakai *J. Food. Sci. 48*:26 (1983).
90. German, J.B., T.E. O'Neill and J.E. Kinsella, *J. Am. Oil Chem. Soc. 62*:1358 (1985).
91. Kim, S.H., and J.E. Kinsella, *J. Food Sci. 52*:128 (1987).
92. Galembeck, F., D.A. Ryan, J.R. Whitaker and R.E. Feeney, *Agric. Food Chem. 25*:238 (1977).
93. Chiba, H., H. Doi, M. Yoshikawas and E. Sugimoto, *Agric. Biol. Chem. 40*:1001 (1976).
94. Morr, C.V., *J. Dairy Sci. 68*:2773 (1985).
95. Li-Chan, E., and S. Nakai, *J. Agric. Food Chem. 29*:1200 (1981).

Chapter Seven

Molecular Properties of Proteins Important in Foams

J.B. German[a] and L.G. Phillips[b]

[a]Department of Food Science and Technology
University of California, Davis
Davis, CA

[b]Department of Food Science
Cornell University
Ithaca, NY

Foams are thermodynamically unstable colloidal systems in which gas is maintained as a distinct dispersed phase in a liquid matrix. In foods the kinetic barrier to bubble coalescence and rupture is typically provided by a protein film surrounding the bubble. In foam formation soluble proteins are subjected to an interfacial exposure/adsorption which alters their structure and allows for subsequent associations with other proteins in the interface. Our goal in understanding this functionality is to determine the role of protein structure upon these interactions. Thus, we need to know how the forces enjoined on soluble proteins during this dynamic interfacial exposure and absorption period alter their structure, and then how these altered structures lead to novel molecular and macromolecular interactions. It is these novel associations which then provide the energetic barrier to resist desorption and collapse. The consensus of recent research is that the ability of proteins to form a gel-like film at the bubble interface is critical for food foam stability(1-3). Some of the research on the structural basis for film formation will be reviewed. Alternatively, it has become clear that the capacity of some proteins to interfere with and even to disrupt the integrity of existing films is dramatic. The structural basis of these net disruptive interactions is not well known, but practically this is important in many food applications, and some recent works on the properties of protein preparations as foam depressants are described.

Interfacial Events

The various molecular events believed to be responsible for the functionality of foaming are complex and not entirely understood. The film formation process can be divided into four steps. The generation of new interface is the necessary, high energy input, initial event. In step two, soluble proteins, maintained in a particular conformation (primarily by entropically favorable solvent (hydrophobic) interactions) arrive at this interface by diffusion. The elimination of solvent energetics from the balance of stability allows the protein to unfold and pursue more energetically favorable associations; first intramolecular (step three), and , as more protein arrives and unfolds at the interface, increasingly intermolecular (step four). Forces and structures which favor intermolecular associations in this step improve foaming ability, while forces and structures which prevent intermolecular associations in this step decrease foaming ability (2,4).

An additional event of interest to most foaming applications, though it is not studied in basic monolayer techniques, is the effect of continuously generating new surface as the events above occur. Whipping is not an instantaneous process. New surface is generated over finite periods. Thus, protein aggregates formed by an initial surface event can be redistributed onto new, clean surfaces. This has not been well studied but the overwhipping phenomenon of egg white as discussed below implies that changes in protein structure produced by interfacial absorption can impact on subsequent interfacial properties.

Finally, foam retardants function by intercalating into the interfacial film network and disrupting existing associations (5). Films provide stability to foams by virture of a continuous kinetic barrier to coalescence, either physically or chemically. Since coalescence requires a sole point nucleation site for rupture to then proceed spontaneously, foam breakers are effective at very low concentrations and are therefore very difficult to investigate structurally (5).

Protein Conformation

The importance of the retention of some native protein conformation and structure in the development of foams is well documented by both model systems and actual food foams (2). This fact is most clearly evident in foods by the frequency of loss of this functionality in response to protein processing. Foaming is perhaps the most sensitive functional property to alterations in the structure of dairy proteins (6). Changes in the structure of specific food proteins which are virtually undetectable by the most modern physicochemical techniques can destroy the ability of the protein to form an acceptable foam (7). The ability to associate in a denaturing environment, i.e., an interface, and yet retain some residual conformation appears to be crucial to the formation of a stable film. This increased structural demand presumably arises from the necessity not only to accumulate at an interface but also to develop a film which must physically resist an energetically favored collapse. The importance of protein conformation, and particularly secondary structure, to this latter phase is suggested by experiments in which proteins are modified and the rate extent of surface adsorption compared with film and foam stability. Extensive modification, including glycosylation, succinylation and enzymatic hydrolysis, reduces secondary structure. While these treatments have frequently increased surface adsorption kinetics and surface pressure, film strength and foam stability were almost invariably decreased (2,4).

Intramolecular Forces

The structure of a protein in solution and the forces which maintain this structure are clearly important to its ability to interact at an interface. Perturbation of native structure significantly alters the interfacial behavior of proteins (8). We have interpreted a variety of studies to suggest that the magnitude of forces maintaining native structure are critical to subsequent film formation at an interface. Those proteins which most owe their solution structure to solvent entropy unfold most readily after leaving the solvent and adsorbing to a surface. The formation of novel intermolecular associations would be more favored, and desorption from the surface less likely. In contrast, those proteins which are stabilized predominantly by intramolecular interactions in solution (ion pairing, disulfide bonds etc.) even though they could spontaneously adsorb to a clean interface, would not as readily alter their native structure and continue to form novel intermolecular associations. Thus, these proteins have a higher desorption tendency. For example, soy 11S globulin is stabilized in solution by both hydrophobic associations and intramolecular disulfide bonds and is relatively resistant to heat and urea denaturation (9). Cleavage of the disulfide linkages does not in itself dissociate the protein but significantly reduces the stability to dissociation in response to a solvent destructuring denaturant. The reduction of these intramolecular disulfide bonds dramatically enhances both film strength and foam stability (3). Alternatively, foam stability of solutions of β-lactoglobulin which readily unfolds at an interface is not affected by the reduction of disulfide bonds (7).

Intermolecular Forces

The forces which favor or retard the interaction of proteins unfolding at an interface are the most critical to foam stability but are also the least understood. This is a technical problem since the stability of protein films presumably requires multiple layer protein associations beyond that which basic studies of monolayers can readily investigate. Although hydrogen bonding and hydrophobic interactions probably play a role here, electrostatic interactions have been more completely studied. The most straightforward evidence to support the requirement for interaction beyond the single protein monolayer in foam formation is provided by studies in which electrostatic interactions are manipulated, typically by altering charge distributions as a result of changing pH. This research is circumstantial but voluminous (1). Proximity to the isoelectric pH is the most widely observed environmental correlate to foaming ability, in contrast to emulsion stability which is maximized away from the isoelectric point (2). Such studies have been widely interpreted to indicate that a strong repulsive force developed as a consequence of adsorption of highly charged proteins to dilute monolayer coverage prevents the approach of additional, like-charged proteins. Furthermore, one would

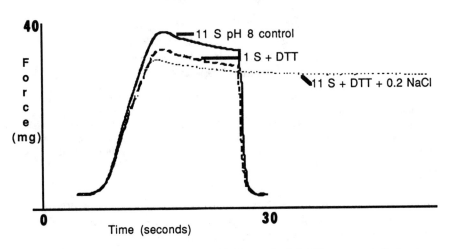

Fig. 7.1. Tensiolaminometric analyses of 0.1% soy glycinin. Films were pulled onto a 1 cm platinum frame from solutions after cleaning the surface by aspiration (3).

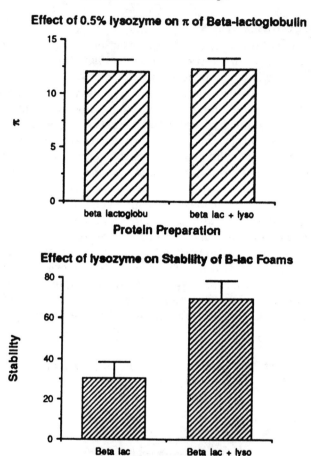

Fig. 7.2. Effect of hen egg lysozyme 0.5% by weight on the surface pressure π and the stability of foams made from 5% solutions of β-lactoglobulin. Surface pressure of the protein solutions was measured by the de Nouy ring method using a Fisher tensiomat. Foam stability was estimated after solutions were whipped five minutes in a double beater mixer, and the stability estimated as the time in minutes to drain 50% of the foam by weight (7).

predict that masking these charges with counterions would reduce electrostatic repulsion and favor protein associations beyond a monolayer. Salts as neutralizing counterions are indeed highly effective protein foam enhancers at pH's distant from the isoelectric point (1) but are actually depressants at the isoelectric point (10). Thus, at the isoelectric point, the forces of repulsion are minimized and the forces of attraction can be observed to contribute to film formation.

A balance of electrostatic interactions are therefore thought to be critical for foaming. Too much, and repulsive forces preclude the formation of protein films. Too little, and film strength is compromised. Again, soy glycinin has proven a useful model protein. The native structure consists of both acidic and basic subunit fragments which interact in the native state through hydrophobic interactions and disulfide bonds. If these disulfide bonds are cleaved and the structure specific hydrophobic associations disrupted (e.g., by heat or urea) the oppositely charged subunits will spontaneously reassociate, but via electrostatic interactions (9,11). The protein fragments on surface absorption would thus exhibit the ability to associate via electrostatic interactions in the developing interfacial film only if the intramolecular disulfide bonds were cleaved. This was found to be true. The sensitivity to electrostatic repulsive forces was demonstrated by the fact that salt as a counterion was still required for the disulfide reduced protein to form stable films far from the isoelectric point of the protein (Fig. 7.1). The ability to utilize basic proteins as foam stabilizers is the most exciting development of this understanding of the forces involved in film integrity. The basic protein lysozyme has long been considered to be important to the whipping properties of egg whites (12). Hen egg lysozyme was added at 0.5% to solutions of 5% β-lactoglobulin, an acidic protein already exhibiting good foam stability. This addition had no significant effect on the surface pressure of the protein solution; however, the foam stability (time to 50% drainage) was more than doubled (Fig. 7.2).

Foam Depressants

The majority of studies in this area have identified polar lipid contaminants as important foam retardants (1). Two other classic food systems illustrate that properties of certain proteins contribute to these effects as well. Egg whites exhibit a well described overwhipping phenomenon. Optimal overrun and stability of egg white foams occur after a short period of whipping. Continued whipping actually reduces the stability of the foam. This has been traced to the development of protein aggregates which specifically depress foaming. Surface denaturation of native egg protein produces a protein conformation which is capable of disrupting surface films and depressing the overrun and stability of egg white foams. The binding of divalent cations, particularly copper, stabilizes conalbumin and effectively delays the development of foam depressing

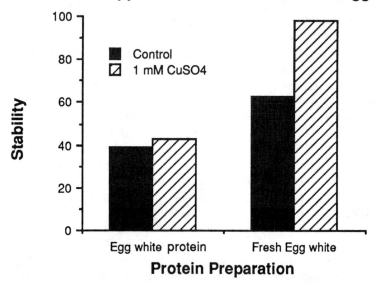

Fig. 7.3. Effect of copper (as 1 mM CuSO4) on the stability of foams whipped from solutions of egg white protein (spray dried egg white protein < 0.1% SDS) and fresh egg white. Solutions were whipped for 10 minutes and foam stability determined as in Figure 7.2.

activity by overwhipping (13). These are clearly structure specific events. The dependence of protein structure is illustrated in Fig. 7.3. Copper enhanced the stability of foams prepared from fresh egg whites; however, it had no significant effect on processed egg white protein, even though both protein preparations exhibited overwhipping. This suggests that the capacity to bind copper and confer increased conformational stability is lost during processing.

Another rather classic foam retardant in the food literature is proteose peptone in whey (14). This proteolytic fragment is thought to be the sole component in whey which characteristically depresses loaf volume in bread. In attempts to further describe this phenomenon we have found a potentially more important component with entirely different properties. This foam depressant activity is associated with a high molecular weight protein. It is heat stable, retained on 100,000 dalton molecular weight cut-off membranes and found in variable quantities in acid and sweet wheys. Although this component has not been fully char-

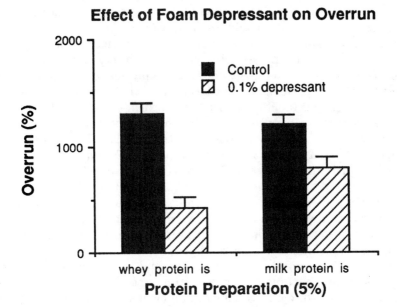

Fig. 7.4. Effect of added foam depressant on the overrun of protein foams. Solutions of whey protein isolate or milk protein isolate 0.5%, pH 7.0 were whipped as above for 15 minutes in the absence or presence of 0.1% by weight of the depressant component in milk (prepared from raw skim milk) (15).

acterized, it is likely a lipoprotein(s) derived from the milk fat globule membrane. This component will yield polar lipids with a fatty acid composition similar to milk fat membranes, but only if exhaustively extracted (7). While the polar lipids may be the actual foam depressant at the level of the interface, their transport and release could be predicted to be a function of the structure of the lipoprotein itself, as well as that of the protein film onto which it adsorbs (Fig. 7.4). This again illustrates that the processing history and structure of proteins will dramatically affect their functional behavior in a foaming application.

Future Objectives

While significant progress has been made, our understanding of the structural basis of protein foaming is still at a formative stage. Accumulating this information, however, promises many potential benefits. The elucidation of specific molecular features of known and commonly used functional proteins, such as egg white, gluten and dairy proteins, would have several immediate implications. First, processing, modification or storage conditions which could be identified as deleterious to the structure and functioning of specific proteins could be avoided. Although this sounds obvious and very straightforward, the utility of most available functional proteins is seriously impaired simply due to the variability in their functional performance resulting from their processing history. Once those critical functional structures have been identified and their relative sensitivity quantified, this disconcerting variability could presumably be eliminated by predictably manipulating processing conditions. Similarly, if molecular features responsible for antifoaming properties could be identified, these fractions could be specifically processed or extracted out of commercial samples. Additionally, if the molecular features of protein structure were understood for individual proteins, the potential advantages for specific applications of mixtures of proteins could be realized. This is likely to be the most productive area in the immediate future for food applications.

In an entirely novel direction, if the specific structural features responsible for this functionality were understood, one could begin to rationally redesign food proteins to add elements of functionality not currently available in any food protein. The potential available from molecular biology to provide a polypeptide with any given sequence implies that one could produce an idealized functional food protein. Unfortunately, the links between sequence, structure and function are currently obscure.

We would predict from studies relating thermal stability to structure at the sequence level, if structural relations in foaming are similar, that there is no single structural feature which provides optimal performance. There is unlikely to be a foam helix or a film pleated sheet. At the same time this does not preclude the discovery of common features of the tertiary structure which can be assigned either to improvement, or conversely, to disruption of film integrity. In this direction the application of site directed mutagenesis techniques will be essential. Of even more interpretive value may be the addition of replacement of entire sequence regions (perhaps intact exons) into proteins as functional modules or cassettes.

Methodology

The fact that very subtle structural differences are important is evident from the effects of copper on overwhipping of egg white protein. The many variables which have impact on these differences emphasize the necessity of adopting standardized, or at least comparable, methodologies for foam measurement. These observations also suggest that comparisons between foaming behaviors of different proteins are unlikely to provide definitive information, due to the profound structural differences between them. Recent developments in model protein systems have elucidated some of the molecular structures which improve and detract from foam formation and stability. Adoption of standardized protein models-at least for comparative purposes-would similarly be very productive.

At the practical level certain descriptive analyses are still lacking. Non-invasive measurements of structure development and collapse are essential. In this regard, novel imaging techniques will prove particularly useful. As an example, analysis of the drainage of egg white foam using Nuclear Magnetic Resonance imaging is illustrated in Fig. 7.5 (16). The magnitude of signal has been used in this application to accurately estimate the water content, in effect the density of the foam, at that point. Taking a planar image of the foam generates a profile of density. Over time this density stratification shown allows us to measure the drainage behavior throughout the foam not simply as it exits at the bottom.

The most obvious descriptive property of foams which needs to be accurately measured is their bubble size distribution. Without this fundamental parameter, mechanistic studies are difficult to interpret. Development of automated analyses for these measurements will advance the field significantly.

From the pragmatic point of view, food foams are almost invariably 'Kugelschaum', i.e., holding a high liquid content. Unfortunately, most fundamental studies address the properties of 'Polyederschaum' foams, those with very low water content and polyhedral bubbles. Since foam drainage is the macroscopic parameter of most immediate interest to stability of foams, the thinning of polyederschaum films is generally considered a critical property of the stability afforded by functional proteins. This is probably not sufficient. Food foams are not stable indefinitely by virtue of the protein film around the bubbles, in most cases the protein film must retain the bubble structure only until other components in the continuous phase are induced to gel. This affords long term stability. The critical question therefore is not related to lamellar thickness, but to film stability. If a protein(s) can maintain a stable interfacial film long enough for bulk phase agents to develop stabilizing interactions, the protein is a good foaming protein. If the film tends to break

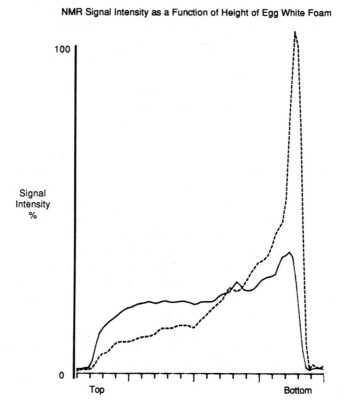

Fig. 7.5. Use of NMR image analysis to monitor drainage of foam. Egg whites were whipped for one minute and a cylindrical core was punched from the center of the foam. This sample container was placed in the imaging coil and planar, and 4 mm thick vertical images recorded. The proton signal was integrated horizontally, generating a vertical signal intensity profile from top to bottom of the foam cylinder. Sequential scans were repeated every minute and scans 2 (—) and 10 (—) minutes after whipping are shown to illustrate the density alterations resulting from drainage within the foam.

initially, or in response to the conditions necessary to induce gelation of the continuous phase (i.e., heating), the protein is not competent for foaming and slight differences in resistance to lamellar thinning are irrelevant. Therefore, the critical parameter to estimate in the practical utility of a foaming protein is its ability to form and maintain a stable film. Methods currently available to estimate this include the tensiolaminometer (3,17) and measures of dynamic dilatational modulus (2). These and other novel methods need to be incorporated into actual foaming studies.

In a more basic sense, the necessary information concerning protein conformation at the interface must be ultimately developed to gain an understanding at the molecular level of the protein structures, associations and their functions. While considerable methodology exists to study protein structure and conformation in solution, similar information of the structures formed at interfaces is needed. In this direction, surface circular dichroism, FTIR, and Total Internal Reflectance Fluorescence spectroscopies and electron tunneling microscopy will prove very useful in the near future.

Studies to provide clues to the type and magnitude of forces acting between proteins, especially at the interface, are critical. Specific alterations in protein sequence will be invaluable, necessary tools to this end.

Acknowledgment

We would like to thank the National Dairy Board for their support.

References

1. Halling, P.J., *CRC Critical Rev., Food Sci. Nutr. 155*:13 (1981).
2. Phillips, M.C., *Food Technol. 35*:50 (1981).
3. German, J.B., T.E. O'Neil and J.E. Kinsella *J. Am. Oil Chem. Soc. 62*:1358 (1985).
4. Kinsella, J.E., and D.M. Whitehead, in *Proteins at Interfaces* ACS Symposium Series 343, Am. Chem. Soc., Washington, DC, 1987.
5. Prins, A., in *Food Emulsions and Foams*, edited by E. Dickinson, Royal Society of Chemistry, Burlington House, London, UK, 1987.
6. Liao, S.Y., and M.E. Mangino, *J. Food Sci. 52*:1033 (1987).
7. Phillips, L., M.S. Thesis, Cornell University, Ithaca, NY, 1988.
8. Song, K.B., and S. Damodaran, *J. Agric. Food Chem. 35*:236 (1987).
9. German, J.B., S. Damodaran, and J.E. Kinsella, *Ibid 30*:807 (1981).
10. Kim, S.H., Ph.D. Thesis, Cornell University, Ithaca, NY 1984.
11. Damodaran, S., and J.E. Kinsella, *J. Agric. Food Chem. 30*:1249 (1981).
12. Poole S., S. West, and C. Walters, *J. Sci. Food Agric. 35*:701 (1984).
13. McGee J., S. Long, and W. Briggs, *Nature 308*:607 (1984).
14. Kinsella, J.E., *Food Chem 7*:273 (1981).
15. Aschaffenburg, R., and J. Drewery, *Biochem 65*:237 (1957).
16. German, J.B., and M. McCarthy, in preparation, 1988.
17. Eydt, A.J., and H.L. Rosano, *J. Am. Oil Chem. Soc. 45*:607 (1968).

Chapter Eight
Lipid-Protein-Emulsifier-Water Interactions in Whippable Emulsions

N.M. Barfod[a], N. Krog[a] and W. Buchheim[b]

[a]Grindsted Products A/S
DK-8220 Brabrand, Denmark

[b]Milk Research Institute
D-2300 Kiel, West Germany

The air bubbles of whipped dairy cream are stabilized by a layer of partly coalesced fat globules adsorbed at the air-water interface. This layer is important for foam texture and firmness (1,2). In contrast to whipped dairy cream we have found that the air-bubbles in whipped topping (imitation cream) are stabilized by adsorbed fat crystals rather than globular fat particles (3). The fat phase of lipid droplets in topping powders with good whippability and foam stiffness is in a supercooled state due to strong lipid-protein interactions (4). At the moment the topping powder comes into contact with cold water, the resulting emulsion starts to destabilize, and the fat phase in the lipid droplets undergoes a spontaneous crystallization. The fat crystals formed during this process are oriented at the air-water interface during whipping and stabilize the air bubbles in whipped topping, thus contributing to foam stiffness (4).

The interplay of protein, lipid, emulsifier and water in toppings has been further elucidated by means of low-resolution pulsed NMR, interfacial tension and X-ray diffraction analysis. The interactions studied may be significant for other oil-in-water emulsions containing both proteins and emulsifiers.

Experimental Procedures

Preparation of topping powders.

Topping powders were made by spray-drying an emulsion containing 25% hydrogenated coconut oil, 5% distilled propylene glycol monostearate (Promodan SP, Grindsted Products, Denmark) as emulsifier, 15% maltodextrin, 0–5% sodium caseinate and 50% water.

The emulsions were prepared by melting coconut fat and emulsifier together at 80°C and dissolving the caseinate and maltodextrin separately in the water phase at 90°C. The two phases were mixed and homogenized on a one-stage high-pressure piston homogenizer with a Rannie liquid whirling-type valve at a pressure of 100 kg/cm^2 at 80°C. The emulsions were then spray-dried using a rotating atomizer (16,000 rpm) with an air inlet temperature of 150°C and an outlet temperature of 90°C. The spray-dried topping powders were cooled to 5°C for one hour and then stored at 20°C or at a lower temperature.

Pulse NMR analysis.

The measurements were taken on a pulse low-resolution NMR-spectrometer (model Minispec PC/20B from Bruker Spectrospin, West Germany) operating at 20 MHz for protons.

Solid fat content was analyzed by a standard program using a 90 degree pulse. Simple dry mixtures of topping powder ingredients were made by mixing a finely ground topping fat phase with powdered maltodextrin and sodium caseinate in the same ratios as in topping powder. However, the dry mixture was not subjected to the homogenization and emulsification process as were the topping powders. The samples were stored at 5°C overnight and equilibrated for 1 hr at 5, 10, 15, 20, 25 and 30°C before pNMR-analysis.

Lipid-protein interaction in dry topping powder and water-binding in reconstituted topping emulsion were studied by multiexponential T_2-relaxation analysis using the spin-echo technique (5). The Bruker Minispec pNMR spectrometer used is equipped with computer facilities to measure the spin-echo amplitudes at time 2τ after a $90°/\tau/180°$ pulse sequence. The time delay τ between the pulses was varied from 0.1 to 50 msec. The delay between two consecutive measurements was 2 sec to allow the nuclear magnetization to return to its equilibrium value. Low bandwidth and diode detection were used. Each relaxation analysis was calculated from 30 to 40 data points. T_2 values were determined from semilogarithmical plots of signal amplitude (M_{obs}) versus pulse spacings τ. Nonexponentiality of spin-echo curves is often found in real food systems. This can be accounted for by the presence of 2, 3 or more recognizable components representing species of hydrogen atoms with different mobility (5). Such curves can be resolved by subtracting from the experimental curve the slowest decaying component, so that a new decay curve appears. Again, this does not show single-exponential decay and the curve may be resolved once more by subtracting the slowest decaying component. In this way, the T_2-relaxation of different types of hydrogen species may be determined step by step. The extrapolated signal amplitude M(O) corresponding to $\tau = 0$ is proportional to the relative abundance of each hydrogen species (5). A program written in BASIC for the resolution of multiexponential T_2-relaxation curves into single exponential decay curves was used in a Hewlett-Packard HP87 computer for data processing.

Interfacial tension analysis.

The interfacial tension of oil-water two-phase systems was measured with a Digital-Tensiometer K10 (Krüss, Hamburg), using the Wilhelmy plate method. Emulsifier was dissolved in oil, and protein was dissolved in water both by heating to about 50°C before allowing contact between the oil and water phases. In order to obtain near equilibrium values, the interfacial tension analyses were started 30 min after contact between oil and water. The interfacial tension and temperature were measured continuously, using a SE 120 Servogor two-channel recorder (Metrawatt, Nürnberg). The systems were analyzed during cooling at a rate of 0.3°C per minute using a Hetotherm thermostat type 04PG623 extended with a gradient unit (Heto, Copenhagen).

X-ray diffraction.

A DPT (diffraction pattern temperature) Model 300 X-ray camera (Incentive Research and Development AB, Sweden) equipped with a Lauda P120 electronic temperature programmer was used.

Results and Discussion

There is generally a lower percentage of solids in topping powder than in a corresponding dry mixture. This difference is due to lipid-protein interactions resulting in a supercooling of the fat phase in the lipid droplets, as shown recently (4). A cold treatment at 5°C for at least 1 hour is essential for obtaining the maximum supercooling and functionality of the product. The amount of protein in the formulation determines the degree of supercooling by controlling the total droplet surface area. This is shown in Fig. 8.1, where increased lipid droplet size is obtained with decreased protein content, as studied by transmission electron microscopy (3). However, with marginally hydrolyzed protein (3% of peptide bonds cleaved), the supercooling is greatly reduced with-

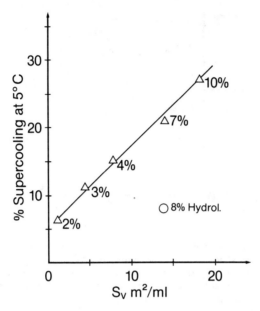

Fig. 8.1. Effect of protein content on particle surface area per unit volume and on supercooling in fat globules of dry topping powders.

out affecting particle size, as seen in Fig. 8.1. This indicates that the emulsifying capacity and the degree of supercooling are not directly related but must be due to two different mechanisms.

The nature of the lipid-protein interaction in dry topping powder was studied by multiexponential T_2-relaxation analysis. Fig. 8.2 shows the relaxation behavior of pure hardened coconut oil, melting point 31°C, and soybean oil at 35°C. Two distinct relaxation lines are observed for both oils. The increased proportion of fast-relaxing hydrogen atoms seen in line II in coconut oil (56%) as compared to that of soybean oil (20%) is probably due to the high content (more than 50%) of shortchain fatty acids, mainly lauric acid.

Since marginally hydrolyzed sodium caseinate greatly reduces the fat supercooling observed below 20°C, it was of interest to study this effect using T_2-relaxation analysis. The results are shown in Table 8.1.

The supercooling in topping powders observed below 20°C is primarily due to an effect on the fast relaxing protons (line II in Table 8.1). These fast relaxing protons represent a much bigger fraction of intact caseinate (60%) than hydrolyzed caseinate (41%) at 15°C. The combination of

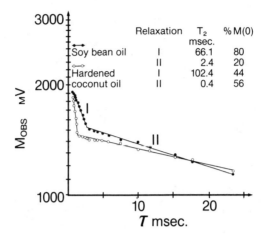

Fig. 8.2. T_2-spin echo relaxation of soybean and hardened coconut oil at 35°C. M_{obs} = Maximum amplitude of the echo signal observed. $M(0)$ = Echo signal amplitude from the single-exponential part of the relaxation curve extrapolated to $\tau = 0$. This signal is proportional to the percentage of hydrogen atoms with the T_2-time corresponding to that line.

TABLE 8.1

T_2—Relaxation of Dry Toppings with Different Sodium Caseinates

		15°C		35°C	
Sample	Relaxation	T_2 msec	$M(0)$ %	T_2 msec	$M(0)$ %
Topping with intact	I	27.1	40	45.0	50
sodium caseinate	II	0.6	60	0.7	50
Topping with hydrolyzed	I	35.0	59	50.3	48
sodium caseinate	II	0.5	41	0.7	52

these results with those obtained from the pure oils indicates that sodium caseinate exerts its supercooling effect on coconut oil by interacting with its short-chain fatty acid residues and inhibiting crystallization. Experiments with other fat types such as butter, partially hardened soybean and fish oil showed that the supercooling effect of sodium caseinate is obtained only with short-chain fatty acid fat such as coconut and palm kernel oil.

The emulsion that results from the reconstitution of topping powder in cold water is unstable because of protein desorption from the fat

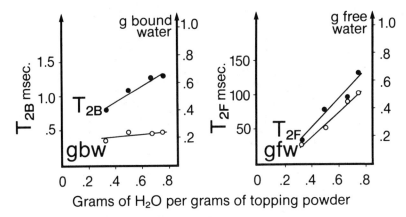

Fig. 8.3. Exchange of free and bound water in topping emulsion at varying water contents at 5°C. T_{2B} = T_2-time of bound fraction, T_{2F} = T_2-time of free fraction, gbw = grams of bound water, gfw = grams of free water.

globules resulting in coalescence and spontaneous fat crystallization (4). The water-binding of the topping emulsion at 5°C after destabilization was also investigated by multiexponential T_2-relaxation analysis. This type of analysis may supply information about water mobility, the relaxation time of slow and fast relaxing protons and the relative amounts of the two types of water (in the following called "free and bound water"). When the ratio of water to powder is varied, it is possible to analyze the exchange process between free and bound water (5). The influence of this variation on relaxation time and on amounts of free and bound water is shown in Fig. 8.3.

With increasing amounts of water, the fraction of bound water remains fairly constant. This might be due to the constant thickness of water layers in a gel phase in the fat phase due to the presence of emulsifier observed by X-ray diffraction analysis (see below). Since T_2 of bound water varies with increasing amounts of water, it is presumed that water is freely exchangeable between the free and bound state. If there were restrictions in water exchange, then the relaxation time of bound water would be independent of water content (5). The increase of the fraction of free water proportional to the amounts of water added is self-evident.

According to the results, it seems appropriate to modify the term "free water," which is normally regarded as a very loosely bound type of water. The mobile water fraction exhibits T_2-values of about 100 msec or less,

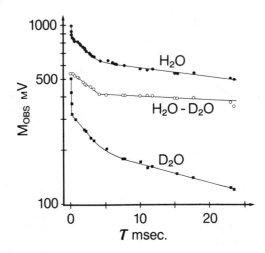

Fig. 8.4. T_2-spin echo relaxation of topping powder in H_2O and D_2O (1:1, w/w) at 5°C. T_2-relaxation of water protons in topping powders at 5°C (H_2O - D_2O).

TABLE 8.2

T_2-Relaxation and Amounts of Free and Bound Water in Topping Emulsion

Type of topping	Relaxation	5°C		30°C	
		T_2 msec	M(0) %	T_2 msec	M(0) %
+ Emulsifier	I Free H_2O	102.2	65	149.7	76
	II Bound H_2O	0.9	35	2.1	24
– Emulsifier	I Free H_2O	110.7	79	200.5	62
	II Bound H_2O	4.2	21	1.2	38

much shorter than the 1000 to 2000 msec for free or bulk water. This implies that the free water measured in the emulsion is relatively immobile when compared to pure water.

The relaxation lines of topping in water are the sum of the relaxation of free and bound water, of water-solubilized sugar and protein protons, and of liquid protons in the coconut oil. It was of interest to study the effect of lipophilic emulsifier on the water-binding in the system. Consequently, an attempt was made to eliminate the signal from oil and solubilized non-water protons. The nature of this investigation is outlined in Fig. 8.4.

The relaxation behavior of the sample is investigated in both H_2O and D_2O. By subtracting the signal amplitude obtained in D_2O from the corresponding amplitude obtained in H_2O, the relaxation behavior of the water component in the system is obtained. The extremely fast relaxing line observed at short τ values in the curves with H_2O or D_2O is due to the relaxation of solid-like protons. These are very difficult to quantify, and fortunately they are eliminated after subtraction. The technique might be called differential multiexponential T_2-relaxation analysis. The results of such an analysis are shown in Table 8.2 for toppings with and without emulsifier. At 30°C the emulsion is relatively stable whereas at 5°C it is totally destabilized.

The emulsifier only has a very small reducing effect on the T_2-time of free water. However, both the fraction and the T_2-time of bound water are influenced to a considerable extent by the presence of the emulsifier. The data suggest that water-binding of bound type (low T_2 and high % $M(O)$) is due to protein at 30°C and emulsifier at 5°C. In other words, at 30°C water-binding is dominated by protein hydration, whereas at 5°C water-binding is primarily due to emulsifier-mediated micellar solubilization of water in the oil phase. This type of water is freely exchangeable with bulk water, suggesting that the water-in-oil solubilization takes place only a short distance from the oil/water interface.

To investigate the strong water-binding mediated by emulsifier, simple two-phase systems in petri dishes were made, consisting of equal volumes of sunflower oil + emulsifier and 0.02 M citrate-phosphate buffer pH 7.0. The emulsifier was melted in the oil at 80°C and carefully poured down onto the water surface (also 80°C). After three days at 4, 20 and 35°C the amount of water absorbed into the oil phase was measured by Karl Fischer titrations (Figs. 8.5 and 8.6).

Increasing amounts of water are absorbed with (a) increasing emulsifier concentration in the oil phase and (b) decreasing temperature (Fig. 8.6). This behavior is quite normal for a nonionic emulsifier. According to Shinoda and Friberg (6), the hydration between water and the hydrophilic moiety of a nonionic emulsifier increases as the temperature is decreased. Upon cooling a two-phase system with hydrocarbon, nonionic emulsifiers and water, a new intermediary phase containing large amounts of water, hydrocarbon and emulsifier concurrently develops (6). After heating the crystalline water-containing fat phase, the water became visible as droplets in the melted fat phase, (w/o emulsion)

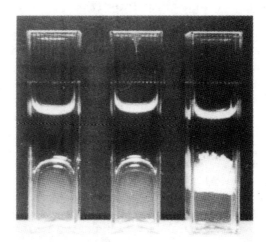

Fig. 8.5. Emulsifier-mediated water-uptake into oil phase using water-oil-emulsifier two-phase systems in plastic cuvettes after three days incubation. Left: Sunflower oil:water at 15°C. Middle: Sunflower oil with 20% emulsifier:water at 30°C. Right: Same as middle, but at 15°C.

when studied by light microscopy. The phenomenon seems to involve an emulsion reversion.

The water absorption in the oil phase was investigated by X-ray diffraction analysis. A melted blend of coconut oil (31°C m.p.) and emulsifier (1:1) was carefully poured onto a water phase containing 1% sodium caseinate in a petri dish, and then allowed to cool to 5°C. After 20 hours of equilibration at 5°C, the solid lipid phase was removed and a thin layer of the interface that had been in contact with the water phase was removed for X-ray diffraction analysis. The results appear in Table 8.3, and show that the long spacings Å of emulsifier in the lipid phase are increased from 48.9 Å in the bulk phase to 56.1 Å in the interfacial layer. This increase of approximately 7 Å can only be due to the penetration of water into the polar regions of the emulsifier due to the so-called hydration force (7). The second long spacing (B) of 35.4 - 36.3 (Å) is from the chain packing of the coconut oil, i.e., methyl-end group distance of the triglycerides in the β'-form. The short spacings are the lateral fatty acid chain distances and correspond to a β'-form of the triglycerides overlapped by the α-form of emulsifier. The appearance of

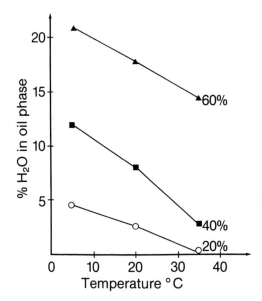

Fig. 8.6. Water content in oil phase of two phase systems in petri dishes after three days. Effect of emulsifier concentration in oil phase (20, 40 and 60% w/w) and of temperature.

TABLE 8.3
X-ray Diffraction of Interfacial Layers from Coconut Oil/Emulsifier/Water Systems

Sample preparation	Long spacings d(Å)		Short spacings d(Å)
	(A)	(B)	
Interfacial layer: Coconut oil: Emulsifier (1:1) + water, temp. 5°C	56.1	− 35.4	4.18 − 3.83
Bulk phase: Coconut oil: Emulsifier (1:1) Melt, cooled to 5°C	48.9	− 36.3	4.29 − 4.16 3.97 − 3.82 − 3.65

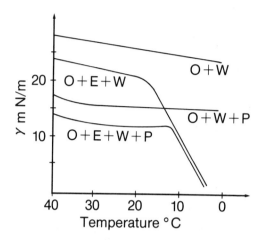

Fig. 8.7. Effect of cooling on interfacial tension of sunflower oil:water two phase systems containing 2% emulsifier (in oil phase) and 0.01% sodium caseinate (in water phase). O = Oil; W = Water; E = Emulsifier; P = Protein. Cooling rate 0.3°C per minute.

two distinct long spacings shows that coconut oil and emulsifier do not form solid solutions, but crystallize in separate fractions.

The interfacial tension of simple two phase systems of emulsifier + soybean oil and water was measured to investigate the effect of emulsifier. Propylene glycol monostearate is not particularly surface active above its melting point (Fig. 8.7). However, a large drop in interfacial tension is observed from 20 to 5°C. This drop may be due to more efficient packing or to the increased density of polar groups of the emulsifier near the crystallization point. Similar observations have been described by Lutton et al. (8), who propose that such a decrease of interfacial tension at low temperatures may be due to either the crystallization of a monomolecular layer of emulsifier adsorbed at the interface and/or micellization near the interface. The results indicate that both emulsifier and protein are located at the oil-water interface above 20°C. Below this temperature, the emulsifier dominates at the interface. From earlier studies we know that protein is actually desorbed effectively during the cooling of a reconstituted topping emulsion (4). This effect might be due to the increased interfacial activity of emulsifier at low temperatures. The increased interfacial hydrophilicity will probably weaken the hydrophobic lipid-protein interaction, resulting in protein desorption.

Lactylated and acetylated monoglycerides as well as propylene glycol monostearate (which is used in this study), are more efficient in desorbing caseinate than monoglycerides, due to a very low affinity between the former surfactants and caseinate (results obtained from studies on model systems). The interfacial tension between coconut oil (without surfactant) and sodium caseinate in water was much lower compared to soybean oil and sodium caseinate in water, and suggests a very high affinity of sodium caseinate for coconut oil surfaces.

From the experiments performed, the following model for lipid-protein-emulsifier-water interactions is suggested.

Lipid-protein interactions.

The strong affinity of sodium caseinate for high lauric acid triglycerides results in the supercooling of the fat phase. A co-operative effect, arising from any binding segments in the polymeric protein molecule to triglycerides at the interface, inhibits the proper packing of fat crystals.

A reduced protein to lipid ratio results in decreased interfacial area and increased droplet size, resulting in reduced supercooling, which shows that the lipid-protein interaction is of relatively short-range character.

Hydrolyzed protein results in reduced supercooling, due to decreased co-operative effect from the reduced number of joining binding segments. However, the emulsifying capacity of intact and hydrolyzed protein is similar (producing identical droplet size distributions), showing that the two types of lipid-protein interaction are due to two different mechanisms.

Above room temperature there are strong lipid-protein interactions, strong protein hydration and low interfacial activity of emulsifier which is present as *multilayers in the oil phase* (Fig. 8.8).

Below room temperature the emulsifier molecules orient their polar groups towards the interface just before the start of crystallization, resulting in increased interfacial activity (Fig. 8.9). The increased hydrophilicity of the oil-water interface weakens the hydrophobic lipid-protein binding, initiating a desorption of protein from the interface. A similar desorption of milk proteins induced by emulsifiers has also been observed in dairy emulsion systems by Oortwijn and Walstra (9), using surface rheological analysis.

Because of the high water-binding capacity of emulsifier at low temperature, water penetrates into the emulsifier multilayers in the fat phase, accelerating the desorption of protein. After the lipid-protein

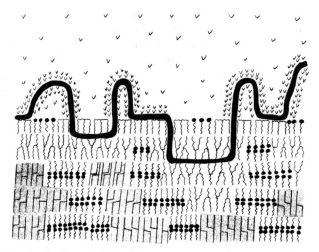

Fig. 8.8. Oil-water interface of reconstituted topping above room temperature.
V = Water molecule.
~ = Protein molecule.
= Crystalline and melted triglyceride.
= Crystalline and melted emulsifier.

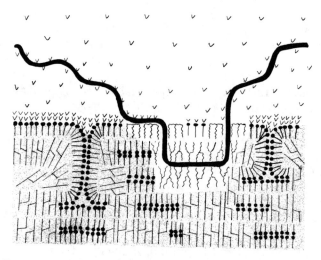

Fig. 8.9. Protein-emulsifier interaction below room temperature at the oil-water interface.

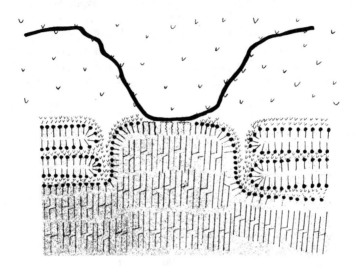

Fig. 8.10. Water-emulsifier interaction and emulsion breakdown below room temperature.

binding has been weakened, the supercooled fat starts to crystallize. After protein desorption, the emulsifier is unable to maintain the fat in an emulsified state on its own, probably due to the decreased steric stabilization, and coalescence sets in, which is further enhanced by spontaneous fat crystallization.

The fat crystals are oriented at the air-water interface of the air bubbles during the whipping of the destabilized emulsion, and thus contribute to foam stiffness. The concentration of emulsifier in the fat phase controls the rate of emulsion destabilization. The surface-denatured desorbed protein molecules are probably transported to the air-water interface during whipping, and contribute to foam formation.

The increase in fat crystallization is probably an important destabilizing factor (Fig. 8.10), as it has been reported that "the stability of an emulsion against coalescence is adversely affected by the formation of crystals in the oil droplets" (10).

In conclusion, a model on lipid-protein-emulsifier-water interactions in an oil-in-water emulsion system (topping) has been proposed. Competition between protein and lipid emulsifier for the oil-water interface is important for the functionality of this food product. At elevated temperatures the protein dominates at the interface and stabilizes the emulsion. At lower temperatures the protein is displaced from the interface by the emulsifier. This will abolish the lipid-protein-induced supercooling of the fat phase and is accompanied by the absorption of water into the fat phase (phase inversion), resulting in fat crystallization and droplet coalescence. The model proposed may be relevant for other emulsified fat-containing foods containing both proteins and lipid emulsifiers.

References

1. Buchheim, W., *Gordian* 6:184 (1978).
2. Schmidt, D.G., and A.C.M. van Hooydonk, *Scanning Electron Microsc. III*:653 (1980).
3. Krog, N., N.M. Barfod, and W. Buchheim, in *Food Emulsions and Foams*, edited by E. Dickenson, Royal Society of Chemistry, London, England, p. 144.
4. Barfod, N.M., and N. Krog, *J. Am. Oil Chem. Soc. 64*:112 (1987).
5. Brosio, E., G. Altobelli, and A. Di Nola, *J. Food Technol. 19*:103 (1984).
6. Shinoda, K., and S. Friberg, *Emulsions and Solubilization*, John Wiley and Sons, New York, NY, 1986.
7. De Nevey, D.M., R.P. Rand, D. Ginger, and U.A. Parsegan, *Science 191*:399 (1976).
8. Lutton, E.S., C.E. Stauffer, J.B. Martin, and A.J. Fehl, *J. Colloid Interface Sci. 30*:283 (1969).
9. Oortwijn, H., and P. Walstra, *Neth. Milk Dairy J. 33*:134 (1979).
10. Van Boekel, M.A.J.S., Ph.D. Thesis, University of Wageningen, Holland, 1980.

Chapter Nine
Molecular Properties and Functionality of Proteins in Food Emulsions: Liquid Food Systems

M.E. Mangino

The Ohio Agricultural and Development Center
Department of Food Science and Nutrition
The Ohio State University
Columbus, Ohio 43210

Emulsions are thermodynamically unstable mixtures of immiscible liquids such as vegetable oil and water. If energy is applied the systems may be dispersed, but increased surface energy causes the phases to rapidly coalesce unless an energy barrier is established. Emulsified droplets can be stabilized against coalescence by the addition of molecules that are partially soluble in both phases. In foods a number of small molecules can serve this function. Proteins capable of unfolding at the interface may also function as emulsifiers. The protein coats the lipid droplet and provides an energy barrier to both particle association and to phase separation.

The study of emulsions is complicated by the interactions that can occur when multiple components are present, and by the fact that conditions that are important in dilute solutions where the systems are easier to study may not apply to conditions likely to be found in foods. Generally it is possible to explain much of how proteins function in emulsions from a knowledge of the forces that are operational during emulsion formation and that are responsible for protein structure.

Once formed, food emulsions may be stored for months or even years. It is important, therefore, that the mechanisms and the nature of the forces responsible for emulsion breakdown be understood. This paper will review the forces involved in emulsion formation and relate these to the forces that govern protein lipid interactions. The mechanism of emulsion destabilization and the forces involved are discussed. Two excellent reviews have recently been published (1,2).

Forces

An understanding of the forces involved in emulsion formation and stabilization requires a discussion of the molecular properties of proteins which aid in the formation and stabilization of liquid emulsions. Generation of an emulsion involves the mixing of two immiscible liquids. The reason that the liquids are immiscible can be related to their relative polarities. When the nonpolar portions of a protein or any other molecule are exposed to the aqueous phase, they tend to spontaneously associate in a manner that minimizes contact with water. Measurements of the enthalpy of hydration of a number of nonpolar molecules yield values that are similar and negative. This suggests that interaction between nonpolar molecules and water should be favorable. When solubility data are examined, however, it is found that nonpolar molecules are only slightly soluble in water. Measurements of the free energy of transfer of nonpolar molecules from organic solvents to water give values that are positive. The negative values for the enthalpies of hydration and the positive free energy of transfer to the aqueous phase suggest that an entropically driven aggregation of nonpolar molecules occurs in an attempt to minimize their contact with water. The reason for this is that the intrusion of a nonpolar molecule interferes with the normal structure of water in such a way as to increase its order (3).

The ordering of water molecules around nonpolar molecules has been hypothesized to result in the formation of cage-like structures called clathrates. This ordering results from an attempt by the water to maintain a maximal number of energetically favorable hydrogen bonds. The decrease in entropy must be balanced by a greater decrease in the enthalpy of the system compared to what would have occurred if the hydrogen bonds had been broken. The net effect is that the least energetic state occurs when the area of contact between water and the nonpolar groups is minimized. The association of hydrophobic molecules thus results in an unfavorable increase in enthalpy that is accompanied by an even greater increase in the total entropy of the system. In addition, the favorable London interactions between hydrophobic groups adds to the stability of the association. This leads to the unusual situation where the associated state is more random than the unassociated state. It also leads to the observation that when the proper values for enthalpy and entropy are inserted into the free energy equation, that hydrophobic associations are weakened as the temperature is lowered.

When a liquid of low polarity, such as fat, is mixed with water there is a strong driving force to limit the contact between the two liquids. This happens when phase separation occurs. To increase the interfacial area, and the energy of the system, requires the input of work. If the liquids are dispersed through the application of work, the system attempts to achieve the conformation of lowest free energy. The total energy can be minimized if the area of contact between the two liquids is kept to a minimum. This can initially be achieved by the formation of spherical particles. Thus, when two immiscible liquids are forced into contact by the application of work, the result will be the formation of a number of spherical droplets within the dispersed phase. Larger spheres have a smaller ratio of surface to volume than do smaller spheres and hence a lower surface energy. If there is no energy barrier to prevent coalescence, the system will continue to lower its total energy content by the formation of larger droplets from smaller ones. Given enough time this leads to the situation of minimum contact and phase separation.

The dispersed system can be stabilized against coalescence and phase separation if another component that is partially soluble in both phases is added. Molecules that are composed of portions that are soluble in water and portions that are soluble in lipids can serve as emulsifiers. Phospholipids are a class of naturally occurring compounds that can serve this function. When mixed with lipid in an aqueous environment, the fatty acid portion of the phospholipid molecule is inserted into the oil phase, while the phosphate ester head group remains in contact with the aqueous phase. Thus, a portion of the molecule is in contact with the lipid phase while another portion of it is in contact with the aqueous phase. More importantly the two immiscible phases are not in contact with each other and the total energy of the system is lower. The head portions of the phospholipid molecules contain like charges and tend to repel each other causing an energy barrier to coalescence and phase separation.

An emulsion formed from a mixture of oil, water and emulsifier is at a higher energy level than the deemulsified system. Thus, an emulsion is thermodynamically unstable, and given enough time, breaks or separates. The goal of the food scientist is to make the energy of activation high enough to give the emulsion a reasonable lifetime.

Emulsion Formation

In order to form an emulsion, energy must be provided in excess of that due to the creation of the new interfacial area of the emulsion. The size of the droplets (and thus their interfacial energy) depends on the amount of work done on the system. As more work is applied, the droplet size becomes smaller. As soon as new interfacial area is created, the system attempts to reach a lower energy state by coalescence of fat globules. The rate of coalescence depends on the energy barrier and the rate of droplet collision. For uncoated fat globules the energy barrier to coalescence is so small that it can be ignored. In most lipid systems the density of the lipid phase is much less than the aqueous phase and the fat droplets tend to rise to the surface. This increases the rate of collisions, the rate of droplet coalescence and phase separation.

The presence of an emulsifier makes the situation more complex. The emulsifier has a portion of the molecule oriented away from the phase that it is dispersed in. In the case of phospholipids in water, a micelle forms with the fatty acid tail groups oriented away from the aqueous phase. When such a micelle approaches the lipid/water interface it tends to reorient itself. The charged groups on the phospholipid resists removal from the aqueous phase. If the micelle approaches a lipid droplet the structure reorients in an attempt to prevent the dehydration of the charged phospholipid head groups. This causes an exposure of the fatty acid tail portions of the molecules. As these come into contact with the lipid phase, the fatty acid portions of the molecules orient into the lipid, while the charged head groups remain in contact with the water. This results in the formation of a monolayer of phospholipid molecules at the surface of the droplet oriented such that their hydrophobic groups are inserted into the lipid phase and their charged head groups are in contact with the aqueous phase. For lipids in water with no emulsifier present, a rapid coalescence of fat globules occurs, and in time, phase separation follows. If emulsifier molecules are present, they diffuse to the fat lipid interface as coalescence is occurring (4).

If the newly created surface could be instantaneously coated with emulsifier molecules, the emulsion would consist of particles having the same size distribution as they did at the moment of homogenization. In real emulsions, the emulsifier molecules require a finite time to diffuse to the interface and to be absorbed in order to provide a barrier to coalescence. The rate of droplet coating by phospholipid is a complex function of the rate of fat droplet coalescence, the rate at which the phospholipid molecules reach the lipid surface and the rate at which the micelles are able to reorient. Small emulsifier molecules are able to diffuse rapidly, and the amount of reorientation required to interact in the surface is small. In general, the use of small emulsifier molecules results in a relatively narrow particle size distribution and in the formation of emulsions with relatively small fat globules (5).

Proteins at Interfaces

Proteins are often included in emulsions to aid in their formation and to increase their stability. Proteins are much larger and more complex than are simple emulsifier molecules and the formation of a protein stabilized emulsion requires that the protein molecule must first reach the water/lipid interface and then unfold so that its hydrophobic groups can contact the lipid phase. To illustrate the forces involved, the situation of a protein molecule approaching a static water/lipid interface will first be considered. In native proteins most of the nonpolar amino acid side chains are located in the interior of the molecules. It has been estimated that the removal of one mole of hydrophobic groups from the surface results in an energy gain of 12 Kj (6). Any hydrophobic groups that remain at the surface increase the total energy of the system. Proteins have charged groups at the surface of the molecule and in contact with water molecules. The favorable interaction of water with surface charges lower the total energy of the protein molecule. In some respects the protein may be envisioned as resembling the micelles of phospholipids in the previous example. The hydrophobic groups are removed from contact with the aqueous phase while charged groups maximize solvent contacts.

As a protein molecule approaches the interface, there is less opportunity for the charged groups to interact with solvent. In the extreme case, charged groups are removed from the aqueous phase and enter the lipid phase. This is energetically unfavorable and these groups are repelled from the interfacial area. If the groups nearing the interface are in a region of the protein molecule that contains some flexibility, then the molecule may begin to unfold. This unfolding causes the exposure of hydrophobic groups to the surface. If these groups are exposed to the aqueous environment, there is an increase in total energy, and random fluctuations in protein structure will cause these groups to return to the interior of the molecule. If the exposure occurs at an interface, the state of lowest free energy depends on the nature of the interface. In the case of a protein unfolding near a lipid, the hydrophobic groups are inserted into the lipid phase. This insertion has a very low energy of activation and proceeds spontaneously (7). For most proteins studied, the size of the hydrophobic region inserted is about six to eight amino acid residues (8). It has been shown that the enthalpy for this step is positive (9) so that the driving force must be an increase in the entropy of the system. This increase in entropy is generally thought to have two components, one due to the conformational entropy of the protein and one due to the structure of water near hydrophobic groups (10). There is an increase in the conformational entropy of the protein as the hydrophobic groups are removed from the interior of the molecule and placed into another nonpolar environment. The original protein had a limited number of ways of arranging its components to attain a low energy state. The partially unfolded molecule has many ways of inserting a hydrophobic group into a nonpolar environment, and once there, the group can assume more conformations than before. The solvent molecules at the interface are arranged in highly ordered structures as was previously noted. The approach of the protein with the insertion of hydrophobic groups into the oil phase will, in essence, coat the nonpolar material and will allow for the release of solvent from the surface. The release of this water is responsible for a significant increase in the entropy of the system.

While the original insertion of a hydrophobic group proceeds spontaneously with a small energy of activation, the reaction is not readily reversible (11). In time, other sections of the protein molecule approach the surface, and if these occur in flexible portions of the protein they too may be inserted into the lipid phase. The protein will unfold at the interface as this continues.

Proteins that become attached by more than one hydrophobic group desorb very slowly from the surface, if at all. Langmuir and Schaeffer (12) calculated that if absorption was completely reversible and the Gibb's absorption equation applied, changes in surface pressure of the magnitude they observed in ovalbumin stabilized emulsions should result in essentially complete desorption of protein from the interface. This does not occur for protein stabilized emulsions, suggesting that a significant energy barrier to protein desorption exists. Removal of hydrophobic groups from the lipid exposes lipids to the aqueous phase as well as the hydrophobic groups that are being removed. Even if the removed hydrophobic groups could be buried in the protein interior, the protein would remain attached to the fat globule at other points and reattachment would be likely. If other hydrophobic molecules are available to cover the exposed lipid area, desorption is easier to achieve. For instance, it has been shown that gelatin molecules can be replaced by more hydrophobic casein molecules from the water/lipid interface (13).

Once a layer of protein has been adsorbed additional protein cannot be added in the same manner because an energy barrier to absorption exists (14). In order for more protein to be absorbed, the protein already at the surface must be compressed to make room. The amount of compression that is possible depends on the rigidity of the protein and also on the amount of residual charge near the surface. At some level of compression, the absorption of more protein will require more energy than can be gained by the insertion of hydrophobic groups into the lipid layer. Further interaction involves the interaction of protein molecules in the bulk phase with those already absorbed to the lipid and the formation of multilayers. This sequence of events can be inferred from an examination of the change in surface concentration with time as protein is added to a system containing an oil/water interface. The changes that occur with time in both surface concentration and surface pressure for β-casein which contains little or no secondary structure is shown in Figure 9.1. There is a rapid initial increase in surface concentration followed by a more gradual decrease and finally a plateau. The initial rapid decrease in Γ results from the absorption of the first molecules to arrive when there is essentially no barrier to absorption. As the surface is nearly covered, new molecules are only slowly absorbed. The values for surface pressure change in parallel with those for surface concentration, suggesting that as the molecules are absorbed at the interface they can exhibit their full effect on interfacial tension. The situation for a protein that contains considerable secondary structure such as β-lactoglobulin is shown in Figure 9.2. There is a rapid increase in surface concentration with time as protein is absorbed at the surface. The change in interfacial pressure shows a lag and continues to decrease with time long after the absorption of new protein molecules has ceased. This can be interpreted as being due to the unfolding of structured areas of the proteins already absorbed at the interface (15).

When proteins are used for the generation of emulsions the system becomes highly complex. Some form of shear is generally responsible for the creation of new surface area. The high energy state is relieved by rapid coalescence of fat globules. For prevention of coalescence, protein molecules need to diffuse to the fat/water interface and then unfold and coat the surface. When enough of the new surface is covered, coalescence ceases. It has been demonstrated with proteins (16) that the rate of diffusion to the interface is a significant variable in the amount of protein that absorbs to the interface during emulsion formation. Anything that tends to decrease the rate of diffusion of the protein molecules decreases the protein load (17).

Once a protein molecule reaches the surface it must be able to unfold enough to expose hydrophobic groups if it is to function as an emulsifier. Molecules like the various caseins are extremely flexible and contain little secondary structure (18). Caseins are excellent emulsifiers because of their ability to easily unfold at interfaces. Molecules that

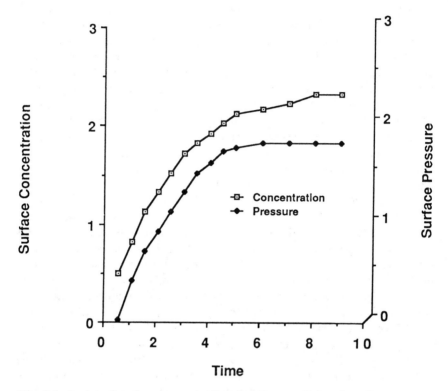

Fig. 9.1. A plot of surface concentration (arbitrary units) and surface pressure (arbitrary units) vs time (arbitrary units) for β—casein (12).

contain crosslinks such as disulfide bonds are more rigid and less able to unfold. Such molecules are less effective in emulsion formation (3). Reduction of disulfide bonds enhances the emulsifying ability of some proteins as long as the molecules do not unfold to the point that there is a large increase in viscosity (19). The content of disulfide bonds has been related to the emulsion capacity of complex mixtures of proteins such as whey protein concentrates (20). Small highly crosslinked protein molecules tend to perform poorly as emulsifiers.

Once protein begins to unfold, there must be hydrophobic groups present to insert into the nonpolar phase. In theory, a measure of the relative hydrophobicity of a protein should be related to its ability to function as an emulsifying agent. In practice, relative hydrophobicity measurements have been difficult to obtain. The early methods gener-

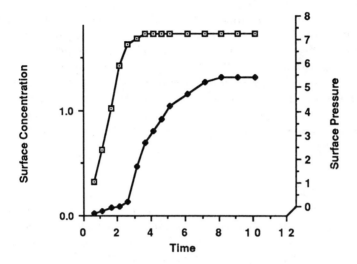

Fig. 9.2. A plot of surface concentration (arbitrary units) and surface pressure (arbitrary units) vs time (arbitrary units) for β-lactoglobulin (12).

ally assigned some relative value to each amino acid and then the value for the protein was calculated from its composition. These procedures have rarely correlated well with functionality because they measure the total potential hydrophobicity of the protein rather than those hydrophobic groups which can actually reach the surface upon unfolding. Recently a number of procedures have been developed which measure what is termed the effective hydrophobicity of proteins. Generally this means obtaining a quantitative measure of those hydrophobic groups that are capable of binding to a selected probe molecule. The groups that are deeply buried in a portion of the protein that does not unfold are not measured. The groups that interact with the probes are generally accessible to the surface and are capable of interacting in emulsions or foams. Both surface hydrophobicity, as determined by the binding of cis-paranaric acid (21), and effective hydrophobicity, as determined by alkane binding (22), have been related to the ability of various proteins to form stable emulsions.

The distribution of hydrophobic groups is also important. In proteins like β-lactoglobulin, the hydrophobic groups are rather evenly distributed throughout the molecule (23). There are no large portions of the molecule where hydrophobic amino acids are grouped, nor are there large sections of the molecule that do not contain charged amino acids. This makes it difficult to find portions of the molecule that are sufficiently hydrophobic or find residues that do not also contain amino acids with charged groups that would resist their removal from the aqueous phase. In molecules like β-casein there are large sections of the protein that contain hydrophobic amino acids without the presence of charged groups (24). The molecule has such an uneven distribution of charge and hydrophobic groups that is amphipathic. It is easy to find portions of this molecule that contain at least six nonpolar amino acids and no charged groups.

Emulsion Stability

Once formed, an emulsion can undergo a number of changes. The most striking change would be phase inversion. In order for phase inversion to occur, the surface of a number of fat globules have to be exposed and allowed to coalesce. With protein stabilized emulsions phase inversion is generally not a problem because when fat globules near each other, the proteins usually provide an effective barrier to coalescence. The removal of protein from the surface of a fat globule is energetically unfavorable and does not occur at any appreciable rate. In order to destabilize the emulsion, large changes in the structure of water in the system, large inputs of energy, or both are required. In food products, fluctuations in temperature are a common cause of emulsion destabilization. As the temperature is lowered, water attains more and more structure. As the water becomes more ordered, there is less of an energy difference between hydrophobic groups exposed to the aqueous phase and those buried in the oil phase. Low temperature alone does not usually cause an emulsion to break, but it can be the deciding factor in the stability of an otherwise poorly emulsified system. Recently an interesting paper has discussed the changes that occur in the relative solubilities of nonionic surfactants with changes in temperature (25). An extension of this treatment to proteins would be of interest.

The largest temperature induced changes to emulsions occur upon freezing and subsequent thawing. Not only is the energy difference between the associated and free state minimized by the low temperature, but the formation of ice crystals can cause physical damage to the emulsion. When the system is thawed, coalescence occurs if the physical damage has been extensive. One of the best ways to minimize this type of damage is to add substances that will modify the size and extent of water crystal formation (26).

A more common defect in food emulsions results from the phenomenon known as creaming. If density differences between the dispersed and continuous phase exist, particles of the dispersed phase either sediment or rise depending on the relative densities. In most emulsions the dispersed phase is less dense than the continuous phase, causing creaming to occur. Given enough time, a depletion of lipid from the bulk aqueous phase occurs with the formation of a compact cream layer containing the majority of the lipid. The rate of creaming is given by Stoke's law:

$$V = \frac{2r^2 g \Delta p}{9\mu}$$

Where v equals the velocity of the fat globule, r is the radius of the fat globule, g is the force of gravity, Δp is the density difference between the two phases and μ is the viscosity of the continuous phase. In theory for an emulsion to have an extended shelf life either the density of the fat globules must be made identical to that of the continuous phase or the viscosity must be high enough so that the yield value is greater than the acceleration due to buoyant differences. Few emulsions can be made totally stable to creaming without the formation of some sort of matrix. Homogenized milk, for example, can be shown to have approximately 1% of the fat globules entering the cream layer for each day of storage (27). Thus a product with a fat globule size distribution similar to that found in homogenized milk would have 90% of its fat globules in the cream layer after only three months of storage. For products with shelf lives approaching 24 months, such as infant formulas, even small differences in density between the dispersed and continuous phases results in the formation of a cream layer during the useful shelf life of the product unless additional measures are taken to increase stability.

It has been recently shown (28) that the addition of polysaccharide stabilizers to emulsions has little effect on the stability of the systems unless they increased the viscosity to the point of imparting a yield value. It has been suggested for some time that carrageenan stabilization of heated milk products is the result of the formation of a network that physically prevents the fat globules from coming into contact (29). This has been likened to the formation of a loose gel matrix within the fluid phase. Evidence for this mechanism can be found in the micrographs shown in Figure 9.3. In Figure 9.3A, a scanning electron micrograph of a 20% fat emulsion stabilized by 2% whey protein concentrate, the fat globules are spherical and free of each other. A similar emulsion that formed with the addition of a polysaccharide stabilizer is shown in Figure 9.3B. Figure 9.3C shows the same sample prepared for microscopy in such a way that the fat globules are removed. The network in which they were entrapped is clearly visible. The fat globules are still spherical, but they appear to be trapped in a network that is not present in the absence of stabilizer. This network may be viewed as a physical means of preventing cream layer formation. An extension of the formation of such a network is the formation of non-fluid emulsions as are discussed in the next chapter. In fluid emulsions the network must be firm enough to retard or prevent fat globule coalescence, but must be weak enough to allow fluid flow when the product is poured. Walstra

9.3A

Fig. 9.3. Scanning electron micrographs of a whey protein stabilized emulsion, (A); as in A with added gum arabic, (B); same sample as B but freeze fractured without fixing the lipid phase (C). The lipid has been removed from (C) showing the matrix it was held in.

9.3B

(30) has suggested that a yield stress of greater than 0.1 Pa would be enough to prevent creaming. If the yield stress were less than about 10 Pa, the gel would be readily reversible and would be fluid when poured. Even if there is not a yield value of more than 0.1 Pa, most emulsions are sufficiently pseudoplastic to exhibit higher than expected viscosities at very low shear rates, and thus creaming is often slower than predicted. Thus, while Stoke's law is important in predicting the rate of emulsion creaming, for most products with any appreciable shelf life other factors, especially viscosity, pseudoplasticity and yield stress must also be considered.

The amount of damage done to a product by the formation of a cream layer depends on the type of product and the tenacity of the formed layer. If the layer can be readily dispersed by shaking, little damage ensues. If on the other hand, the proximity of the particles in the cream layer leads to coalescence or the layer is somehow crosslinked to an extent that prevents it from being readily redispersed, considerable economic loss can occur.

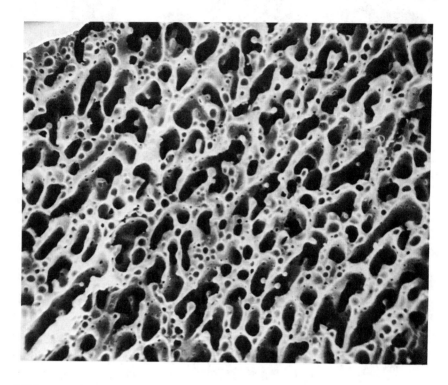

9.3C

Formation of emulsions with food proteins

There has been a considerable amount of research concerning emulsion formation with food proteins. In this section a few of the major studies are discussed and the methods utilized to study emulsions in food products reviewed with suggestions for future research.

Tornberg (16,17,31) studied the formation of emulsions with soy protein, caseinates and whey protein concentrates. The emulsions formed with deionized water and in salt solutions generally contained 2.5% protein in 40% soybean oil. In some studies different types of homogenizers were utilized. In these studies the energy input into the emulsions was kept controlled so that different levels of energy input did not become an uncontrolled variable. The amount of protein deposited at the interface was assumed to be equal to the initial protein concentration, minus that in free solution following formation of the emulsion. The surface areas of the emulsions were determined by light scattering. The surface area of the emulsions formed increased with power consumption up to some maximal level. Beyond this point additional passes through the homogenizer caused a destabilization of the already emulsified droplets. The amount of protein per unit of fat globule surface was fairly constant for a given protein. Differences in protein load were most pronounced at very low or very high surface areas. Generally, the slope of the surface protein concentration versus $t^{1/2}$ was linear and it was suggested that this indicated that protein absorption was a diffusion controlled process. Others have argued that the situation is very complex and the time scale utilized to measure the rate of absorption may not be appropriate for emulsion formation. They have suggested that further evidence will be necessary before it can be concluded that the rate of emulsion formation is truly diffusion controlled. It is unlikely that the rate of protein diffusion is the only factor that affects the size distribution of emulsions, but it is undoubtedly an important factor.

Graham and Phillips (14, 15, 32) studied the foaming and emulsifying properties of β-casein, lysozyme and bovine serum albumin. Casein was utilized because it is a molecule with little secondary strucutre and considerable flexibility. Lysozyme was considered as a small rigid molecule and serum albumin was a model of a globular protein which is easily unfolded. The authors conclude that at low protein concentrations, β-casein molecules were completely unfolded at the interface, while lysozyme still maintained considerable residual structure. As the surface concentration approached monolayer saturation, the β-casein molecules were able to alter their conformations. They ceased to exist in a totally spread layer and started to form a series of loops. Lysozyme molecules are rigid and their conformation does not change with increased protein concentration. The flexible β-casein molecules result in a surface with very low surface viscosities, while the rigid lysozyme molecules form surfaces that are more viscous. The proteins that formed more viscous films were found to be more resistant to phase separation. The authors related the surface and viscous properties of the resulting emulsions to the strucutres of the original proteins. Further detail is not warranted here, but these papers are highly recommended for a more detailed discussion of the relationship of protein structure to emulsion formation and the resulting rheological behavior of the emulsions. However, the applicability of the data to food emulsions is limited because the differences noted in behavior were less prominent at concentrations approaching those commonly utilized in foods.

Haque and Kinsella (19) investigated the emulsion characteristics of bovine serum albumin in a valve homogenizer as described by Tornberg and Lundh (31). The homogenizer was able to form emulsions usually formed at known energy inputs. The surface area was determined by an optical method (33). The protein concentration, oil to water ratio, energy input, salt concentration and pH were related to the properties of the emulsions formed. The results were related to the molecular properties of serum albumin and were generally in agreement with previous literature and theoretical considerations. The effects of pH and energy input on emulsion surface area are of particular interest because they demonstrated that both variables should be considered when emulsions are studied. The conditions of protein and oil concentration selected in this study were in the range generally used in food emulsions and thus the data should be applicable to foods.

The determination of meaningful emulsion data with complex food products is difficult. Much of the experimental work with model systems has been done in very dilute solutions. The surface pressure or interfacial tension is often the quantity measured. With a food product the relevant information is concerned with the question: How much lipid can be emulsified and how long will it be stable to coalescence and/or creaming? The situation in food products is also complicated by the presence of other surface active molecules in addition to the proteins present.

Three main types of test have been devised to give an indication of the efficiency of proteins to serve as emulsifiers in food products. A number of tests measure emulsion capacity (34-36). These generally involve adding lipids to an aqueous solution of the protein to be tested. The addition of lipids is continued until phase inversion occurs. Thus, the test measures the capacity of the protein to emulsify fat at very high lipid to protein ratios. Values for emulsion capacity are commonly in the range of a few hundred milliliters of oil emulsified per gram of protein. The test does not measure the stability of the emulsion formed. While these values can predict which protein of a group will emulsify the most fat or what conditions of pH, salt content, etc., lead to maximal emulsion capacity, the conditions do not usually resemble those found in most food systems. Extrapolation of these values to determine the stability of a food emulsion with storage are difficult at best.

Another means of estimating emulsion stability is to form an emulsion under conditions that resemble those in the product. The emulsion is then allowed to separate either under the influence of gravity or after exposure to a centrifugal field. The change in lipid distribution throughout the sample with time can be measured and the phase separation with time noted. Within a centrifugal field the fat globules are compacted into a cream layer and an aqueous layer devoid of fat is formed. The ratio of either the cream layer or the aqueous layer formed to the volume of the initial emulsion is often utilized as an indicator of emulsion stability. The emulsion volume index (37) is an example of such a test. Generally, forces of many hundreds or thousands times that of gravity are employed to obtain separation in fairly short periods of time. These high forces tend to distort fat globules and to overcome forces of repulsion that would probably still be operative in a product stored at 1g for an extended period of time. Studies that have measured differences in distribution of lipids with time in emulsions have been generally limited to times in the order of hours.

The size distribution of the particles in an emulsion can also be utilized as an indicator of the effectiveness of the emulsifier. In general, the more efficient the homogenization, the smaller the particles. Determination of the size distribution can be tedious and may involve the introduction of artifacts due to sample preparation. This is especially true when electron microscopy is used to measure particle diameters. A method based on light scattering (EA) (33) has become very popular recently. This method is easy and requires only small amounts of sample. The procedure as described makes a number of assumptions, well discussed by the authors, that may not be valid for a number of samples.

When comparing results in the literature it is important that the method of evaluation be considered since the different methods of emulsion evaluation may be more important than the variables being examined to explain differences in results obtained by workers. Recently it has been reported that none of the above methods gave results that were able to predict the shelf life of either a coffee whitener or a salad dressing (38). A method has been reported which may be applicable for both of these food systems (22). The original formulations were modified so that a test product could be produced that was not as stable as the commercial product, but for which the stability was strongly correlated with the commercial product. Such an approach is difficult and a new test formulation must be applied for each commercial product investigated. If properly done, however, a system can be developed that is capable of predicting in a few weeks the shelf life of a product that is measured in months or years.

In the future, more work will have to be directed toward model systems that more closely resemble foods if the results are to have meaning to the food industry. Darling and Birkett (5) have recently suggested that one of the greatest challenges facing those who study food emulsion is to make their model systems and the data collected relevant to real food systems. We would like to echo this sentiment and suggest that the complexity and uncertainty of working with systems that resemble food may make interpretation of the data more difficult and ambiguous, but the data obtained from such studies are necessary. One goal of research into emulsion systems is to improve our understanding of the physical chemistry involved; another goal is to translate such understanding into meaningful recommendations for the food industry.

References

1. Dickinson, E., *Food Emulsions*, The Royal Society of Chemistry, London, England, 1987.
2. Dickinson, E., and G. Stainsby, *Colloids in Food, Applied Science*, London, England, 1982.
3. Tanford, C., *The Hydrophobic Effect: Formation of Micelles and Biological Membranes*, John Wiley and Sons, New York, NY, 1980.
4. Mangino, M.E., *J. Dairy Sci. 67*:2711 (1984).
5. Darling, D.F., and R.J. Birkett, in *Food Emulsions and Foam* edited by E. Dickinson, The Royal Society of Chemistry, London, England, 1987.
6. Kinsella, J.E., in *Food Proteins*, edited by P.F. Fox and J.J. Cowden, *Applied Science*, London, England, 1982 pp. 51-103.
7. Tanford, C., *Adv. Prot. Chem 24*:1 (1970).
8. MacRitchie, F., and A.E. Alexander, *J. Colloid Interface Sci. 18*:453 (1963).
9. Stainsby, G., in *Food Macromolecules*, edited by J.R. Mitchell and D.A. Ledward, Elsevier, London, England, 1985.
10. MacRitchie, F., *Adv. Prot. Chem. 32*:283 (1978).
11. MacRitchie, F., in *Protein at Interfaces*, edited by J.L. Brash and T.A. Horbett, Amer. Chem. Soc., Washington, DC, 1987.
12. Langmuir, I., and V.J. Schaeffer, *Chem Rev. 24*:181 (1939).
13. Musselwhite, P.R., *J. Colloid Interface Sci. 21*:99 (1966).
14. Phillips, M.C., *Food Technol. 35*:50 (1981).
15. Graham, D.E., and M.C. Phillips. *J. Colloid Interface Sci. 70*:403 (1979).
16. Tornberg, E., in *Functionality and Protein Structure*, edited by A. Pour-El, Am. Chem. Soc., Washington, DC, 1979.
17. Tornberg, E., *J. Sci. Food Agric. 29*:762 (1978).
18. Farrell, H.M., *J. Dairy Sci. 56*:1195 (1974).

19. Haque, A., and J.E. Kinsella, *J. Food Sci. 53*:416 (1988).
20. Peltonen-Shalaby, R., and M.E. Mangino, *Ibid 51*:91 (1986).
21. Kato, A., and S. Nakai, *Biochim. Biophys. Acta. 576*:269 (1980).
22. Mangino, M.E., D.A. Fritsch, A.M. Fayerman, and W.J. Harper, *New Zealand, J. Dairy Sci. Technol. 20*:103 (1985).
23. McKenzie, H.A., in *Milk Proteins: Chemistry and Molecular Biology*, Vol. 2, pp. 255–330, Academic Press, New York, NY, 1971.
24. Eigel, W.N., J.E. Butler, C.A. Ernstron, H.A. Farrell, V.R. Harwalker, R. Jennes and R.M. Whitney, *J. Dairy Sci. 67*:1599 (1984).
25. Kahlweit, M., *Science 240*:617 (1988).
26. Courts, A., in *Applied Protein Chemistry*, edited by R.A. Grant, *Applied Science*, London, England, 1980 pp. 1–29.
27. Walstra, P., and R. Jennes, in *Dairy Chemistry and Physics*, John Wiley and Sons, New York, NY 1984.
28. Howe, A.M., A.R. Mackie, P. Richmond, and M.M. Robins, in *Gums and Stabilizers for the Food Industry*, Vol. 3, edited by C.O. Philips, D.J. Wedlock and P.A. Williams, Elsevier Applied Science Publishers, London, England, 1985.
29. Hansen, P.M.T., *Prog. Food Nutr. Sci. 6*:127 (1982).
30. Walstra, P., in *Food Emulsions and Foams*, edited by E. Dickinson, The Royal Society of Chemistry, London, England, 1987 pp. 242–257.
31. Tornberg, E., and G. Lundh, *J. Food Sci. 5*:1553 (1978).
32. Graham, D.E., and M.C. Phillips, *J. Colloid Interface Sci. 70*:427 (1979).
33. Pearce, K.N., and J.E. Kinsella, *J. Agric. Food Chem. 26*:716 (1978).
34. Swift, C.E., C. Lockett, A.J. Fryar, *Food Technol. 15*:468 (1961).
35. Webb, N.B., F.J. Ivey, H.B. Craig, V.A. Jones and R.J. Monroe, *J. Food Sci. 35*:501 (1970).
36. Acton, J.A., and R.L. Saffle, *Ibid 37*:904 (1972).
37. Harper, W.J., R.I. Peltonen, and J. Hayes, *Food Prod. Dev. 14*:52 (1980).
38. Harper, W.J., *J. Dairy Sci. 67*:2745 (1984).

Chapter Ten
Are Comminuted Meat Products Emulsions or a Gel Matrix?

Joe M. Regenstein

Department of Poultry and Avian Sciences
and Institute of Food Science
Cornell University
Ithaca, NY 14853-5601

Recent studies suggest that the structural elements of muscle (i.e., the highly organized myofibrillar, sarcoplasmic reticulum system, etc.) in meat batters may be more important than the behavior of proteins at the oil/water interface. In this paper some of the research by our laboratory group to develop methods to study this problem and some of the more interesting results of this work are reviewed.

The interest in meat batters was stimulated initially by the work of Tsai et al. (1). They studied individual muscle proteins using the classical emulsion capacity test, i.e., the amount of oil that had to be added to a fixed quantity of protein in aqueous solution until the emulsion either broke or inverted. The results suggested that the emulsion capacity was essentially independent of the particular protein used. This was difficult to accept because of the vastly different structures of the various proteins studied, ranging from the water soluble, supposedly non-emulsion participating globular sarcoplasmic proteins to the uniquely structured myosin. In fact, the hyperbolic emulsion capacity function (ml oil emulsified/g protein vs mg protein) obtained suggested that emulsion capacity measures a constant amount of oil emulsified with an ever changing denominator of protein content (2). This is explained by the fact that the amount of oil emulsified remains relatively constant but the amount of soluble protein changes (with and without the addition of ATP to identical samples) so the apparent emulsion capacity values change (2). On the other hand, if the initial protein concentrations (soluble plus suspended material) were used, the various samples gave the same emulsion capacity. The small differences in the amount of oil emulsified with different proteins and protein concentrations may reflect the effects of viscosity on the efficiency of the equipment rather than protein differences. For example, actomyosin is much more viscous than the sarcoplasmic proteins, and therefore the EC test might not reach an endpoint until slightly more oil had been added. Other reports in the literature had already suggested that the technique had a number of limitations because of problems with controlling various experimental variables. At best, the test is a complicated way to measure solubility.

The emulsion capacity test is used in the meat industry as part of the calculations leading to the bind value, i.e., a factor that supposedly reflects the solubility and the emulsion capacity of various meat components which can then be used in obtaining least-cost formulations for producing comminuted meats. The bind value is based on the emulsion capacity of the soluble protein fraction, and no measure of emulsion stability is incorporated into the measurements. The numbers are tabulated by meat cut (with no consideration of batch-to-batch variation). The formulator then specifies the desired total bind value for the final product, and the least-cost software insures that the formulation meets this standard. For most meat products this constraint was rarely invoked by the computer in formulating a product. Apparently other constraints, such as the regulatory requirements for such formulations, controlled the formulation selected. However, at least one company is trying to produce a true 'least-cost' product, and they have found that the computer generated formulations result in a surprisingly large percentage of product failures.

Therefore, studies were initiated to develop a test system that would allow us to accomplish two objectives; to develop a practical system which is convenient and easy to use in a plant environment, and to provide some insights into the emulsion process in meat products to enable the testing of some of the hypotheses that developed with the technology. In this paper the focus is on timed emulsification and the subsequent emulsion stability measurement, and how it addresses such questions as the importance of protein solubility in meat emulsions and the possible role of muscle structure in comminuted meat products.

Timed emulsification (TE) uses a fixed oil to protein solution ratio (2) i.e., 6 ml oil and 3 ml water. The original test conditions were optimized using myosin. The ratio of oil to water is greater than that used in real meat products where the legal requirement is usually a maximum of 30% fat.

The muscle sample is blended for various time periods in an Omnimixer (Dupont Sorvall, Norwalk, CT) at an intermediate blade speed. (The same equipment can be used at higher speed settings followed by centrifugation to remove materials in the same solvent prior to testing so that the additional solubilization of material during the actual TE experiment is minimized.) Sharp blades are essential for good results. An ice-water bath is used to maintain temperature, which may still heat up to 10°C over the course of a three minute blending. Because of the temperature increase, the blending is now usually a maximum of three minutes. The Omnimixer used centrifuge tubes that can be centrifuged without requiring a quantitative transfer of viscous emulsions. The sample is then creamed in the centrifuge at 30,000 x g (16,000 rpm in the Sorvall RC2B centrifuge with the SS-34 rotor). The various fractions can then be analyzed using a variety of techniques.

The major measurement following timed emulsification is the loss of protein (usually by Lowry) from the aqueous layer. If any pellet phase remains, the protein content can also be measured. The actual composition of the proteins remaining in the aqueous (or pellet) phase may be determined by sodium dodecyl sulfate-polyacrylamide gel electrophoresis (SDS-PAGE). Working with both natural actomyosin (i.e., soluble thick and thin filaments) or with glycerinated myofibrils (i.e., the insoluble components of muscle tissue), it was found that the presence of ATP (or a food analog polyphosphate such as pyrophosphate) resulted in the selective removal of myosin to the cream layer or the selective retention of actin in the aqueous phase (3,4). Thus, the myosin is the important "emulsifying" component in "actomyosin" systems. This was the first example of this newer methodology detecting effects that could not have been determined using the emulsion capacity test.

By measuring the amount (weight or height) of the cream layer formed, the relative amount of oil incorporated into the cream layer was determined. Only in a few marginal cases (e.g., with very low protein concentrations) was all of the oil not incorporated. Furthermore, the cream layer formed during the centrifugation step had sufficient mechanical strength that it could be removed from the centrifuge tube and used to measure the emulsion stability.

The emulsion stability (5,6) is measured by placing the cream layer in the center and on top of three to four layers of filter paper (in a petri dish) topped with a polyester mesh (to allow easier handling of the cream layer). A similar series of materials are placed above the cream layer and the petri dish is covered. The entire sample is placed in the cooler (0-1°C). Weight loss of the pellet, which mainly reflects the loss of oil, although some moisture is also lost, is measured every day for four days and the filter paper is replaced each day. Thus, the stability measurement is made at a little greater than one atmosphere pressure; the filter paper provides some capillary action and the top layer of the filter paper also provides a small amount of extra pressure.

There are three different stages where heat may be applied in the timed emulsification test (7,8). The first is while the proteins are still in the aqueous solution. The second is when the samples have been emulsified but not creamed. And the third is when the sample has been creamed. To a first approximation, these three points should mimic the three major steps of the meat emulsion/sausage/batter preparation procedures. That is, first the meat is chopped with ice and water (heating the protein in aqueous solution). Second, the fat is added and chopped to a slightly higher temperature (heating the emulsion after the Omnimixer step but before centrifugation to form a cream layer). And third, the batter is cooked (heating the cream layer).

One of the issues in dispute by meat scientists is whether the final maximum chopping temperature is determined by the behavior of the fat or the protein. Based on the differences in melting point temperature between the fat from various meat sources, i.e., beef vs chicken, and the results on the protein phase (i.e., very little changes in the timed emulsification or the emulsion stability between 0°C and 30°C), changes in the protein at low temperature seem to be unimportant. The only temperature effects observed were with myosin, NAM (natural actomyosin), or exhaustively washed muscle when the muscle was heated to above 40°C in the aqueous solution before emulsification. The denaturing temperature in the presence of structurally intact muscle (i.e., exhaustively washed muscle which is presumed to still have myofibrils, including thin and thick filaments) was higher (60-75°C) compared to NAM (40-60°C), which has approximately the same proteins, but in solution.

Exhaustively washed muscle is prepared first by washing the muscle tissue two times at low salt (0.15 M NaCl which is used to remove the "sarcoplasmic" proteins), and then four times at high salt (0.6 M which is used to remove the soluble "myofibrillar" proteins such as NAM). The Omnimixer speed is much higher than that used for the timed emulsification studies. Preliminary evidence from SDS-PAGE suggests that not all of the actomyosin is removed from the material. However, that which remains after exhaustive washing is probably still a part of the myofibrillar structure (insoluble) and is in 'rigor' linkage.

Current ideas about emulsions suggest that protein solubility is an important property in emulsion formation. Proteins migrate to the interface between the aqueous and oil phases and then undergo conformational changes leading to the hydrophobic groups entering the oil phase while the hydrophilic regions remain in the aqueous phase (9). Using this model, exhaustively washed muscle should not form a good emulsion. The results, however, showed a stable cream layer was formed with insoluble protein, with and without heating up to 75°C, which is higher than the temperature used for meat processing.

The highly varied, essentially "globular" sarcoplasmic proteins ought to be good emulsifiers if surface denaturation at interfaces was important. Other globular proteins are believed to be able to unfold easily at these interfaces in order to minimize their free energy (10). Unfortunately, both the results from our laboratory and from the classical meat processing wisdom suggest that these proteins are in fact poor participants in meat batter systems, possibly because they are not able to form a gel structure and/or they interfere with the gel structure of myofibrillar proteins (5).

By using lower amounts of protein to determine the minimum amount of protein needed for forming relatively stable emulsions, the amount of protein needed for stable cream layers of exhaustively washed muscle was only twice that of myosin (11). The droplet sizes were measured by taking pictures and analyzing the pictures. The greatest error in the droplet surface measurements would occur with any smaller droplets that would be missed in the pictures. (These would have a great deal of surface area in comparison to the larger droplets.) Calculations of initial protein concentration divided by droplet surface area showed that there was less than the 2-3 mg of protein per mm of interfall surface area that is required to give a protein monolayer to emulsified droplets in emulsions (12,13), strongly suggesting that the stable cream layers formed by myosin did not depend on surface droplet coating. Both myosin, an aggregated, coiled-coil α-helical protein for most of its length and insoluble exhaustively washed muscle were not likely to be easily denatured at the oil/water interface. (Unfortunately, it has not been possible to recover any protein participating in the cream layer in order to test whether it is denatured or not.) These experiments suggest a need for new explanations of the meat emulsion/batter process that take the role of structure into account.

The current hypothesis is that the meat proteins set up a structure in the aqueous phase, i.e., a gel matrix based on the myofibrillar proteins. This matrix is different from that of gelatin since it is known that gelatin interferes with the formation of hot dog-like products.

Food additives such as soy and milk proteins in processed meats may then function in one or more of three ways: to supplement the muscle proteins in the formation of the protein matrix in the aqueous phase by directly participating in the myofibrillar matrix, to improve water retention and/or to complement the activity of the aqueous phase muscle proteins by acting as surface active material at the oil/water interface.

The ability of proteins, either from meat or an additive, to aggregate in the appropriate way to form the aqueous matrix may be more important than their solubility (13).

How can the structural properties of proteins (both meat and non-meat) in meat batters be measured? Two tests may be relevant. The first is the measurement of the textural properties of cooked batters using the General Foods Texture Profile Analysis on the Instron or similar equipment to characterize the actual textural properties of (a) an experimental system such as the cream layers, (b) a model emulsion product such as that developed by Whiting and Miller (14) or (c) an actual meat-batter product. Comparisons between the three types of products should permit critical testing of various hypotheses (15).

The second test may be derived from studies of the water retention properties of muscle/meat systems. Water holding capacity measurements in the literature often are of two distinct types: either (a) some force is used to remove the "free" water, which will now be referred to as an "expressible moisture" measurement or (b) the ability of a sample's non-soluble component to incorporate additional water, which will now be referred to as a "water binding potential" measurement (16,17). An examination of the pH profile and the effect of various salt ions on these two types of measurements has established that, although they both may be called "water holding capacity," they are very different measurements. The latter is, in reality, a measure of the swelling capacity of muscle under various conditions. Offer and Trinick (18) showed that muscle tissue in the presence of salts will show an increase in the thin to thick filament distance due to swelling of the sarcomere, presumably by adsorbing additional moisture as sarcomere length does not change. Of particular interest is the effect of NaI on this system. This salt, at about 0.25 M, will cause a significant swelling of muscle, but only after the resolution of rigor has begun (19,20). At this NaI concentration, the myofibrillar proteins are not solubilized. Thus, it becomes possible under these conditions to separate swelling effects from solubilization effects. Preliminary results using this salt in the timed emulsification system with whole muscle (unwashed) suggest that the behavior of 0.25 M NaI samples is very similar to that of 0.6 M NaCl samples (where both swelling and protein solubility occur) supporting the concept that swelling (structural changes) may be more important than solubility (21). Allen Foegeding (personal communication) has asked the important question, "Can one make a hot-dog with 0.25 M NaI?"

In summary, the use of these newer methods to examine meat batters has led to a reevaluation of the mechanisms involved in meat batter products. The results suggest a major role for the aqueous phase protein matrix compared to the traditional emphasis on reactions at the interface. The tests are simple enough to run that they may also prove practical in determining the functional parameters for use in least-cost formulations on individual lots of meat.

Acknowledgment

The work summarized here represents the contributions of a number of graduate students over the years including Steve Galluzzo, Marilyn Gaska, Michele Perchonok, Donna Gerwig Huber, Tehmasp Gorimar, Zwi Weinberg, Carlos Jauregui and Dong-Hoon Shin, as well as collaborators JoAnn Rank Stamm, Peter Lillford and Peter Wilding. We thank Allen Foegeding for reviewing the manuscript.

References

1. Tsai, R., R.G. Cassens and E.J. Briskey, *J. Food Sci. 37*:286 (1972).
2. Galluzzo, S.J., and J.M. Regenstein, *Ibid. 43*:1757 (1978).
3. Galluzzo, S.J., and J.M. Regenstein, *Ibid. 43*:1761 (1978).
4. Galluzzo, S.J., and J.M. Regenstein, *Ibid. 43*:1766 (1978).
5. Gaska, M.T., and J.M. Regenstein, *Ibid. 47*:1460 (1982).
6. Gaska, M.T., and J.M. Regenstein, *Ibid. 47*:1438 (1982).
7. Perchonok, M.H., and J.M. Regenstein, *Meat Sci. 16*:17 (1986).
8. Perchonok, M.H., and J.M. Regenstein, *Ibid. 16*:31 (1986).
9. Cante, C.J., R.W. Franzen and F.Z. Saleeb, *J. Am. Oil Chem. Soc. 56*:71A (1979).
10. Halling, P.J., *Crit. Rev. Food Sci. Nutr. 12*:155 (1981).
11. Gerwig Huber, D., and J.M. Regenstein, *J. Food Sci. 53*:1282 (1988).
12. Graham, D.E., and M.C. Phillips, *J. Colloid Interface Sci. 70*:415 (1979).
13. Graham, D.E., and M.C. Phillips, *Ibid. 70*:427 (1979).
14. Whiting, R.C., and A.J. Miller, *J. Food Sci. 49*:1222 (1984).
15. Toro-V., J.F., and J.M. Regenstein, Emulsion Activity and Stability of Protein Additives Designated for Comminuted Meat Products. *IFT* Abstract (1988).
16. Jauregui, C.A., J.M. Regenstein and R.C. Baker, *J. Food Sci. 46*:1271 (1981).
17. Regenstein, J.M., C.A. Jauregui and R.C. Baker, *J. Food Biochem. 8*:123 (1984).
18. Offer, G., and J. Trinick, *Meat Sci. 8*:245 (1983).
19. Weinberg, Z.G., J.M. Regenstein and R.C. Baker, *J. Food Biochem. 8*:215 (1984).
20. Lillford, P.J., J.M. Regenstein and P. Wilding, *J. Sci. Food Agric. 34*:1021 (1983).
21. Shin, D.H., and J.M. Regenstein, Effect of Heating on Emulsion Stability of Chicken Breast Muscle. *IFT* Abstract (1988).

Chapter Eleven
Molecular Properties and Functionality of Proteins in Food Gels

E. Allen Foegeding

Department of Food Science
North Carolina State University
Raleigh, NC 27695-7624

The manufacturing of processed foods often requires ingredients which gel and provide a desirable texture and mouthfeel. Proteins form gels by polymerizing into a three-dimensional matrix which converts a viscous liquid into a viscoelastic solid. The polymerization process can be caused by heat denaturation or by any other process that converts proteins to a state that favors intermolecular protein interactions. This review will be confined to gelation of heat denatured proteins, where polymerization occurs as the temperature is increased. This mechanism is different than the gelation of gelatin, where polymerization is due to interactions of molecules refolding upon cooling. The functional properties of gelling proteins are related to several factors: (a) the lowest concentration of protein required to form a gel and desired gel texture, (b) the conditions required for gelation (temperature, pH, ions) and (c) the bonding, geometry and physical properties of the gel matrix.

The chemical properties of the protein determine the polymerization process which forms the gel matrix; gel texture (rheological properties) is related to the microstructure of the gel matrix. The chemistry/rheology link can be addressed by asking "what chemical principles regulate unfolding and subsequent interactions between proteins and how do various aggregated states translate into product texture?"

Gelation Models

Flory (1) described the process of gelation as formation of an infinite network by aggregation of trifunctional and bifunctional units (Fig. 11.1). One key element of gelation is that a simple end to end polymerization would not produce gelation; however, it would increase the viscosity of the solution. A gel matrix can be represented as consisting of interconnected cage-like unit structures, with the solvent continuous

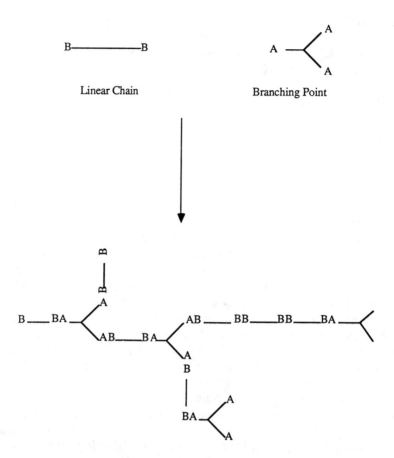

Fig. 11.1. Formation of an infinite gel matrix. Adapted from Flory (1).

throughout the matrix (Fig. 11.2). The matrix geometry, flexibility of the polymer, and strength of the junctions (chemical nature and extent of protein-protein interactions) determine rheological characteristics.

Gelation of denatured proteins is a two step process of activation and association (2). The activation is caused when heat produces a change in protein structure so that interactions can occur intermolecularly. The subsequent polymerization process is determined by the protein surface available for intermolecular interactions. The simple model of a protein polymerizing to form a homogeneous network is not valid when there is more than one protein present. Multicomponent gels can be separated into categories of filled gels, complex gels and mixed gels (3).

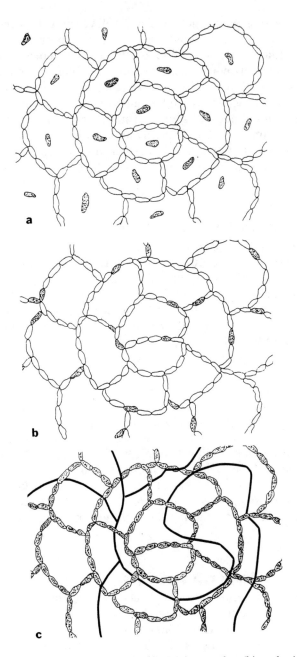

FIG. 11.2. Schematic representation of a filled (a), complex (b) and mixed (c) gel.

In filled gels, one macromolecule is forming the gel matrix while the other molecules are acting as fillers within the interstitial spaces (Fig. 11.2a). The filler molecules (for example, nongelling proteins) can affect certain textural properties and/or water binding. In complex gels, the network is formed by interactions among the components (Fig. 11.2b). Mixed gels are those in which the macromolecules form two or more three-dimensional networks without interactions among the polymers (Fig. 11.2c). From the limited data available, it was suggested that mixed and complex gels have the potential to produce textural characteristics which cannot be achieved individually with either component (3).

Heat Denaturation (Activation)

When the temperature of a protein is increased, structural changes occur. The denaturation temperature (T_d) is the point at which the extent of the reaction is equal to 0.5 and [Native] = [Denatured] (4,5). Under this definition, any analytical method that detects changes in secondary, tertiary or quaternary structure can be used to study denaturation, though the degree of structural change detected may vary with specificity and sensitivity of techniques.

There is evidence that only partial unfolding, with slight changes in secondary structure, is needed for gelation of some proteins (6). Thus, depending on the degree of unfolding prior to association, there could be multiple denatured states available to associate into a gel matrix. The degree of complexity this adds to the interpretation of results can be illustrated by the results of Ishioroshi *et al.* (7), who reported that rabbit myosin formed more rigid gels at pH 6 than at pH 7.5. The initiation of gelation was at 35° and 45°C in buffers at pH 6 and pH 7.5, respectively. From the data, it is not possible to discern if the variation is caused by the gel matrix formed by association of different denatured states of myosin or if the pH is affecting the bonding within the gel matrix.

Rheological Properties of Gels

A definition that describes protein gels was given by Flory (1) : "the one feature identified almost universally as an essential characteristic of a gel is its solid-like behavior. When deformed its response is that of an elastic body." Most food gels are viscoelastic; a property which can be subjected to rigorous investigation by the techniques of rheology. The rheological and textural properties of gels can be determined by numerous techniques. In general, they can be separated into analysis at small, non-destructive strains (sample deformation) and techniques which strain the sample to failure (i.e., rupture). Rheological analysis at low, non-destructive strains permits a dynamic measurement of rheological transitions. Changes in rigidity or shear modulus (stress/strain), storage modulus (describes the elastic nature of the material), and loss modulus (describes the viscous nature of the material) can be measured as a function of time and/or temperature (8). These are fundamental physical properties of the material that are not dependent on sample size or shape and can be determined by a variety of methods (8). While the evaluation of rheological properties at non-destructive strains is beneficial in understanding gelation, the results are not always pertinent to textural properties perceived by sensory analysis. This is because sensory perception is related to how the sample ruptures.

Rheological properties determined at failure have been shown to correlate with sensory perception of texture (9). Shear stress at failure (shear force/area of shear surface) and shear strain at failure (shear deformation/thickness of layer sheared) are physical properties of gels which can be determined by mathematical analysis of samples of various simple shapes in compression, torsion or tension (10). Conversely, Instron texture profile analysis parameters (force to fracture, hardness) and other properties such as gel strength, are empirical and dependent on sample size and shape. Furthermore, determination of shear stress and strain gives a more complete description of the texture than single property evaluations. For example, a soft, deformable gel may have the same failure stress (hardness) as a rigid, brittle gel; however, the failure strain (deformation) could vary greatly between the samples. Methods which determine the force required for rupture of gels (hardness or gel strength) tend to be correlated with failure stress but give no information on failure strain.

Rheological analysis of failure stress and strain can be used to distinguish between filled gels and mixed/complex gels. Fillers tend to increase the failure stress but not the strain; whereas mixed/complex gels would be indicated by changes in stress and strain (11,12).

The lowest concentration of protein to form a gel, sometimes designated critical concentration (C_o) (6), can be used as a point of comparison among gelling proteins. This value can be determined by direct measurement [least concentration end-point, (13)], or by extrapolating to the concentration where a rheological parameter goes to zero (14).

Gelation of Single Proteins
(Bovine Serum Albumin and Myosin)

Bovine serum albumin (BSA) is a single polypeptide of 582 amino acids and 66,267 dalton molecular weight (15). Gelation of BSA is due to an ordered aggregation of partially unfolded molecules (6,16). Molecules heated at 70°C aggregate into chains, indicating that protein-protein interactions are restricted to specific surface locations (16). By contrast, heating BSA to 100°C causes extensive unfolding of the molecule and increased surface area for interactions. The result is random aggregation to form spherical particles and coagulation rather than gelation. Factors which influence order/unordered aggregation are reflected in gel structure.

Gel properties, such as gel time and rigidity modulus, are dependent on protein concentration, pH, ionic strength and heating temperature (17,18). When the concentration of BSA is held constant, the gelation time (time required to form a gel) decreases as temperature increases. A break in the gel time vs temperature plot at 57°C was thought to indicate that the rate limiting step was protein unfolding below 57°C and aggregation above the break temperature. The storage modulus increases with protein concentration (17). By combining variables of pH, protein concentration and NaCl concentration, it was shown that the greater the difference between protein isoelectric point (pI) and solution pH, the more sensitive the storage modulus is to changes in NaCl concentration (18). A similar pH/ionic strength dependence in failure properties (deformation to failure) was shown by Hegg (19). The concentration of NaCl required to form a stable gel increased as the pH moved away from the pI. From the previous discussion, it is clear that gelation is a balance of attractive and repulsive forces which vary with the physical/chemical properties of the protein.

Proteins which are single polypeptides, such as BSA, are the least complex and gelation results from changes in the secondary and/or tertiary structure. Many food proteins are composed of subunits (for example, soy proteins and muscle proteins) and thus can have alterations in quaternary structure, such as subunit dissociation, cause gelation. Myosin is a muscle protein which consists of six subunits (two heavy chains and four light chains), with a combined molecular weight of approximately 480,000 daltons (20). Skeletal muscle myosin exists *in vivo* as thick filaments, which can be formed *in vitro* by reducing the KCl concentration of a myosin suspension from 0.6-0.15 M (21). In most investigations on myosin, the state of the suspended protein is best viewed as an associated⇌monomeric equilibrium, dependent on pH, ionic strength and age of the preparation. Myosin filaments (0.2 M KCl) form more rigid gels than myosin (0.6 M KCl) (22). Gels formed at low ionic strength (0.25 M KCl) have a fine stranded microstructure; the microstructure is coarsely aggregated in gels formed at high ionic strength (0.6 M KCl) (23). This indicates that native protein-protein interactions of myosin (filaments vs monomers) can affect the type of gel formed.

Gelation of Mixed Proteins

Most foods contain many different proteins. The effect/function of each protein in a mixture which is heated to form a gel depends on unique properties of individual proteins, concentration, solution environment (pH, ionic strength, specific ions), and stability to the heating process.

One system is that of a mixture of gelling and non-gelling protein, such as actin and myosin. Actin, which forms a gel by polymerization of native molecules, does not gel when heated (24). Addition of actin increases the rigidity of myosin gels formed by heating for 20 min at 65°C in a pH 6 buffer containing 0.6 M KCl. The optimum condition for the effect of actin on myosin gel rigidity is a myosin to actin weight ratio of 15 (24). These findings suggest that the actin-myosin interaction orients myosin so that the association process forms a more rigid gel network. When the buffer is changed to favor filamentous myosin the results are quite different. The addition of actin decreases rigidity of myosin gels formed in a pH 6, 0.2 M KCl buffer (22).

Bovine serum albumin (BSA), a protein which can form a gel, is found in cheese whey, along with the non-gelling protein α-lactalbumin. When α-lactalbumin is added to BSA, the gelling temperature is increased (25). The limited data just reviewed indicates that non-gelling proteins can have a negative or positive influence on gelling proteins.

Since proteins denature and gel at different temperatures, a question arises as to the effect of denaturation temperature in a mixture of gelling protein. This problem was addressed using myosin-BSA and myosin-fibrinogen mixtures (26). Under the conditions used, myosin, fibrinogen and BSA gelled at 50°C, 50°C and 85°C, respectively. The myosin-fibrinogen mixture formed gels which were stronger (determined by failure analysis) than the sum of singly formed myosin and fibrinogen gels. An interaction between myosin and fibrinogen was suggested by enhanced gelation of myosin in mixtures where fibrinogen was $<C_o$. In a subsequent investigation a disulfide myosin-fibrinogen inter-

action was shown (27), indicating that the myosin-fibrinogen mixture formed a complex gel. Myosin forms a gel network between 50-80°C which is unaffected by the presence of non-gelled BSA (26). The gelation of BSA at 85 and 90°C does not change myosin gelation; however, the decrease in myosin gel strength at 95°C appears to be prevented during mixed gelation of BSA and myosin. The results suggest that in the case of myosin-BSA, the myosin gel network is formed prior to BSA gelation so that the processes are independent and this mixture may form a mixed gel.

Gelation in Food Systems

Investigations with purified proteins allows for close control of variables and analysis of molecular events. By contrast, most foods are complex mixtures which contain lipids, carbohydrates and minerals which can exist in suspension or as particles.

The effect of gums and fat content on rheological properties of meat batters was the subject of investigations in our laboratory (11). All meat batters contained 13% protein and at least 10% fat. The treatments consisted of replacing a portion of the water with fat or hydrocolloid-water mixtures. Since all treatments had equal protein content, the results shown in Table 11.1 can be viewed as the effects of substituting either fat, ι-carrageenan, κ-carrageenan or xanthan gum for water contained by a protein gel matrix. Using control-fat as a basis for comparison, fat, ι-carrageenan (0.5%) and κ-carrageenan (0.5 and 1%) had filler

TABLE 11.1
Textural Properties of Meat Gels

Treatment	Treatment Level	Fat	Shear stress (kPa)	Shear strain
Control fat	—	25.5%	63.1[a]	1.19[b]
Control water	—	10	38.5[c]	1.16[b]
ϕ-carrageenan	0.5%	10	39.0[c]	1.17[b]
	1.0%	10	54.6[b]	1.33[a]
κ-carrageenan	0.5%	10	41.8[c]	1.17[b]
	1.0%	10	51.7[b]	1.21[b]
Xanthan gum	0.5%	10	5.3[d]	0.76[c]
	1.0%	10	1.8[d]	0.57[d]

[a]Values in columns with different superscripts are significantly different (P<0.001). All gels contained 13% protein. Data from Foegeding and Ramsey (11).

effects; shown by an increase in shear stress and no strain effect. Xanthan gum (0.5 and 1%) and ι-carrageenan at 1% had an effect on the gel matrix, as indicated by changes in shear strain.

Future Research

The gelation of proteins involves events on a molecular level that ultimately determine the texture and water-holding ability of foods. While analysis of individual proteins produces information on individual molecular events, it is sometimes difficult to extrapolate those results to final product texture. Likewise, rheological analysis of products does not give information on chemical events. The greatest amount of insight will be gained by studies that use rheological analysis on individual proteins, final products and various mixtures of product components. Furthermore, the methods used for rheological/textural analysis require further development. Studies which investigate the relationship between rheological properties (determined at failure and non-destructively) and sensory perception of texture are needed so that standardized methods can be established.

References

1. Flory, P.J., *Chem. Soc. London, Faraday Discuss. 57*:7 (1974).
2. Ferry, J.D., *Adv. Protein Chem. 4*:1 (1948).
3. Tolstoguzov, V.B., and E.E. Braudo, *J. Texture Stud. 14*:183 (1983).
4. Edsall, J.T., and H. Gutfreund, in *Bio-Thermodynamics: The Study of Biochemical Processes at Equilibrium*, John Wiley and Sons, 1983, p. 248.
5. Privalov, P.L., and N.N. Khechinashvili, *J. Mol. Biol. 86*:665 (1974).
6. Clark, A.H., and C.D. Lee-Tuffnell, in *Functional Properties of Food Macromolecules*, J.R. Mitchell and D.A. Ledward, (eds.), Elsevier Applied Sci. Publishers, New York, NY, 1986, pp. 203-272.
7. Ishioroshi, M., K. Samejima and T. Yasui, *J. Food Sci. 44*:1280 (1979).
8. Hamann, D.D., *Food Technol. 41*:100 (1987).
9. Montejano, J.G., D.D. Hamann and T.C. Lanier, *J. Texture Stud. 16*:403 (1985).
10. Hamann, D.D., in *Physical Properties of Foods*, E.B. Bagley and M. Peleg, (eds.), AVI Publishing Co., Westport, CT, 1983, pp. 351-383.
11. Foegeding, E.A., and S.R. Ramsey, *J. Food Sci. 52*:549 (1987).
12. Brownsey, G.J., H.S. Ellis, M.J. Ridout and S.G. Ring, *J. Rheology 31*:635 (1987).
13. Trautman, J.C., *J. Food Sci. 31*:409 (1966).
14. Bikbow, T.M., V.Y. Grinberg, Y.A. Antonov, V.B. Tolstoguzov and H. Schmandke, *Polymer Bull. 1*:865 (1979).

15. Peters, P.T., Jr., *Adv. Protein Chem. 37*:161 (1985).
16. Tombs, M.P., in *Proteins as Human Food*, R.A. Lawrie, (ed.), Avi Publishing Co., Westport, CT, 1970, pp. 126-138.
17. Richardson, R.K., and S.B. Ross-Murphy, *Int. J. Biol. Macromol. 3*:315 (1981).
18. Richardson, R.K., and S.B. Ross-Murphy, *Br. Polym. J. 13*:11 (1981).
19. Hegg, P-O., *J. Food Sci. 47*:1241 (1982).
20. Harrington, W.F., and M.E. Rodgers, *Ann. Rev. Biochem. 53*:35 (1984).
21. Pepe, F.A., B. Drucker and P.K. Chowrashi, *Preparative Biochem. 16*:99 (1986).
22. Ishioroshi, M., K. Samejima and T. Yasui, *Agric. Biol. Chem. 47*:2809 (1983).
23. Hermansson, A-M., O. Harbitz and M. Langton, *J. Sci. Food Agric. 37*:69 (1986).
24. Yasui, T., M. Ishioroshi and K. Samejima, *J. Food Biochem. 4*:61 (1980).
25. Paulsson, M., P. Hegg and H.B. Castberg, *J. Food Sci. 51*:87 (1986).
26. Foegeding, E.A., W.R. Dayton and C.E. Allen, *Ibid. 51*:109 (1986).
27. Foegeding, E.A., W.R. Dayton and C.E. Allen, *J. Agric. Food Chem. 35*:559 (1987).

Chapter Twelve

Functional Roles of Heat Induced Protein Gelation in Processed Meat

James C. Acton and Rhoda L. Dick

Department of Food Science
South Carolina Agricultural Experiment Station
Clemson University
Clemson, South Carolina 29634-0371

Stabilization of comminuted meat matrices involves three principal physiochemical events that occur in the protein fraction during heat processing. Functionally these events are (a) protein-water interaction, (b) protein-fat interaction, and (c) protein-protein interaction. The responses of the meat tissue's myofibrillar proteins, primarily myosin and actomyosin, determine the ultimate development of the internal gel structure in heat processed products. The functional roles of myosin and actomyosin in water binding and fat stabilization are dependent on the protein sol-to-gel transformation during the continuous input of thermal energy associated with cooking. In the 30°C to 70°C temperature zone, molecular transition temperatures of myofibrillar proteins provide valuable evidence for the sequence of events in protein-protein interactions and gel matrix formation. The protein-protein interactions associated with gelation of myosin and the myosin constituent of actomyosin during thermal heating are presented in this report.

Functional properties have been described as measures of the descriptive actions of performance attributes imparted from inherent or added ingredients used in the processing of meat products (1,2). The myofibrillar proteins myosin and actomyosin are most directly involved in the protein-protein interactions that lead to the formation of an ordered structural matrix. One of the best examples of the protein-protein interaction reactions in meat processing is the development of "binding" at the surface junction between meat pieces and the recognition that binding strength is dependent on the nature of the gelation properties of the proteins lying within the junction (3-7). Protein-protein interactions are also recognized as being important to processing of comminuted meat batters (8-11), where development of internal structure during heat processing contributes to textural aspects and stabilization of dispersed fat particles (12,13).

Each of the physical and chemical factors which affect proteins in general are important to the functional behavior of myosin and the actomyosin protein complex. The background to understanding the nature of protein-protein interactions in processed meats, the involvement of successive steps of thermal energy input for denaturation of proteins during meat processing, and the sequence of myosin and actomyosin molecular transitions important to gelation are presented and discussed. Experimental methods which have been used for measuring protein transition temperatures (Tm's) are included for confirming that two major thermal stages or zones for myofibrillar protein gelation occur during heat processing of meat products.

Meat Tissue Factors and Protein Aggregation

Tissue pH, salt concentration, thermal energy input, protein concentration (in sol state), and even time are some of the factors affecting protein-protein interactions. In processed meats, the aqueous phase pH is generally that attained in two cases: prerigor meat normally in the 6.8-7.0 pH range, and postrigor meat in the 5.4-6.0 pH range. Salt concentration (as NaCl) is an important variable in protein-protein interactions. With a normal level of 2.0-2.5% addition of NaCl to meat tissues having a moisture content of 55% to 65%, the effective ionic molarity of the aqueous phase ranges from 0.67-0.93 when the approximately 0.15 ionic molarity inherent to the tissues is included. At temperatures below 15.6-21.1°C, there appears to be very limited protein-protein interaction. However, protein-protein interactions would occur if the system were held for any length of time in the 15-21°C zone, due to the accumulation of energy. The practice in meat processing operations has always been either the "keep it cold" or "move it through" method of processing. The majority of the protein-protein interactions occur in heat processing of meat mixtures and batters with temperatures spanning the approximate 15-21°C (60-70°F) to 65-71°C (150-170°F) region.

Aggregation is a general term which has been used to describe many types of protein-protein interactions. However, a distinct difference exists between protein aggregations termed "coagulation" and "gelation." Coagulation is the random protein-protein interaction of denatured protein molecules and the clue is that randomness does not lead to an orderly structural assembly of the final aggregate. By comparison, gelation is the orderly interaction of proteins, which may or may not be denatured, and which leads to formation of a three-dimensional well-ordered structural matrix (14).

Since denaturation is involved in these definitions, it is evident that native protein structure is altered when a "coagulated mass" or a "gelled protein matrix" is formed. For myosin and the actomyosin complex, transition from the native to denatured state is a continuous process of native protein structural changes involving secondary, tertiary, and/or quarternary structures. Hydrogen bonding, hydrophobic interactions and electrostatic linkages are altered during transition to the denatured state(15). Once aggregation occurs there generally will not be a return of molecular structure to the native state. Thus, with matrix formation by myosin or actomyosin the course is irreversible.

Bond energies that contribute to native structure and maintenance of a protein's conformation are given in Table 12.1. These bonds can also be viewed as important in protein that is denatured. Once a new structure (gel) is formed, the same types of bondings contribute to its structure. Thermal energy input to a protein system is known to drive denaturation reactions, i.e., sufficient energy is provided to drive the reaction at a given rate through a particular path of bond breaking and altered bond reformation stages in the new structure. With electrostatic and hydrogen bonding decreases, and with conformational change, there is greater tendency for more interchain hydrophobic interactions to occur.

The gelled matrix formation by myosin and actomyosin has been hypothesized to involve an initial interaction of myosin globular heads which is followed by cross-linking of the rod or tail sections of the molecules [Fig. 12.1; Ziegler and Acton (19)]. This heat-induced two-step process is consistent with Ferry's (20) proposed reaction sequence for formation of protein gels:

Native Protein → Denatured Protein → Aggregated Protein

Two reactions, occurring in separate temperature zones, were reported by Acton *et al.* (21) for actomyosin in dilute solutions and in solutions at gelling concentrations. The overall reaction follows first-order kinetics with an activation energy (E_a) of 24.3 kcal/mole at pH 6.0 (22). Smith *et al.* (23) reported a similar denaturation E_a for postrigor myofibrillar protein during gelation.

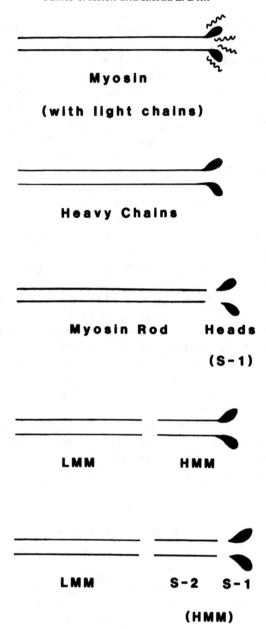

Fig. 12.1. Schematic of the myosin molecule and various segments or fragments obtained by enzymatic digestion. (See Goll *et al.* (18) for a detailed description of myosin and its segments and subfragments.)

TABLE 12.1
Structural Bondings Found in Proteins

Type of bond[a]	Mechanism	Energy kcal/mole	Distance of interaction (Ångstroms)
Covalent bond[b]	Electron sharing	30 - 150	1 - 2
Electrostatic (Ionic) bond	Coulombic attraction between oppositely charged groups	10 - 20	2 - 3
Hydrogen bonding	Hydrogen sharing between two electronegative atoms (directional)	1 - 5	2 - 3
Hydrophobic interaction (Van der Waals attractive)	Induction of dipole moments in apolar groups	1 - 3	3 - 5
Electrostatic repulsive force	Coulombic repulsion between similarly charged groups	$\dfrac{q_1 q_2}{r^2}$	
Van der Walls repulsive force	Repulsion between apolar groups in close proximity	$\dfrac{1}{r^{1/2}}$	

[a]Adapted from Jones (16) and Whitaker (17).
[b]The bond energy for the disulfide bond as occurs in proteins, important to protein-protein interactions discussed in this report, was not found by the author in searching available literature.

Heat-Induced Interaction Sequence

In applying a two-step reaction sequence for myosin and actomyosin to processed meats, it is obvious that the two stages will involve distinct segments of the myosin molecule shown in Fig. 12.1. The molecular transformations occur in separate temperature zones between 30–50°C and above 50°C. The sequence of these changes are separated in the following sections using these temperature zones.

The 30-50°C heating stage.

The first stage involves aggregation of the globular head regions of the molecule. It is an irreversible reaction, assuming that heating will be continuous with continuous temperature elevation of the system. Through studies with the S-1 fraction, HMM segment, myosin and PCMB (p-chloromercuribenzoate) treated myosin, the aggregation is thought to be dependent on oxidation of -SH groups which are found predominantly in the globular head region (24-26). While -SH group reduction (moles -SH/mole segment) progressively increases from 20-70°C, considerable reduction of -SH content occurs in the early temperature range of 20-50°C for myosin or segments containing a globular head portion. No -SH reduction has been reported for myosin rod or the LMM segment (both devoid of globular heads) over the 20-50°C range. The changes in measurable free -SH groups for myosin and its segments are given in Table 12.2.

It is possible that the free -SH group reductions are reflective of possible disulfide bond participation or of structural alterations which renders them inaccessible to measurement. Hamm and Hoffmann (27) reported an increase in free NEM (N-ethylmaleimide) reacting -SH groups when myofibril preparations were heated from 30-70°C, although the total -SH group content reactive with $AgNO_3$ did not change. They concluded that heat denaturation was not accompanied by new disulfide bond formation, at least up to 70°C of heating. However, the myofibrils were not in an environment of or equivalent to 0.6 M KCl. Actomyosin, isolated from Atlantic croaker, shows a small decrease of total -SH groups when heated from approximately 30-60°C, with a 50% decrease in heating from 60-70°C (28). Liu et al. (28) concluded on the basis of ease of solubilization (by 2% SDS) of actomyosin gels heated to 48°C, that hydrophobic interactions were the predominant force in actomyosin aggregation below 50°C. Solubilization of myosin gels heated to 50°C by guanidine hydrochloride and urea, as reported by Foegeding et al. (29), also implies that hydrophobic and hydrogen bonding are as important as -SH group reduction for gel stabilization.

Several workers have reported increased binding and enhancement of fluorescent intensity for actomyosin (30) and myosin (31,32) upon heating in the presence of ANS (8-anilino-1-napthalene sulfonate). ANS is a fluorescent probe capable of binding with hydrophobic regions of proteins when conformational changes allow reaction with nonpolar residues. The binding of ANS is initiated at temperatures beginning in the 27-33°C range and increases to a maximum at 41-49°C (32). The

TABLE 12.2
Net Reduction in Free Sulfhydryl Content for Myosin and Components from <20°C (Initial) through Heating to 70°C[a]

Component	Reduction in –SH content (moles –SH/mole component)
HMM	8.6
S-1	3.9
Myosin	1.4
Myosin rod	0
LMM	0

[a]Adaped from Samejima et al. (25) and Ishioroshi et al. (26).

exposure of hydrophobic residues facilitates hydrophobic interactions and thus increases the potential for globular head cross-linking.

It is reasonable to assume that myosin free –SH content decreases are due to the globular head conformational changes in the temperature region of 30–50°C and that the conformational alterations render –SH groups unavailable to detection by reagents used for their measurement. Whether covalent disulfide bonding is formed or not is not critical in this low temperature region. Further considerations of gel rigidity and thermal transition data (in the following sections) indicate that these early conformational changes in the globular myosin head segment satisfy the first step in the overall reaction sequence for myosin and actomyosin gelation.

The heating stage above 50°C.

The second stage is associated with structural change of the helical rod segment of myosin which culminates in network formation through cross-linking of these segments. The data in Table 12.3 (25,26) for myosin and its segments, show the temperatures at which: (a) gel formation is initiated; (b) maximum gel rigidity occurs; and (c) whether further heating results in the loss of gel rigidity or maintenance at a fairly constant level (plateau). Gel rigidity results are based on shear modulus of the segments in solutions heated over the temperature range of 20–70°C.

If we assume that the initial temperature zone of 30–35°C denotes transit into the first aggregation stage, it is evident (Table 12.3) that this initiation is shared by S-1, HMM and myosin, each of which contains the

TABLE 12.3
Gel Rigidity Aspects of Solutions of Myosin, HMM, S-1, Myosin Rod and LMM in the Temperature Range of 20°C to 70°C[a]

Component	Rigidity initiation °C (Range)	Maximum rigidity °C	Stability after max rigidity
S-1	30-35	50-55	decline
HMM	30-35	60	slight loss at 70°C
Rod	35-40	60	plateau
LMM	35-40	60	plateau
Myosin[b]	30-35	60	plateau

[a]Estimated from original Figure 2 in Samejima et al. (25) and original Figure 2 in Ishioroshi et al. (26).
[b]Actomyosin formed from myosin and F-actin at apparent mole ratios of 1.5 or 2.7 myosin to F-actin follow the same gel rigidity-temperature profile as myosin (33). However, natural actomyosin has a mole ratio ranging from 0.21 - 0.26 (dependent on extraction) which suggests free myosin in the high mole ratio combinations. Comparable data for natural actomyosin not available.

globular head (or heads). The delayed transit of the rod and LMM segments, both devoid of heads, at 35-40°C rather than at 30-35°C, may indicate that the first stage or a parallel to its equivalent is absent; then these segments may only be initiating an early phase of the second stage. It is myosin, the full component with head and tail, that shows both early initiation of rigidity development and a plateau attainment at 60°C (Table 12.3). The plateau of rigidity is absent for S-1 which contains nothing of a tail segment. In fact, S-2 rapidly loses gel rigidity at temperatures exceeding 55°C. HMM (intact S-1 + S-2) by comparison, with its S-2 segment of the tail region, behaves similar to myosin. HMM does lose rigidity slightly at 70°C, which is well beyond the 60°C plateau. The molecular transitions associated with gel rigidity development in addition to studies of other thermally-induced structural changes of myosin and actomyosin are presented in the following section.

Molecular Transition Temperatures (Tm's) for Myosin and Actomyosin

The molecular response of myosin and actomyosin, measured using

TABLE 12.4
Transition Temperatures (Tm's) of Rabbit Muscle Myosin, Myosin Segments and Actomyosin from Studies Utilizing Various Techniques for Thermal Alteration

Protein or segment	Method[a]	Solution conditions		Tm (C)[b]			Ref.
		pH	M or Γ	Tm¹	Tm²	(Tm³)	
Myosin	GR	6.0	0.6M KCl	43	55		(25,26)
HMM				45	54		
S-1				43	none		
LMM				none	53		
Rod				none	53		
Myosin	GR	6.5	0.6M KCl	48			(32)
LMM	OR + CD	6.0	0.6M KCl	none	50-55		(25)
Rod	OR + CD	6.0	0.6M KCl	43	54		(26)
Rod	OR	7.0	0.5M KCl	44	55		(36)
LMM				42	52		
S-2				39	50		
Myosin	ANS-B	6.5	0.6M KCl	44	—		(32)
Myosin	PB	6.0	0.5M KCl	41 est	—		(37)
HMM				41 est	—		
LMM				41 est	—		
S-1				40 est	—		
Rod				43 est	—		
S-2				---->	52 est		
S-2 (short)	DSC	7.2	0.6M KCl	none	50 est		(38)
S-2 (long)				42	50 est		
Myosin	DSC	6.0	1.0Γ	43	49.5	60.5	(39)
Actomyosin		7.0	0.5Γ	51.5	60	73	

[a]Method abbreviation: GR = gel rigidity; OR = optical rotation; CD = circular dichroism; ANS-B = 8-anilino-1-napthalene sulfonate binding; PB = proton binding; DSC = differential scanning calorimetry.
[b]None denotes no Tm present, dash denotes technique does not measure and "est" indicates the Tm was estimated from figures in the reference.

experimental techniques such as optical rotation, circular dichroism, proton and ANS binding, gel rigidity and light scatter over the temperature range of 30-70°C, can be obtained by using derivative plots of change in response with a change in temperature (dR/dT). The deriva-

TABLE 12.5

Thermal Transitions (Tm's) of Myosin and Actomyosin of Various Muscle Foods

Protein	Method[a]	Solution conditions		Tm (C)[b]			Ref.
		pH	M or Γ	Tm^1	Tm^2	Tm^3	
Myosin							
fish	GR	6.5	0.6 M KCl	43	49	55	(32)
poultry white				49			(32)
Myosin							
fish	ANS-B	6.5	0.6 M KCl	37	—		(32)
fish				39	—		(32)
poultry white				44	—		(32)
Actomyosin							
bovine	LS	6.0	0.6 M KCl	48.5	57.5		(22)
poultry white				49.2	60.2		(40)
poultry red				52.6	57.9		(40)

[a]Method abbreviation: GR = gel rigidity; ANS-B = 8-anilino-1-napthalene sulfonate binding; LS = light scatter.
[b]Blank space denotes no Tm present; dash denotes technique does not measure.

tive procedure is extensively used in protein research and yields fairly accurate assignments of the transition temperatures (Tm's) at which each part of an overall process occurs under the conditions and methods by which the response is measured (19). The technique is not an application of strict derivatives, but follows the smoothing method of Savitzky and Golay (34) of a sliding fit of data points equally spaced over selected temperature intervals. What is important here is that the transition temperature or temperatures are associated with "cooperativeness" — meaning that the change or transition is a major, exact event which occurs within a narrow temperature range [adapted from meaning of "cooperativeness" in protein denaturation as given by Tanford (35)]. No derivative plots are necessary with the technique of differential scanning calorimetry.

Transition temperatures from the gel rigidity responses cited in Table 12.3 and from other protein responses through application of experimental techniques where continuous heating was utilized are summarized in Tables 12.4 and 12.5. As pointed out by Ziegler and Acton (19) "...exact Tm value comparisons must be done with caution because of the differences in methods utilized...". The comparisons given in Tables

12.4 and 12.5 correspond, where possible, to protein solution conditions near those of interest, i.e., pH range of 6.0 or above and an ionic strength of 0.5 or higher (exceptions are present in Table 12.4). Some techniques do not allow observation of the second temperature region (>50°C).

The Tm's shown in Tables 12.4 and 12.5 do confirm the existence of the two stages of interaction previously discussed and can be differentiated on the basis of:
 (a) the presence or absence of the head region and/or the tail segment; and
 (b) the first aggregation region from 30–50°C versus the second aggregation region at temperatures >50°C.

Two myosin segments, LMM and S-2, have been found to yield an apparent Tm in the lower temperature region (36–38) although neither possesses the globular head (S-1) portion. From gel rigidity data (26), LMM shows only a single transition at approximately 53°C. This technique (and others in Tables 12.4 and 12.5) may not be sensitive enough to ascertain that another Tm in the 40–45°C range is present.

The molecular region linking LMM and S-2, as found in the myosin rod, is referred to as the "hinge" region (36). When limited enzymatic digestion is utilized in preparing the respective segments, it is possible that parts of the "belt" spanning the hinge region may be possessed by LMM and/or S-2, and if present in either fragment, could be responsible for the lower Tm occasionally observed. The lower Tm has been identified in the myosin rod where the hinge region remains intact (25,36). From the data of Burke et al. (36), who used papain-derived LMM for α-helix fractional alteration measured by optical rotatory dispersion, LMM showed two cooperative transitions, the first at approximately 42°C and the second at 52°C (Table 12.4). The chymotryptic prepared LMM of Ishioroshi et al. (26), when analyzed by optical rotatory dispersion and circular dichroism, yielded only one apparent Tm in the 50–55°C region.

Since cleavage at the "hinge" region (or "belt") is also used in producing S-2, additional inferences from this region were obtained by Swenson and Ritchie (38). They prepared "short S-2" and "long S-2" using chymotryptic digestion (two steps) for long S-2, and subsequent trypsin digestion of long S-2 to obtain short S-2. Since the primary difference between these two segments is the absence of the hinge region in short S-2, their examination by differential scanning calorimetry (Table 12.4) showed that short S-2 yielded only one Tm at approximately 50°C (0.6M KCl, pH 7.2), whereas long S-2 (same conditions) yielded two Tm's, the first at approximately 42°C and the second at 50°C.

The presence or absence of a portion of the hinge region may be responsible for the overlapping of apparent "major" events occurring in the first stage (30-50°C) discussed earlier. While the globular head interaction predominates in the first stage, there is also apparent early disruption of the α-helix at the hinge region in moving from the coiled-coil (α-helix) to a random coil type structure in the same lower temperature region. Further helical disruption of the "tail" portions requires a higher energy input, thus these helical alterations predominate in the second stage at temperatures >50°C.

The coiled-coil to random conformational change in the tail region is extremely important to aggregation occurring in the second stage of events. For LMM (26) and myosin rod (25), ultraviolet absorption difference spectra at 285 nm, where the aromatic side chains of the protein absorb, confirmed that absorption increases as both segments are heated from 20°C to 65°C. The early exposure of hydrophobic residues in the myosin globular head region, which is based on ANS binding (32), and the subsequent continuous exposure of nonpolar aromatic residues in the helical rod region, which are based on UV absorption spectra (25,26), strongly indicate that hydrophobic interactions are involved in head-to-head and tail-to-tail cross-linking in establishing the final ordered gel matrix.

Applied Sol-to-Gel Conversion in Meat Systems

By mechanically shearing intact muscle tissue cells, by the addition of limited free water, by addition of salt and through means of dispersive actions, a suspension of macroparticles exists in a continuous aqueous phase. The term "sol" appears to apply to the aqueous phase of macromolecular hydrated myofibrillar proteins. Even the sheared tips of suspended myofibrils, partly solvated in the high ionic environment, are free to interact when heat is applied to the total system. One notable feature of the "sol" of myofibrillar proteins is apparent — its protein concentration is high, and further solubilizing effects by water addition are limited (legally as applied to several meat products). A sol-to-gel transformation occurs in such meat tissue minces due to continuous energy input (detected by temperature) during heat processing.

The tacky-sol transformation of myofibrillar proteins to the gelled state results in the formation of the ultimate three-dimensional interlinked protein network. This protein network both physically (due to capillarity) and chemically (due to noncovalent bondings) stabilizes water, and physically or structurally restrains dispersed fat (in comminuted meats) from rendering. Reviews of applied research in this area were given by Schmidt et al. (a) and Acton et al. (10). Payne and Rizvi (41) and Burge and Acton (42) reported pseudoplastic behavior for comminuted meat batters and related alterations of apparent viscosities to fat melting characteristics and protein-protein interactions. Further studies are needed for the rheological behavior of dispersed meat systems spanning the temperature zone where a flowing system changes to a nonflowing solid product.

The temperature region of importance in gelation has been given here as initiation in the 30-35°C zone and termination near 65-70°C. From all of the gelation and meat binding studies, the temperature region above 50°C appears most critical. Gels do not reach appreciable strength until the myosin tail portion has undergone helix-coil transformation and subsequent cross-linking. Of course, the myosin head region is important since from ultrastructure studies, these appear to form the initial "super-junctions" upon which the "super-thick" filament or rod network interlinks (5, 25). While similar studies of gelation by natural actomyosin have begun to appear (8,12,21,22,28), the ultrastructure of actomyosin gels is one of thinner filamentous strands with larger pore size distribution and a different cross-linked appearance when compared to myosin gels [see scanning electron micrographs presented by Yasui et al. (43)]. Binding strength development at the junction between adjacent cut meat surfaces shows temperature dependence, particularly above 50°C (44), and has been previously reviewed (3).

In summary, protein-protein interaction is a functional event that can be related to the structural integrity of meat products through orderly heat-induced aggregations. These aggregations are two-fold, involving the head portion(s) of myosin at temperatures between 30-50°C and the rod segment in the temperature region above 50°C. Some initiation of conformational change occurs in the rod segment in the lower temperature zone due to the sensitivity of the "hinge" region connecting the S-2 and LMM segments. The complete myosin molecule is necessary for attaining appreciable continuity and strength in the protein matrix. From ultrastructure studies, there are morphological differences between myosin and actomyosin aggregates, implying that in processed meats, differences in textural attributes between prerigor and postrigor raw materials may emerge in the finished products.

Acknowledgments

Technical contribution 2315B of the South Carolina Agricultural Experiment Station, Clemson, University, Clemson, SC 29634 and publication support from the American Meat Science Association.

References

1. Acton, J.C., in *Proceedings of 21st Ann. Meat Sci. Inst.*, Vol. 21, University of Georgia, Athens, GA 1979, pp. 14-22.
2. Whiting, R.C., *Food Technol. 42*:104 (1988).
3. Theno, D.M., D.G. Siegel, and G.R. Schmidt, in *Proceedings of Meat Ind. Res. Conf.*, Am. Meat Inst. Foundation, Arlington, VA, 1977, p. 53.
4. Addis, P.B., and E.S. Schanus, *Food Technol. 33*:36 (1979).
5. Siegel, D.G., and G.R. Schmidt, *J. Food Sci. 44*:1686 (1979).
6. Turner, R.H., P.N. Jones and J.J. MacFarlane, *Ibid. 44*:1443 (1979).
7. Solomon, L.W., and G.R. Schmidt, *Ibid. 45*:283 (1980).
8. Deng, J., R.T. Toledo and D.A. Lillard, *Ibid. 41*:273 (1976).
9. Schmidt, G.R., R.F. Mawson, and D.G. Siegel, *Food Technol. 35*:235 (1981).
10. Acton, J.C., G.R. Ziegler and D.L. Burge, *CRC Crit. Rev. Food Sci. Nutr. 18*:99 (1983).
11. Smith, D.M., *Food Technol. 42*:116 (1988).
12. Acton, J.C., and R.L. Dick, *Meat Ind. 31*:32 (1985).
13. Asghar, A.,K. Samejima and T. Yasui, *CRC Crit. Rev. Food Sci. Nutr. 22*:27 (1985).
14. Hermansson, A.M., *J. Texture Stud. 9*:35 (1978).
15. Anglemier, A.F., and M.W. Montgomery, in *Principles of Food Science, Part 1: Food Chemistry*, edited by O.R. Fennema, Marcel Dekker, Inc., New York, NY, 1976, p. 238.
16. Jones, R.T., in *Proteins and Their Reactions*, edited by H.W. Schultz and A.F. Anglemier, AVI Publishing Co., Inc., Westport, CT, 1964, p. 34.
17. Whitaker, J.R., in *Food Proteins*, edited by J.R. Whitaker and S.R. Tannenbaum, AVI Publishing Col, Inc. Westport, CT, 1977, pp. 14-49.
18. Goll, D.E., R.M. Robson and M.H. Stromer, in *Food Proteins*, edited by J.R. Whitaker and S.R. Tannenbaum, AVI Publishing Co., Inc., Westport, CT, 1977, pp. 121-174.
19. Ziegler, G.R., and J.C. Acton, *Food Technol. 38*:67 (1984).
20. Ferry, J.D., *Adv. Prot. Chem. 3*:1 (1948).
21. Acton, J.C., M.A. Hanna and L.S. Satterlee, *J. Food Biochem. 5*:101 (1981).
22. Ziegler, G.R., and J.C. Acton, *Ibid. 8*:25 (1984).
23. Smith, D.M., V.B. Alvarez and R.G. Morgan, *J. Food Sci. 53*:359 (1988).
24. Ishioroshi, M., K. Samejima, Y. Arie, and T. Yasui, *Agric. Biol. Chem 44*:2185 (1980).
25. Samejima K., M. Ishioroshi and T. Yasui, *J. Food Sci. 46*:1412 (1981).

26. Ishioroshi, M., K. Samejima and T. Yasui, *Ibid.* *47*:114 (1982).
27. Hamm, R., and K. Hoffmann, *Nature* *207*:1269 (1965).
28. Liu, Y.M., T.S. Lin and T.C. Lanier, *J. Food Sci.* *47*:1916 (1982).
29. Foegeding, E.A., C.E. Allen and W.R. Dayton, *36th Ann. Reciprocal Meat Conf.* *36*:190, Abstract (1983).
30. Niwa, E., *Bull. Jpn. Soc. Sci. Fish.* *41*:907 (1975).
31. Lim, S.T., and J. Botts, *Arch. Biochem. Biophys.* *122*:153 (1967).
32. Wicker, L., T.C. Lanier, D.D. Hamann, and T. Akahane, *J. Food Sci.* *51*:1540 (1986).
33. Yasui, T., M. Ishioroshi and K. Samejima, *J. Food Biochem.* *4*:61 (1980).
34. Savitzky, A., and M.J.E. Golay, *Anal. Chem.* *36*:1627 (1964).
35. Tanford, C., *Adv. Protein Chem.* *23*:121 (1968).
36. Burke, M., S. Himmelfarb and W.F. Harrington, *Biochemistry* *23*:701 (1973).
37. Goodno, C.C., T.A. Harris and C.A. Swenson, *Ibid.* *15*:5157 (1976).
38. Swenson, C.A., and P.A. Ritchie, *Ibid.* *19*:5371 (1980).
39. Wright, D.J., I.B. Leach and P. Wilding, *J. Sci. Food Agric.* *28*:557 (1977).
40. Acton, J.C., and R.L. Dick, *Poultry Sci.* *65*:2051 (1986).
41. Payne, N.N., and S.S.H. Rizvi, *J. Food Sci.* *53*:70 (1988).
42. Burge, D.L., Jr., and J.C. Acton, *J. Food Technol.* *19*:719 (1984).
43. Yasui, T., M. Ishioroshi and K. Samejima, *Agric. Biol. Chem* *46*:1049 (1982).
44. Acton, J.C., *J. Food Sci.* *37*:244 (1972).

Chapter Thirteen

Effects of Medium Composition, Preheating, and Chemical Modification upon Thermal Behavior of Oat Globulin and β-Lactoglobulin

V.R. Harwalkar and C.-Y. Ma

Food Research Center
Agriculture Canada
Ottawa, Ontario
Canada, K1A OC6

Functional properties of food proteins are closely related to their molecular structure and have been the subject of several reviews and a monograph (1-4). To gain some insight into the molecular basis of protein functionality, an assessment of protein structure and conformation is essential, particularly under various processing conditions. Most spectroscopic techniques for monitoring protein conformation require protein to be soluble and are limited to highly purified protein preparation in dilute solutions. Most foods contain proteins in a complex system and some proteins, particularly plant storage proteins, have limited solubility. Thermoanalytical techniques such as differential scanning calorimetry (DSC) are ideally suited for studying food proteins since concentrated solutions (or dispersions) or solid samples can be examined, and relatively crude samples (e.g., protein concentrates) can be analyzed.

In this study, DSC was used to monitor conformational changes in oat globulin and β-lactoglobulin subjected to alteration by manipulating medium composition, heat treatment, and by chemical modification. Many food processing steps (e.g., pH adjustment and salt variation) can be simulated by changing the medium composition. The addition of reagents that modify protein structure, such as urea, sodium sulfate, polyols and reducing agents, can provide information on the chemical forces involved in the stabilization of protein structure. Heat treatments are commonly applied to foods during processing and DSC can easily be used to simulate different heating processes. Chemical modification is used to alter the physicochemical and functional properties of food proteins, and DSC data can provide information concerning the structural changes associated with modification. Some of the data have been reported elsewhere (5-9).

Materials and Methods

Materials.

Oats (Sentinel or Hinoat variety) were grown at the Central Experimental Farm, Ottawa, Canada. Dehulled groats were ground in a pin mill (Alpine model 160 Z) and defatted by Soxhlet extraction with hexane. β-Lactoglobulin (βLG) was purchased from Sigma Chemical Company, St. Louis, MO, and was used without further purification.

Heat treatments.

Oat globulin was heated at either 100 or 110°C in 0.01 M phosphate buffer (pH 7.4) containing 1.0 M NaCl. Protein samples (about 1 mg dispersed in 10 μL buffer) were heated in a sealed DSC pan on the reference or sample platform of a DSC cell preset at 100 or 110°C. After heating for the prescribed time, the pans were cooled rapidly in an ice bath and reequilibrated to room temperature.

To study the nature of heat aggregated protein, 1% oat globulin was prepared in the phosphate buffer, heated in 500 ML Erlenmeyer flasks covered with aluminum foil in either a boiling water bath or in a temperature-controlled autoclave at 110°C. The heated samples were centrifuged at 10,000 × g for 20 min. and the supernatants were dialyzed against distilled water at 4°C. Both the residues and supernatants were freeze-dried to recover the proteins.

To study the DSC characteristics of heat-induced oat globulin gels, 10% protein dispersions were prepared in 0.2 M NaCl, and the pH was adjusted to 9.7 with 1 M NaOH. Sealed DSC pans containing 10 μL of the dispersions were heated in a boiling water bath for 20 min., then cooled and equilibrated to room temperature. Under such conditions, a solid gel matrix was formed.

Chemical modification.

Succinylation. Oat globulin and βLG were succinylated according to the procedure described by Groninger and Miller (10). Succinic anhydride was added at levels of 0.1 and 0.2 g per g protein for oat globulin and 0.02 and 0.1 g per g for βLG. The extent of succinylation was determined from the free amino contents by the method of Concon (11).

Fatty acid modification. Succinimidyl esters of fatty acids were prepared as described by Paquet (12), and used for acylating the free amino groups of βLG according to the method used by Paquet (12, 13). Solutions of βLG (60 mg/mL) in saturated sodium bicarbonate were added to 7 or 15 equivalents of succinimidyl esters of the appropriate fatty acid in dioxane and stirred for five hours. The reaction mixture was extensively dialyzed and freeze-dried. The dried samples were washed with chloroform to remove unreacted fatty acids.

DSC.

The thermal characteristics of oat globulin and βLG were examined by DSC using either a Perkin Elmer DSC-II differential scanning calorimeter or a DuPont 1090 Thermal Analyzer equipped with a 910 DSC cell base and a high pressure cell. Liquid samples (10 μL) were pipetted onto the pan, or solid samples (about 1 mg) were weighed onto the pan and 10 μL of buffer was added. A sealed empty pan was used as reference and Indium standards were used for temperature and energy calibration. The onset temperature (T_m), peak or denaturation temperature (T_d or T_D), width at half peak height ($\Delta T_{1/2}$), and the heat of transition or enthalpy (ΔH) were computed from the thermograms as previously described (6).

The denaturation kinetics of oat globulin were studied by DSC using the heat evolution method described by Borchardt and Daniels (14). The method assumes that the reaction obeys the relationship:

$$d\alpha/dt = k(1-\alpha)^n$$

Where α = fractional conversion; k = rate constant (sec $^{-1}$); and n = reaction order.

This method also assumes that the temperature dependence of the reaction rate follows the Arrhenius expression:

$$k = Z \cdot e^{-E_a/RT}$$

Where Z = preexponential factor (sec $^{-1}$); E_a = activation energy (J/mol); R = gas constant; and T = absolute temperature. Triplicate samples were evaluated by DSC; coefficients of variation ranged from 0.3-0.6% for T_m, T_d and $\Delta T_{1/2}$, 4-15% for ΔH, and 8-10% for kinetic parameters E_a and Z.

Results and Discussion
Effects of medium composition on thermal characteristics.

pH. To study the effect of pH on DSC characteristics of oat globulin and βLG, protein dispersions with desired pH were prepared by the addition of acid or alkali. Figure 13.1 shows that both T_d and ΔH of oat globulin were decreased at extreme acidic or alkaline conditions, indicating decreased thermal stability and partial denaturation. Similar changes were observed for βLG (Table 13.1). The highest T_d and ΔH values of oat globulin were observed at pH ranges near the isoelectric point of the protein, and many proteins have maximum thermal stability near their isoelectric point (15). βLG showed a maximum T_d between pH 2 and 4, similar to that observed by other workers (16,17). At a pH far from the isoelectric point, proteins were unfolded due to intramolecular charge repulsion, leading to rupture of hydrogen bonds and of hydrophobic interactions. Enthalpy is correlated with the content of ordered secondary structure of a protein (18). pH-induced unfolding of oat globulin and βLG decreased the content of ordered secondary structure and caused a reduction in enthalpy.

Extreme acidic or basic pH also led to a broadening of the peak with an increase in $\Delta T_{1/2}$ values in both oat globulin (Fig. 13.2) and βLG (data not shown). Width at half peak height has been used to evaluate the cooperativity of protein unfolding (19). If denaturation occurred within a narrow range of temperature (low $\Delta T_{1/2}$ value), the transition was considered highly cooperative. The result indicates that the partially denatured oat globulin was a less cooperative system than the native protein.

Salts. Although salts are normally present in or added to foods in small amounts, they have significant impact on the functional performance of proteins by influencing the conformation and interaction of proteins.

The effect of NaCl and $CaCl_2$ on the thermal transition characteristics of oat globulin and βLG are summarized in Table 13.2. The T_d of oat globulin was increased by increasing the NaCl concentration, but lowered by the presence of $CaCl_2$ in the medium. The enthalpy was unchanged by NaCl, but decreased by increasing the $CaCl_2$ concentration. For βLG, T_d was increased by both salts, whereas ΔH was not significantly affected. The type and concentration of anion in the medium greatly influenced the thermal stability of oat globulin (Fig. 13.3). T_d was increased with a larger concentration of Cl⁻ and Br⁻ ions, but was decreased in the presence of an equivalent concentration of

TABLE 13.1
Effects of pH on DSC Characteristics of β-Lactoglobulin

pH	$T_d(°C)$[a]	$\Delta H (J/g)$[b]
1.4	47.3	18.3
2.0	80.3	18.8
2.5	83.8	15.4
3.0	86.5	15.1
4.0	86.7	16.5
6.2	80.3	18.4
8.0	75.3	16.0
9.0	73.4	17.0
10.0	72.6	3.6

[a]Denaturation temperature.
[b]Enthalpy.

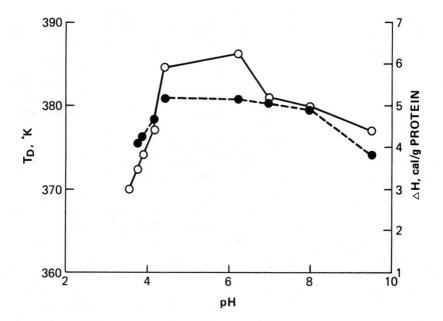

Fig. 13.1. Effect of pH on DSC characteristics of oat globulin. ●, denaturation temperature (T_d); o, enthalpy (ΔH).

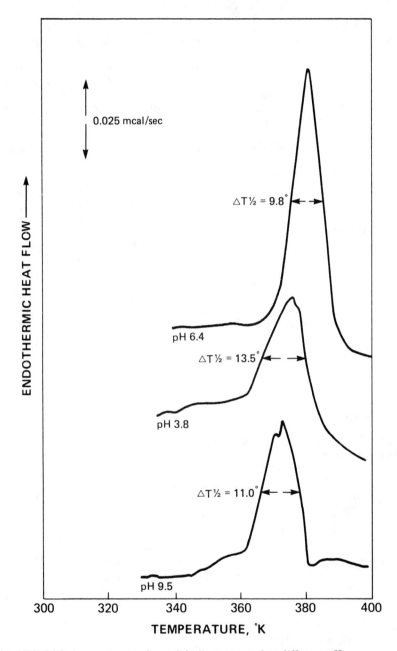

Fig. 13.2. DSC thermograms of oat globulin measured at different pH.

TABLE 13.2

Effect of Salt Concentration on DSC Characteristics of Oat Globulin and β-Lactoglobulin

Protein	Salt conc. (M)	NaCl T_d (°C)[a]	NaCl ΔH (J/g)[b]	CaCl$_2$ T_d (°C)[a]	CaCl$_2$ ΔH (J/g)[b]
Oat	0	108.0	26.3	108.0	26.3
Globulin	0.1	108.0	25.7	104.2	26.4
	0.3	109.7	26.3	n.d.	n.d.
	0.5	n.d.[c]	n.d.	99.4	21.0
	0.7	111.7	26.2	97.3	19.8
	1.0	113.5	26.4	96.0	19.7
β-Lactoglobulin	0	81.5	14.4	81.5	14.4
	0.1	n.d.	n.d.	83.5	13.2
	0.3	84.2	13.1	84.0	13.2
	0.5	85.4	15.2	85.0	13.1
	0.75	86.3	16.3	n.d.	n.d.
	1.0	87.7	15.2	84.6	13.5

[a]Denaturation temperature.
[b]Enthalpy.
[c]Not determined.

TABLE 13.3

Effect of Anions on DSC Characteristics of Oat Globulin and β-Lactoglobulin

Protein	Anion[a]	T_d (°C)[b]	ΔH (J/g)[c]
Oat globulin	Cl$^-$	113.5	26.4
	Br$^-$	109.4	21.8
	I$^-$	104.0	19.2
	SCN$^-$	98.0	16.6
β-Lactoglobulin	Cl$^-$	89.4	11.7
	Br$^-$	87.2	13.6
	I$^-$	83.9	12.5
	SCN$^-$	80.3	13.8

[a]Sodium salts at 1.0 M concentration were used.
[b]Denaturation temperature.
[c]Enthalpy.

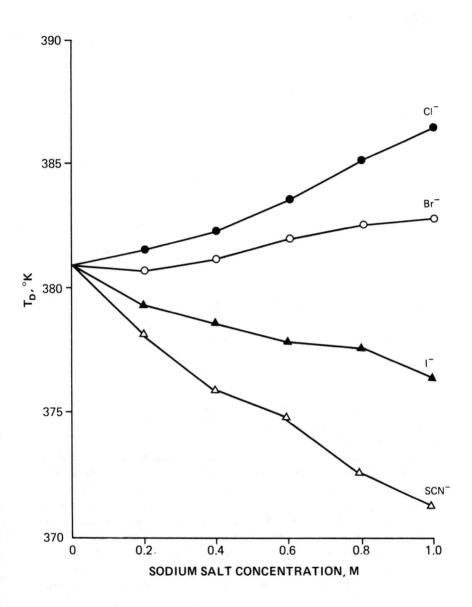

Fig. 13.3. Effect of different sodium anions on denaturation temperature (T_d) of oat globulin. ●, Cl⁻; o, Br⁻; ▲, I⁻; △, SCN⁻.

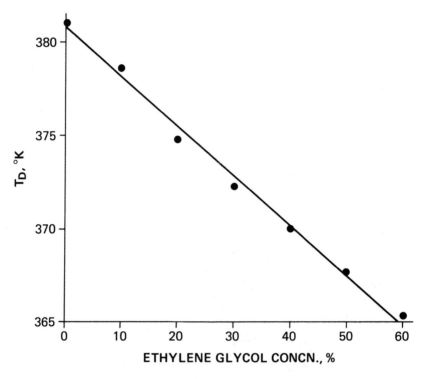

Fig. 13.4. Effect of ethylene glycol on denaturation temperature (T_D) of oat globulin.

I- and SCN- ions (Fig. 13.3). At 1.0 M concentration, both T_d and ΔH of oat globulin and βLG were progressively decreased following the order Cl-, Br-, I- and SCN- (Table 13.3).

The heat stability of proteins is controlled by the balance of polar and nonpolar residues (20), such that the higher the proportion of nonpolar residues, the greater the stability to heat. Protein conformation can be perturbed by the addition of salts which influence the electrostatic interaction with the charged groups and polar groups, and affect hydrophobic interaction via a modification of the water structure (21,22). The degree to which water structure is affected depends on the nature of cations or anions and follow the lyotropic series (23). Cations and anions at the higher order of the series (e.g., Ca^{++} and SCN-) could reduce the energy required to transfer the nonpolar groups into water, and could weaken intramolecular hydrophobic interaction and enhance the unfolding tendency of protein (24), thus lowering both T_d and ΔH. The

TABLE 13.4
Effect of Sugars and Polyols on DSC Characteristics of β-Lactoglobulin

Polyol conc. (%)	Sucrose		Glucose		Glycerol		Ethylene glycol	
	$\Delta T_d(°C)^a$	$\Delta H(J/g)^b$	$\Delta T_d(°C)^a$	$\Delta H(J/g)^b$	$\Delta T_d(°C)^a$	$\Delta H(J/g)^b$	$\Delta T_d(°C)^a$	$\Delta H(J/g)^b$
10	0.9	10.6	1.05	11.35	0.05	16.0	-2.1	13.5
20	1.15	12.3	1.55	7.9	1.1	16.6	-4.2	14.1
30	2.9	10.7	4.4	9.9	1.6	15.4	-7.6	10.9
40	3.9	11.8	4.0	11.5	2.8	14.7	-11.5	9.5
50	5.0	13.0	9.3	—	3.6	16.5	-14.9	11.0

[a] $\Delta T_d = T_d(\beta\text{-LG} + \text{polyols}) - T_d(\beta\text{-LG})$.
[b] Enthalpy.

data suggest that hydrophobic interaction plays an important role in the stabilization of protein structure in oat globulin and βLG.

Protein structure modifying agents. Several chemical reagents known to modify protein conformation were used to assess their effects on thermal characteristics of oat globulin and βLG.

Ethylene glycol caused a marked decrease in T_d of oat globulin (Fig. 13.4). Table 13.4 shows the effect of some polyols on the thermal transition characteristics of βLG. The thermal stability (T_d) was enhanced by these additives, except for ethylene glycol, which lowered T_d of βLG, similar to that of oat globulin. The enthalpy was changed slightly but did not follow any particular pattern. The width at half-peak height remained unchanged (data not shown) indicating that the cooperativity of the thermal transition was not affected. The protective effect of sugars on proteins against thermal denaturation has been discussed (9,25-27). The extent of stabilization (increased T_d) of these sugars or polyols depends upon both the type of polyols and the nature of proteins (25). The stabilization seems to arise from the effect of polyols on water structure which in turn determines the strength of the hydrophobic interaction. Polyols reduce the driving force for transfer of hydrophobic groups from aqueous to nonpolar environment (25). Ethylene glycol, although a polyol, lowers the dielectric constant of the medium and interacts with nonpolar side chains of proteins, weakening hydrophobic interaction and thereby lowering thermal stability (26).

Both T_d and ΔH of oat globulin decreased progressively with increasing concentration of urea (Fig. 13.5). Urea had a similar effect on βLG (Table 13.5) except that ΔH was not significantly decreased at urea concentrations below 4 M. Urea facilitates protein unfolding by weaken-

TABLE 13.5
Effect of Urea on DSC Characteristics of β-Lactoglobulin

Urea conc. (M)	$T_d(°C)$[a]	$ΔH(J/g)$[b]	$ΔT_{½}(°C)$[c]
0	81.6	13.6	5.0
1	80.1	12.7	5.0
3	75.8	12.9	6.3
6	66.7	9.6	7.5
8	——————————NO RESPONSE ——————————		

[a]Denaturation temperature.
[b]Enthalpy.
[c]Width at half-peak height.

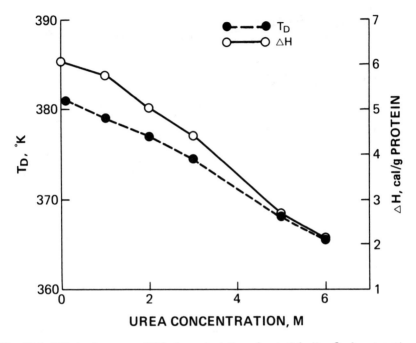

Fig. 13.5. Effect of urea on DSC characteristics of oat globulin. ●, denaturation temperature (T_D); ○, enthalpy (ΔH).

ing hydrophobic interactions (28). Urea also increases the "permitivity" of water (29) for the apolar residues causing loss of protein structure and heat stability.

Some fatty acid salts caused a marked decrease in the thermal stability of oat globulin (Fig. 13.6). Sodium dodecyl sulfate (SDS) did not affect thermal stability but caused a significant decrease in enthalpy of oat globulin (Fig. 13.7). SDS also caused a decrease in enthalpy in βLG, but unlike oat globulin, T_d was increased at lower SDS concentrations and decreased at concentration above 10 mM (Table 13.6). The $\Delta T_{1/2}$ of βLG was significantly increased in the presence of SDS indicating loss of cooperativity. Similar effect of SDS on transition characteristics of βLG were reported by Hegg (16). These detergents are generally regarded as protein denaturants with the sulfates being more potent (30). The slight increase in enthalpy or T_d at low concentration of SDS shows that SDS may stabilize proteins against denaturation by highly specific interactions between proteins and detergent, presumably between cationic

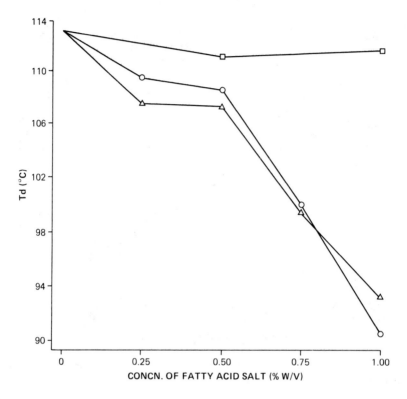

Fig. 13.6. Effect of fatty acid salts on denaturation temperature (T_D) of oat globulin. △, sodium undecanoate; ○, sodium laurate; □, sodium heptadecanoate.

TABLE 13.6
Effect of Sodium Dodecyl Sulfate (SDS) on DCS Characteristics of β-Lactoglobulin

SDS conc. (mM)	T_d (°C)[a]	ΔH (J/g)[b]	$\Delta T_{1/2}$(°C)[c]
0	81.5	14.4	5.0
1	82.3	12.5	5.0
5	84.2	10.1	4.3
10	80.5	11.8	9.0
20	76.4	7.4	13.0

[a]Denaturation temperature.
[b]Enthalpy.
[c]Width at half-peak height.

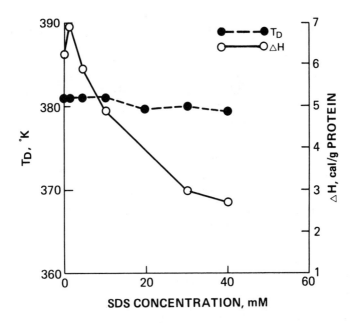

Fig. 13.7. Effect of sodium dodecyl sulfate (SDS) on DSC characteristics of oat globulin. ●, denaturation temperature (T_D); ○, enthalpy (ΔH).

groups of proteins and anionic groups of SDS (31). The resulting hydrophobic side chains of SDS may be entropically transferred to the interior of the protein molecules, leading to the formation of additional hydrogen bonds, and hence, an increase in enthalpy or thermal stability. At higher detergent concentrations, the binding seems to be nonspecific and causes charge repulsion between protein chains leading to either unfolding or lowering in thermal stability.

Both oat globulin and βLG are oligomeric proteins containing disulfide bonds. Oat globulin is composed of six subunits, each containing an acidic and a basic polypeptide linked by disulfide bonds (32). βLG exists as a dimer in milk and as an octamer near isoelectric pH, and in these polymeric states, the monomers are linked by noncovalent bonds. At extreme pH (below pH 3.5 and above pH 8), βLG exists as a monomer which contains two disulfide bonds and a sulfhydryl group (33). To assess the contribution of disulfide bonds to the thermal properties of these proteins, dithiothreitol (DTT), a reducing agent, was added to the medium either with or without SDS. DTT did not cause a marked

TABLE 13.7
Effect of Reducing Agent on DSC Characteristics of Oat Globulin and β-Lactoglobulin

Protein	Additive	T_d (°C)[a]	ΔH (J/g)[b]
Oat globulin	No additive	108.0	26.5
	10 mM DTT[c]	108.6	23.1
	20 mM SDS[d]	106.6	16.0
	10 mM DDT + 20 mM SDS	107.2	13.5
β-Lactoglobulin	No additive	81.5	14.4
	10 mM DTT	81.3	13.3
	10 mM SDS	80.5	11.8
	10 mM DTT + 10 mM SDS	80.9	11.5

[a]Denaturation temperature.
[b]Enthalpy.
[c]Dithiothreitol.
[d]Sodium dodecyl sulfate.

decrease in T_d or ΔH of either protein in the absence or presence of SDS (Table 13.7). DTT reduces disulfide bonds, particularly in the presence of SDS. The present data contrast with the view that disulfide bonds contribute to heat stability since proteins containing disulfide bonds show higher temperatures and enthalpies of denaturation compared to proteins without disulfide linkages (28).

Effects of heat treatments on thermal characteristics of oat globulin.

Transition and kinetic characteristics of oat globulin preheated under nongelling conditions. When oat globulin was preheated near neutral pH at 100 and 110°C, the transition temperatures, T_m and T_d, increased progressively while half peak width and enthalpy decreased (Table 13.8). The changes were much more pronounced at higher temperature. The data show that heating below T_d (114°C in 1.0 M NaCl) caused a progressive unfolding of oat protein and increases in both thermal stability and cooperativity. Oat globulin oligomers can undergo dissociation (into monomers) and aggregation when heated near neutral pH (34). Protein aggregation is an exothermic reaction (35,36) and would lower the net endothermic contribution and enthalpy value. The in-

TABLE 13.8
Effect of Heat Treatments on Thermal Transition Characteristics of Oat Globulin

Heating temperature (°C)	Heating time (min)	T_m^a (°C)	T_d^b (°C)	$\Delta T_{½}^c$ (°C)	ΔH^d (J/g)
100	0	105.2	114.2	10.0	26.4
	5	105.4	114.2	9.9	25.9
	10	105.4	114.4	9.8	24.3
	30	105.9	114.5	9.6	22.5
	60	106.9	114.7	9.0	18.0
	120	108.0	115.1	8.5	14.2
110	1	107.8	114.7	9.0	19.5
	2	109.8	115.5	6.8	12.0
	5	112.1	116.4	5.5	7.9
	10	112.5	116.6	5.3	6.7
	30	113.0	116.8	4.8	4.0
	60	114.5	117.7	4.7	2.2

[a]Onset temperature.
[b]Denaturation temperature.
[c]Width at half-peak height.
[d]Enthalpy.

creases in thermal stability may be due to rearrangement of oat protein to assume a more compact conformation or association into a complex, ordered network structure which has higher thermal stability than unassociated protein and would denature in a highly cooperative fashion.

Heating under nongelling conditions also led to marked changes in DSC kinetic parameters of oat globulin. The activation energy (E_a) and preexponential factor (log Z) were increased rapidly at the initial stage of heating, levelling off after 30 min at 100°C and after 10 min at 110°C (Fig. 13.8). Such changes may again be attributed to the conversion of preheated protein to a more compact conformation.

Transition characteristics of heat aggregated oat globulin. When heat aggregated oat globulin was separated into buffer-soluble and insoluble fractions and analyzed by DSC, the soluble fraction exhibited endotherms with slightly higher enthalpy and lower T_m and T_d values than the control with no significant change in $\Delta T_{1/2}$ (Fig. 13.9). The insoluble fractions did not show any endothermic response. The results

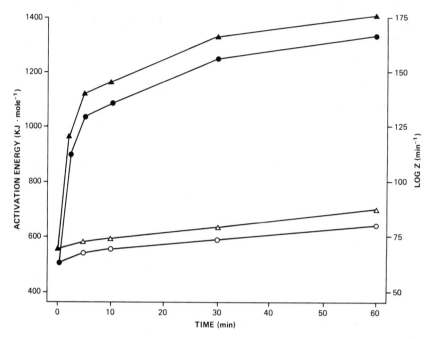

Fig. 13.8. Effect of heat treatments on denaturation kinetic constants of oat globulin. Protein samples (10%) were pre-heated at 100°C (open symbols) or 110°C (solid symbols), ○, ●, activation energy; △, ▲, preexponential factor (Z).

TABLE 13.9
Thermal Transition Characteristics of Oat Globulin Gels

Treatment[a]	T_m[b] (°C)	T_d[c] (°C)	$\Delta T_{1/2}$[d] (°C)	ΔH[e] (J/g)
No pH adjustment	100.8	111.0	10.1	25.6
pH adjusted to 9.7	97.5	109.6	11.3	18.3
pH 9.7, preheated at 100°C for 20 min	104.5	111.1	8.2	14.7

[a]Protein samples (10%) were prepared in 0.2 M NaCl.
[b]Onset temperature.
[c]Denaturation temperature.
[d]Width at half-peak height.
[e]Enthalpy.

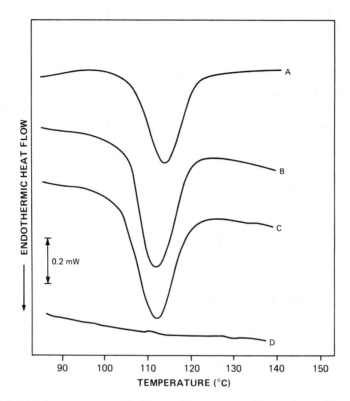

Fig. 13.9. DSC thermograms of buffer-soluble and insoluble fractions of heated oat globulin. A, unheated; B, 100°C heated soluble fraction; C, 110°C heated soluble fraction; D, 100°C heated insoluble fraction.

suggest that the soluble fractions contained essentially undenatured globulin, while the insoluble fractions were made up of extensively denatured protein. The data are consistent with the generally accepted view that aggregation is preceded by denaturation following the scheme: N ⇌ D → A, where N denotes native protein; D, denatured molecule; and A, the aggregate or gel (37).

Transition characteristics of thermally-induced oat globulin gels. When oat globulin was heated at concentrations above 5% at alkaline pH and in the presence of salt (>0.1 M), a self supporting gel matrix can be obtained. Alkaline pH caused a decrease in T_d and ΔH and an increase in $\Delta T_{1/2}$ (Table 13.9). When oat globulin was preheated at alkaline pH, cooled, and reanalyzed by DSC, there was a further decrease in ΔH, but

T_d increased to a value close to that of the native protein, and T_m and $\Delta T_{1/2}$ values also increased markedly. The changes in these transition characteristics are explained by the fact that protein in a gel matrix has a more compact and ordered structure than protein in dispersion. The data also show that the gels retained considerable native structure, suggesting that gelation did not require extensive protein denaturation. This seems to contradict the above model for protein denaturation and aggregation. However, it has been suggested that for oligomeric proteins with complex quaternary structures such as soya glycinin, heat may cause the association/dissociation of the oligomer to induce aggregation, and disruption of the quaternary structure alone can lead to coagulation/gelation (38).

Effects of chemical modification on thermal properties of oat globulin and β-lactoglobulin.

Succinylation. Succinylation caused a progressive decrease in both T_d and ΔH in oat globulin, whereas T_d was increased in the mildly succinylated (18%) βLG although enthalpy was greatly reduced (Table 13.10). At higher levels of modification (72%), βLG did not show any endothermic response indicating extensive denaturation. Succinylation changes the net charge of proteins and may lead to protein unfolding through intramolecular charge repulsion (39), and hence a reduction in enthalpy. The data indicate that oat globulin is more resistant to succinylation-induced changes than βLG, probably because of its more compact structure. The changes in T_d in the succinylated proteins suggest a modification of the conformation leading to enhanced or lower thermal stability.

Fatty acid modification. Modification of proteins, in which fatty acids are covalently attached to free amino groups, is another chemical modification used to improve functionality of proteins by deliberately introducing lipophilicity, making a protein more amphiphilic (40, 41). The effect of lipophilization on DSC characteristics of βLG is summarized in Table 13.11. The use of organic solvent causes significant protein denaturation as indicated by an almost complete disappearance of the endothermic peak. Acylation with some fatty acids restored some protein structure as indicated by reappearance of an endothermic peak. The extent of increase in ΔH (over the solvent-treated control) and the shift in T_d (increase or decrease) varied with the chain length of fatty acids and the extent of acylation. Fatty acid modifications also caused a marked sharpening of the endothermic peak and decrease in $\Delta T_{1/2}$, and

TABLE 13.10
Effect of Succinylation on DSC Characteristics of Oat Globulin and β-Lactoglobulin

Protein	% Succinylated	T_d (°C)[a]	ΔH (J/g)[b]
Oat globulin	0	110.1	26.4
Oat globulin	31.2	106.4	10.8
Oat globulin	58.4	105.0	6.8
β-Lactoglobulin	0	81.5	14.4
β-Lactoglobulin	18.0	88.4	4.94
β-Lactoglobulin	72.1	———————No Response———————	

[a]Denaturation temperature.
[b]Enthalpy.

TABLE 13.11
Effect of Fatty Acid Modification on DSC Characteristics of β-Lactoglobulin

Treatment	T_d (°C)[a]	ΔH (J/g)[b]	$\Delta T_{1/2}$(°C)[c]
Control	81.5	14.4	5.0
Solvent treated	84.8	1.3	11.5
C_{12}(7 eq) modified	53.2	2.2	9.0
C_{14}(7 eq) modified	————NO RESPONSE————		
C_{14}(15 eq) modified	80.2	3.6	2.8
C_{16}(7 eq) modified	83.7, 99.5	1.9, 2.2	2.8, 2.5
C_{18}(7 eq) modified	88.4	5.5	2.5

[a]Denaturation temperature.
[b]Enthalpy.
[c]Width at half-peak height.

the C_{16} modified protein exhibited two endothermic peaks (Table 13.11). The introduction of hydrophobic fatty acid side chain to βLG might enhance a compact conformation through hydrophobic interactions, and may promote the folding of proteins and the formation of additional internal hydrogen bonds, thus restoring the endothermic peak. It is not known why an additional peak appeared in some lipophilized βLG.

Acknowledgments

Technical assistance was provided by B. Boutin-Muma and G. Khanzada. This is Contribution No. 785, Food Research Center, Agriculture Canada, Ottawa.

References

1. Kinsella, J.E., *Crit. Rev. Food Sci. Nutr.* 7:219 (1976).
2. *Protein Functionality in Foods*, ACS Symposium Series No. 147, J.P. Cherry (ed) American Chemical Society, Washington, DC, 1979.
3. Nakai, A., *J. Agric. Food Chem. 31*:676 (1983).
4. Kilara, A., and T.Y. Sharkasi, *CRC Crit. Rev. Food Sci. Nutr. 23*:323 (1986).
5. Harwalkar, V.R., and C.-Y. Ma, *J. Food Sci. 52*:394 (1987).
6. Ma, C.-Y., and V.R. Harwalkar, *Ibid, 53*:531 (1988).
7. Ma, C.-Y., and V.R. Harwalkar, *J. Agric. Food Chem. 36*:27 (1988).
8. Harwalkar, V.R., *Proc. 14th American Thermal Analysis Society Conference* 1985 p. 334.
9. Harwalkar, V.R., *J. Dairy Sci. 69*:84 (1986).
10. Groninger, H.S., and R. Miller, *J. Agric. Food Chem. 27*:949 (1979).
11. Concon, J.M., *Anal. Biochem. 66*:460 (1975).
12. Paquet, A., *Can. J. Chem. 54*:733 (1976).
13. Paquet, A., *Can. J. Biochem. 58*:573 (1980).
14. Borchardt, H.J., and F. Daniels, *J. Am. Chem. Soc. 79*:41 (1957).
15. Privalov, P.L., and N.N. Khechinashvili, *J. Mol. Biol. 86*:665 (1974).
16. Hegg, P. -O., *Acta Agric. Scand. 30*:401 (1980).
17. De Wit, J.N., and G. Klarenbeek, *J. Dairy Res. 48*:293 (1981).
18. Kosiyama, I., M. Hamano and D. Fukushima, *Food Chem 6*:309 (1981).
19. Privalov, P.L., N.N. Khechinashvili and B.P. Atanaasov, *Fed. Proc., Fed. Am. Soc. Exp. Biol. 52*:159 (1979).
20. Bigelow, C.C., *J. Theor. Biol. 16*:187 (1967).
21. von Hippel, P.H., and T. Schleich, in *Structure and Stability of Macromolecules*, Vol. 2, S.N. Timsheff and G.D. Farman (eds), Marcel-Dekker, New York, NY, 1969, p. 187.
22. Damodaran, S., and J.E. Kinsella, in *Food Protein Deterioration, Mechanisms and Functionality*, ACS Symposium Series No. 206, J.P. Cherry (ed), American Chemical Society, Washington, DC, 1982, p. 327.
23. Hatefi, Y., and W.G. Hanstein, *Proc. Natl. Acad. Sci. U.S.A. 62*:1129 (1969).
24. von Hippel, P.H., and K.Y. Wong, *J. Biol. Chem. 240*:3909 (1965).
25. Beck, J.F., D. Ookenfull and M.B. Smith, *Biochemistry 18*:5191 (1979).
26. Gerlsma, S.Y., and E.J. Stuur, *Int. J. Peptide Protein Res. 4*:377 (1972).
27. Wright, D.J., in *Developments in Food Proteins* Vol. 1. edited by B.J.F. Hudson, Applied Science Publishers, London, England, 1982, p. 67.

28. Kinsella, J.E., in *Food Proteins*, P.F. Fox and J.J. Cowden (eds), Applied Science Publisher, London, England, 1982, p. 51.
29. Franks, F., and D. England, *CRC Crit. Rev. Biochem.* *3*:165 (1975).
30. Lapanje, S., *Physicochemical Aspects of Protein Denaturation*, John Wiley and Sons, New York, NY, 1978, p. 156.
31. Tanford, C., *Adv. Protein Chem.* *24*:1 (1970).
32. Brinegar, A.C., and D.M. Peterson, *Arch. Biochem. Biophys.* *219*:71 (1982).
33. Swaisgood, H.E., in *Developments in Dairy Chemistry*, Vol. 1, P.F. Fox (ed), Applied Science Publishers, London, England, 1982, p. 1.
34. Ma, C.-Y., and V.R. Harwalkar, *Cereal Chem.* *64*:212 (1978).
35. Jackson, W.M., and J.F. Brandts, *Biochim. Biophys. Acta* *9*:2294 (1970).
36. Arntfield, S.D., and E.D. Murray, *Can. Inst. Food. Sci. Technol. J.* *14*:289 (1981).
37. Ferry, J., *Adv. Protein Chem.* *4*:1 (1948).
38. German, B., S. Damodaran and J.E. Kinsella, *J. Agric. Food Chem.* *30*:807 (1982).
39. Kinsella, J.E., and K.J. Shetty, in *Functionality and Protein Structure*, ACS Symposium Series No. 92, A. Pouri-El (ed), American Chemical Society, Washington, DC, 1979, p. 37.
40. Haque, Z., and M. Kito, *J. Agric Food Chem.* *31*:1225 (1983).
41. Haque, Z., and M. Kito, *Ibid*, *31*:1231 (1983).

Chapter Fourteen
Effect of Molecular Changes (SH Groups and Hydrophobicity) of Food Proteins on Their Functionality

E. Li-Chan and S. Nakai

Department of Food Science
University of British Columbia
6650 N.W. Marine Dr.
Vancouver, Canada V6T 1W5

Elucidation of the structure-functionality relationship is an ultimate goal of many protein chemists. In this regard, hydrophobic, steric and electrical parameters have been identified as three important categories of variables which may be useful to predict functionality (1,2). Through classical techniques of protein chemistry as well as more recent recombinant DNA and protein engineering technology, great strides are being made on characterizing molecular properties which confer specific biological functions. The native three-dimensional structure of proteins such as enzymes has been well-recognized to be crucial for biological activity. However, in the case of food protein functionality, one must consider not only the structural properties of the undenatured molecule, but also the flexibility of the molecule during processing treatments, which may change the characteristic of the molecular surface by enabling unfolding and exposure of functional groups to the surface.

In this paper, an overview of the structure-functionality relationship of food proteins is first presented. Since properties of the molecular surface such as surface hydrophobicity and net charge are key parameters influencing functionality, steric factors affecting conformational stability and molecular flexibility must also be considered. Proteins such as lysozyme which are held in a rigid globular conformation by intramolecular disulfide bridges may be made more flexible by treatment with reducing agents. Data are presented to illustrate the effects of partial reduction of disulfide bonds of lysozyme on the exposure of hydrophobic regions and the accompanying changes in gelling, emulsifying and foaming properties.

Review of the Literature

The three major categories of protein molecular properties affecting their structure and function are hydrophobic, electrical and steric properties (Table 14.1). The impact of hydrophobic interactions of food proteins on their functional properties has probably received the most attention within the last decade (3). In particular, the incorporation of protein hydrophobicity as a significant variable to predict functionality has enabled development of more accurate equations describing the quantitative structure-activity relationship (QSAR).

For example, significant correlations were found between the decreasing interfacial tension or increasing emulsifying activity index of various proteins and increasing hydrophobicity of the protein surface, measured by fluorescence probes (4). Insolubility of milk and soy proteins was correlated to parameters of hydrophobicity measured by hydrophobic interaction chromatography on phenyl-Sepharose (PSC), and net charge measured as zeta potential (ZP) (Fig. 14.1) (5). This relationship was quantitatively described by the equation:

$$\% \text{ insoluble protein} = 3.43 PSC - 0.057 PSC.ZP - 0.042 ZP^2 + 4.4$$
$$(n=189, R^2=0.61, P<0.001)$$

Similarly, the effects of surface hydrophobicity (S_0) measured by a fluorescence probe and solubility (s) on the emulsifying capacity (EC) of salt-extractable proteins from beef muscle (Fig. 14.2) could be quantitatively described by the equation:

$$EC = -35.3 + 6.46s - 0.0013s.S_0 - 0.55s^2 \quad (n=26, R^2=0.92, P<0.001)$$

TABLE 14.1
Structure-function Relationships of Food Proteins

Structural properties	Functional properties
Hydrophobicity	solubility
surface or exposed	emulsification
Electrical parameters	foam formation
net or surface charge	gel formation
Steric parameters	coagulation
molecular size	viscosity
molecular-flexibility	water binding
sulfhydryldisulfide	fat binding

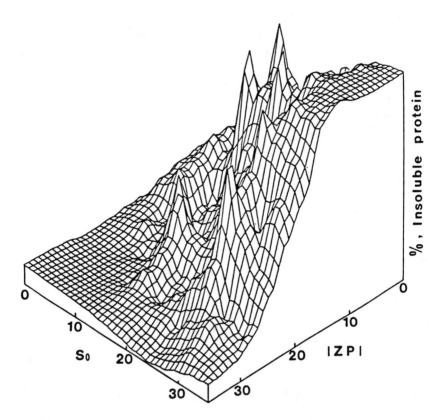

Fig. 14.1. Three-dimensional plot of the relationship between hydrophobicity measured by phenyl-Sepharose chromatography (PSC), net charge measured by zeta potential (ZP) and the % insolubility of milk and soy proteins. [From Hayakawa and Nakai (5)]

Surface hydrophobicity was the key parameter to predict emulsifying properties of samples with over 50% solubility, whereas solubility parameters were influential for samples with lower solubility (6).

In these studies on solubility, interfacial tension and emulsifying properties of proteins, the importance of "surface" hydrophobicity was emphasized, in contrast to earlier studies which dealt with "total" hydrophobicity, such as Bigelow's hydrophobicity parameter, $H(\phi)_{ave}$, computed as the sum of the side chain hydrophobicities of constituent amino acids (7,8). However, attempts to correlate surface hydrophobicity of proteins with their foaming capacity were largely unsuccessful (9).

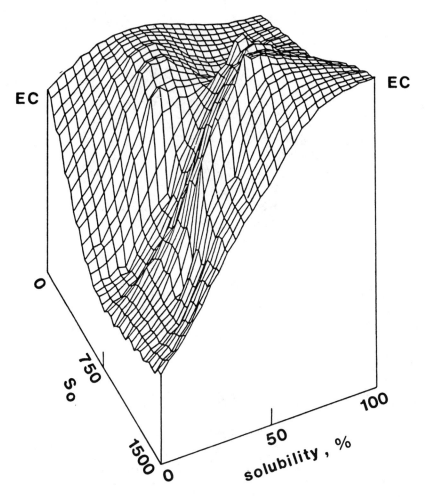

Fig. 14.2. Three-dimensional plot of the relationship between surface hydrophobicity (S_0), solubility and emulsifying capacity (EC) of salt-extractable proteins from beef muscle. [From Li-Chan et al (6)]

Instead, Bigelow's average hydrophobicity values were significantly correlated with foaming capacity. An experimental parameter to approximate the calculated $H(\phi)_{ave}$ values of Bigelow was established by measuring hydrophobicity by the fluorescence probe method, of protein solutions which had been heated at 100°C for 10 min in the presence of sodium dodecyl sulfate. The resulting S_e or "exposed" hydrophobicity

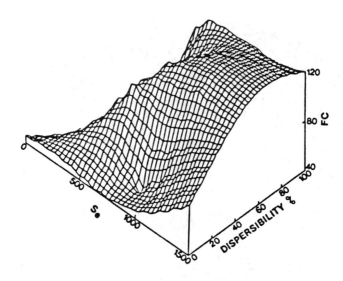

Fig. 14.3. Three-dimensional plot of the relationship between exposed hydrophobicity (S_E), dispersibility and the foaming capacity (FC) of various food proteins. [From Townsend and Nakai (9)]

TABLE 14.2

Relationships Between Hydrophobicity and Sulfhydryl Group Content on Thermal Functional Properties of Eight Food Proteins

Gelation = $-6.54 + 0.0065\ S_e + 0.012\ SH^2$	($R^2 = 0.74$, $P < 0.05$)
Thickening = $-202.7 + 0.75\ S_e + 0.016\ S_e \cdot SH - 0.001\ S_e^2$	($R^2 = 0.96$, $P < 0.01$)
Heat coagulation = $-5.62 - 2.50\ SH + 0.009\ S_e \cdot SH$	($R^2 = 0.74$, $P < 0.05$)

S_e is the exposed hydrophobicity determined by fluorescence probe, and SH is the sulfhydryl group content by Ellman's reagent. [From Voutsinas *et al.* (10)]

values, together with dispersibility parameters, showed significant correlations with foaming capacity (Fig. 14.3) (9). Thermal functional properties such as gelation, thickening and coagulation were also more accurately described using exposed hydrophobicity rather than surface hydrophobicity parameters, together with sulfhydryl group content (10) (Table 14.2).

It is clear that although hydrophobicity is an important parameter to predict functionality, the relationship between steric factors and hydrophobicity must also be considered. The critical parameters affecting functionality are not only those properties at the surface of the native protein molecule, but also the potential surface characteristics which may be exposed upon denaturation. Conformational properties, including flexibility, should be important in adsorption of proteins at air-water or oil-water interfaces. Whey proteins adsorbed on emulsified fat globule surfaces were more susceptible to protease digestion than native whey proteins, indicating drastic conformational changes had occurred at the oil-water interface during the emulsification process (11).

Molecular flexibility or stability may be affected by extrinsic factors, such as the ionic environment, pH and temperature, as well as by intrinsic protein characteristics such as disulfide bond crosslinks which may stabilize the protein molecule against extrinsic perturbations. Amphiphilic proteins possessing a high degree of flexibility may be encouraged to be adsorbed at interfaces by partial denaturation, such as that brought about by mild heating, which may result in exposure of previously inaccessible hydrophobic or hydrophilic regions. For example, Kato *et al.* (12-15) reported increases in protein surface hydrophobicity by heat denaturation when not accompanied by coagulation, which were linearly related to improved emulsifying properties, and curvilinearly related to improved foaming properties. On the other hand, introduction of intramolecular crosslinks to bovine serum albumin greatly reduced the foaming power and foam stability (16). It was suggested that protein flexibility and surface hydrophobicity may be the main governing factors for foaming properties and emulsifying properties, respectively.

Conalbumin and lysozyme from chicken egg white are examples of rigid globular proteins whose conformation is stabilized by intramolecular disulfide bridges (17). Recently, the thiol-induced gelation of egg white at low temperatures was reported (18). This phenomenon was attributed to the conalbumin component of egg white (19,20). The native conalbumin molecule contains 15 disulfide linkages but no free sulfhydryl groups. Following addition of mercaptoethanol and incubation at 35°C, there was an almost immediate increase in sulfhydryl groups as well as in surface hydrophobicity. After a lag of about 20 minutes, turbidity of the solution increased, followed by gradual formation of a gel after an hour. It was suggested that intermolecular hydrophobic interactions were involved in gel formation.

Lysozyme is a relatively stable enzyme consisting of a single polypeptide chain of 129 amino acid residues crosslinked by four disulfide bridges (21). The molecule is quite stable to thermal denaturation, a denaturation temperature (T_D) of 75°C being determined for lysozyme at pH 7 by differential scanning calorimetry (22). Structural and chemical characteristics of lysozyme determined both in solution (by nuclear magnetic resonance) and in crystal state (by X-ray diffraction) were reviewed by Blake *et al.* (23). The polypeptide chain is folded on itself so that the first 40 residues from the N-terminal end form a compact globular domain with a hydrophobic core trapped between two α-helices. A more hydrophilic domain, consisting of residues 40–85, forms one side of the active site cleft. The rest of the polypeptide chain partially fills the gap between the two domains and lines the active site cleft with hydrophobic residues.

Native lysozyme exhibits poor emulsifying and foaming properties (12,13) and does not gel even by heating at 80°C (24). Thermally-induced gel formation of lysozyme at 80°C was achieved by the addition of 7.0 mM dithiothreitol and 46.9 mM NaCl (24). It was suggested that an ion-shielding effect by salt and disruption of disulfide bonds were responsible for causing exposure of hydrophobic bonds upon heating, followed by disulfide-sulfhydryl interchanges which resulted in gel formation. However, no data were presented to support these hypotheses.

The objectives of the present study were to investigate the changes in sulfhydryl groups and surface-exposed hydrophobic regions of lysozyme by addition of dithiothreitol, followed by incubation at 80°C or 37°C. The effects of these molecular changes on functional properties such as gel formation, emulsifying and foaming properties were studied. The forces involved in maintaining the gel structures were also investigated by quantitating the solubilization of gels in different dissociating media.

Experimental Procedures

Lysozyme (E.C.3.2.1.17) from chicken egg white, DL-dithiothreitol (DTT or Cleland's reagent) and 5,5'-dithiobis(2-nitrobenzoic acid) (DTNB or Ellman's reagent) were purchased from Sigma Chemical Co., St. Louis, MO (Product No. L-6876, D-0632 and D-8130, respectively). The magnesium salt of 1-anilinonaphthalene-8-sulfonic acid (ANS) was prepared from the technical grade sodium salt (Eastman Kodak Co., Rochester, NY) according to the method of Weber and Young (25), while *cis*-parinaric acid (CPA) was from Molecular Probes, Inc. (Eugene, OR).

The flowchart for preparation of reduced and/or heated lysozyme solutions and corresponding non-reduced controls is shown in Figure 14.4. Weighed portions of solid DTT giving final concentrations of between 0.7–7.0 mM DTT were added to 0.5 or 5.0 % lysozyme solutions

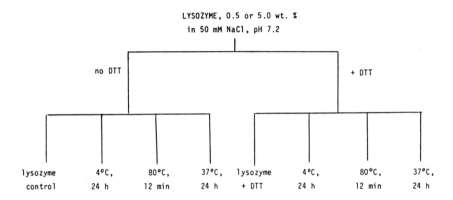

Fig. 14.4. Flowchart of treatment of lysozyme solutions by various reducing agents and/or temperature-time incubation.

in 50 mM NaCl, pH adjusted to 7.2, containing 0.01% sodium azide. After standing for 10–15 minutes at room temperature, the solutions were heated by immersing in water baths at 37°C or 80°C for the specified time-intervals, followed by cooling in an ice-water bath.

Turbidity measurements were carried out by determining absorbance at 600 nm of 0.5% solutions, using a Cary 210 spectrophotometer (Varian Associates, Palo Alto, CA). Hydrophobicity (S_0, %$^{-1}$) was determined with the fluorescence probes, ANS and CPA, as described by Li-Chan et al. (26). Sulfhydryl group content of the samples was determined by the colorimetric reaction with Ellman's reagent, after first removing excess DTT by repeated precipitation of protein with 12% trichloroacetic acid and centrifugation, according to the method of Beveridge et al. (27).

Emulsifying activity index (EAI, m^2/g) and emulsion stability index (ESI, min) were determined by the method of Pearce and Kinsella (28), as modified by Li-Chan et al. (6,26). Foam formation and stability over time were determined by a modification of the shaking method of Graham and Phillips (29). The protein solution — 5 ml of 0.1% protein solution in 0.1M NaCl at pH 7.0 — was placed in a 10 ml graduated test tube, which was then sealed with parafilm and a rubber stopper. Instead of prolonged shaking on a mechanical shaker, the contents were shaken by hand to give 30 inversions in 30 seconds. The initial foam volume and the volume with respect to time were recorded. The foams were transferred to Petri plates for photography under a microscope, using a Wild

M3 stereomicroscope with attached 35mm magazine camera semiphotomat (Wild MPS15/11).

Solubilization of gels, formed by incubation of 5% lysozyme solutions containing 7 mM DTT in 50 mM NaCl, pH 7.2 at 80°C for 12 minutes or 37°C for 24 hours, was determined by modification of the procedures described by Hatta et al. (30) and Utsumi and Kinsella (31). To 0.20 grams of gel weighed into 15 ml Corex centrifuge tubes was added 2.0 ml of one of the following: (a) 0.1M sodium phosphate buffer at pH 7.5; (b) buffer + 1% SDS; (c) buffer + 8M urea; (d) buffer + 1% β-mercaptoethanol; (e) buffer + 1M NaCl; (f) buffer + 1% SDS + 1% β-mercaptoethanol; (g) buffer + 8M urea + 1% β-mercaptoethanol; (h) buffer + 1M NaCl + 1% β-mercaptoethanol. After 22 hours at 22°C, with occasional agitation, the samples were centrifuged at 3000 × g for 20 minutes. Protein concentration in the supernatant was quantitated by A_{280} measurement ($E_{280}^{1\%}$ = 26) (32) or by the biuret-phenol spectrophotometric method using the Folin-Ciocalteu reagent (33). The percentage of solubilization was calculated as (% protein in supernatant)/(% protein before centrifugation) × 100%.

Results
Sulfhydryl group content.
Table 14.3 shows the sulfhydryl content of 0.5% lysozyme solutions after various treatments with dithiothreitol (DTT) and/or temperature-time incubation. In the absence of added DTT, lysozyme had only trace quantities of free sulfhydryl groups. Since the lysozyme molecule reportedly

TABLE 14.3

Sulfhydryl Content(μM/g protein) of 0.5% Lysozyme Solutions with Various Reducing Agents and Temperature-time Treatments

Treatment	Sulfhydryl groups μM/g protein
Lysozyme control	0.7
80°C, 12 min	0.7
37°C, 24 h	0.7
0.7 mM DTT, 22°C, 15 min	0.9
5.0 mM DTT, 22°C, 15 min	5.8
0.7 mM DTT, 4°C, 24 h	21.4
0.7 mM DTT, 80°C, 12 min	420.0
0.7 mM DTT, 37°C, 24 h	240.0

contains only disulfide groups (21), the small quantities of sulfhydryl groups measured here may have been due to impurities of the commercial protein source, a 3x crystallized, dialyzed and lyophilized powder containing 90% protein, with buffer salts as sodium acetate and sodium chloride. No changes in sulfhydryl groups were detected after heating of 0.5% lysozyme solutions at 80°C for up to 12 minutes or at 37°C for up to 24 hours.

A slight increase in free sulfhydryl groups was detected after only 15 minutes at 22°C with 0.7 or 5.0 mM DTT, and a more significant increase was seen upon prolonged incubation for 24h at 4°C. Very dramatic increases in sulfhydryl group content resulted after incubation of lysozyme at 80 or 37°C in the presence of DTT (Table 14.3), corresponding to approximately 6.0 and 3.4 moles of sulfhydryl groups per mole of protein, respectively.

Surface hydrophobicity.

Figure 14.5 shows the changes in surface hydrophobicity of lysozyme solutions heated at 80°C in the absence or presence of DTT. The trends

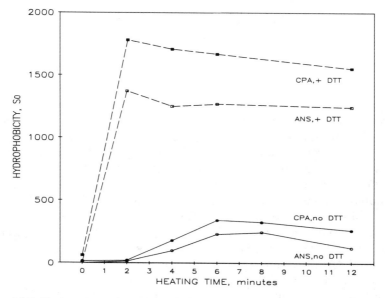

Fig. 14.5. Changes in surface hydrophobicity of 0.5% lysozyme solutions heated at 80°C in the absence or presence of 0.7 mM DTT. (AN'S and CPA are 1-anilinonaphthalene-8-sulfonate and *cis*-parinarate, respectively, for hydrophobicity determination.)

observed by both fluorescence probes, ANS and CPA, are similar. Hydrophobicity values of unheated lysozyme solutions in the presence of DTT were slightly higher than in the absence of the reducing agent. Upon heating lysozyme at 80°C, only a modest increase in surface hydrophobicity occurred in the absence of DTT, with maximum S_0 values found at about 6 to 8 minutes of heating. In contrast, very large increases in surface hydrophobicity were detected within two minutes of heating at 80°C in the presence of DTT. Similarly, high values of surface hydrophobicity (730 and 1340 for S_0ANS and S_0CPA, respectively) were measured for lysozyme after incubation at 37°C for 24 hours in the presence of DTT, but not in its absence. These data strongly suggest that the exposure of buried hydrophobic regions in lysozyme cannot be brought about simply by heating at temperatures of 80°C or lower. The increase in molecular flexibility allowing exposure of these regions must be related to the increase in free sulfhydryl groups which results from incubating lysozyme in the presence of DTT for short periods of time at high temperatures (e.g., 80°C) or for longer periods of time at lower temperature (e.g., 37°C).

Turbidity and gel formation.

Changes in the intermolecular associations of lysozyme after the addition of DTT were implied by the increasing turbidity of these solutions. Figure 14.6 illustrates the more rapid and extensive turbidity development with respect to time of 0.5% lysozyme solutions heated at 80°C in the presence of 0.7 mM DTT, as opposed to its absence. These differences were also significant for solutions incubated at 37°C. After 24 hours at 37°C, the turbidity indicated by A_{600} was 0.315 and 0.014 for 0.5% lysozyme in the presence and absence of DTT, respectively. The 0.5% solutions of lysozyme which had been heated at 37°C or 80°C in the presence of DTT tended to become insoluble when left at room temperature. The insolubilization was prevented or could be reversed by keeping the solutions at 4°C or in an ice-water bath. The temperature dependence of this insolubilization phenomenon is strongly suggestive of hydrophobic interactions in the DTT-incubated lysozyme samples.

The significant differences in intermolecular interactions between native and partially reduced lysozyme molecules were most evident upon heating 5% solutions of these samples (Fig. 14.7). A firm, opaque white gel was formed from lysozyme after 12 min at 80°C or 24 hours at 37°C, in the presence of 7.0 mM DTT, whereas the lysozyme solutions incubated in the absence of DTT only formed slightly turbid solutions.

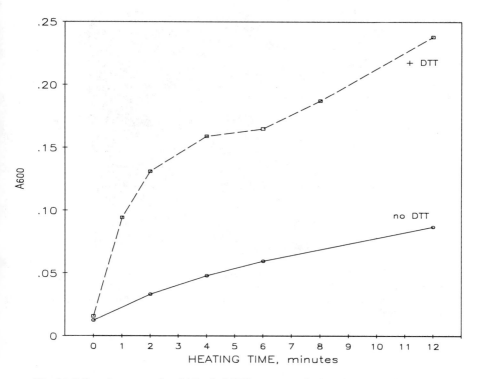

Fig. 14.6. Development of turbidity in 0.5% lysozyme solutions heated at 80°C in the absence or presence of 0.7 mM DTT.

Solubilization of the gels formed from partially reduced lysozyme was investigated using various dissociating media. The results shown in Figure 14.8 suggest that hydrophobic interactions and hydrogen bonding may have been the predominating forces in maintaining gel structure, as both 80°C and 37°C gels could be completely solubilized by 8M urea. The 37°C gel could also be completely solubilized in 1% SDS. Although mercaptoethanol and sodium chloride solubilized 71% and 38%, respectively, of the proteins from the 37°C gel, neither reagent was effective alone in complete solubilization. However, SDS or urea alone was able to effect complete solubilization. It may be inferred that disulfide cross-links and electrostatic interactions were not major forces responsible for the gel structure.

Weaker intermolecular interactions in the structure of the 37°C gel compared to the 80°C gel were indicated by the moderate levels (45%) of

Fig. 14.7. Effect of DTT absence (−) or presence (+) in 5% lysozyme solutions on the gel formation after heating at 80°C for 12 minutes, or 37°C for 24 hours. (Tubes containing the opaque white gels formed in the presence of DTT could be inverted without breaking the gel structure.)

solubilization achieved by simply diluting the 37°C gel in 0.1M sodium phosphate buffer at pH 7.5. It was also observed that the 80°C gel remained stable upon prolonged storage (>4 weeks) at 4°C, whereas the 37°C gel exhibited syneresis within one week at 4°C. These observations indicate that while a firm gel could be obtained by incubating lysozyme with DTT at either 37°C or 80°C, a more stable structure was formed at the higher temperature.

Emulsifying properties.

A significant increase in emulsifying activity index (EAI) was observed for lysozyme solution after incubation with 0.7 mM DTT for 24 hours at 4°C, with EAI value of 10.0 m^2/g, compared to only 1.5 m^2/g for the control lysozyme solution (Fig. 14.9). However, no significant difference was observed in emulsion stability index (ESI) for the DTT-lysozyme compared to control lysozyme. Although the changes in free sulfhydryl groups and surface hydrophobicity brought about by DTT addition alone without subsequent heating were only slight (Table 14.3 and Fig. 14.5), it is likely the mechanical energy of the emulsification process

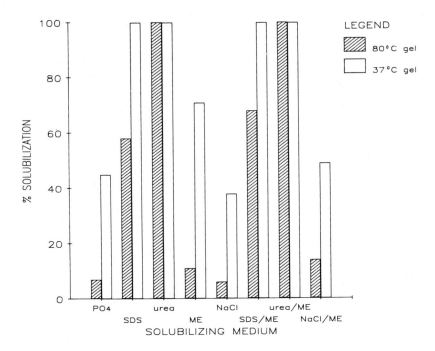

Fig. 14.8. Solubility in different dissociating media of the gels formed from DTT-lysozyme heated at 80°C for 12 minutes or 37°C for 24 hours. Details of the dissociating media are given in the experimental procedures section.

itself was sufficient to enable unfolding of the partially reduced lysozyme molecule at the oil-water interface. In the absence of oil, the blending process itself resulted in significantly higher surface hydrophobicity values for partially reduced lysozyme compared to lysozyme (S_oANS of 60 versus 10 and S_oCPA of 180 versus 40).

Input of energy in the form of heating of DTT-lysozyme at 80°C further improved EAI (to 16.6 m^2/g) and ESI (to 25 minutes, compared to 2 minutes in the control). These improvements were also apparent for DTT-lysozyme incubated at 37°C for 24 hours. However, only slight improvements in EAI and ESI were observed for lysozyme heated at 80°C in the absence of DTT, while EAI and ESI decreased slightly after incubation at 37°C in the absence of DTT.

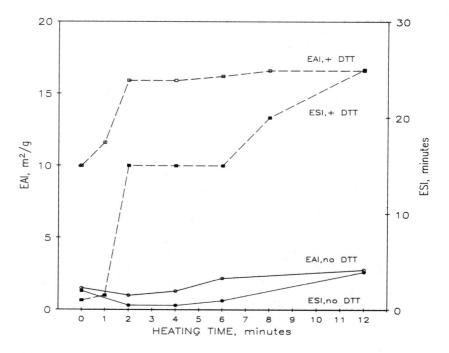

Fig. 14.9. Emulsifying properties of 0.5% lysozyme solutions as a function of time of heating at 80°C, in the absence or presence of 0.7 mM DTT. (EAI= emulsifying activity index; ESI= emulsion stability index.)

It should be noted that the improvements in emulsifying properties of heated DTT-lysozyme were greatest for samples heated at low protein concentrations (e.g., 0.5%) which prevented insolubilization. The EAI values of gelled samples formed by 37°C or 80°C incubation at 5% protein concentration were 12.8 and 8.3 m^2/g, respectively, which represent more moderate improvements than those obtained in non-gelled samples. Their ESI value of 2 minutes was not significantly different from that of the control lysozyme.

Foaming properties.

Figure 14.10 illustrates the foam volume and stability of 0.1% solutions of lysozyme obtained after various DTT and/or time-temperature treat-

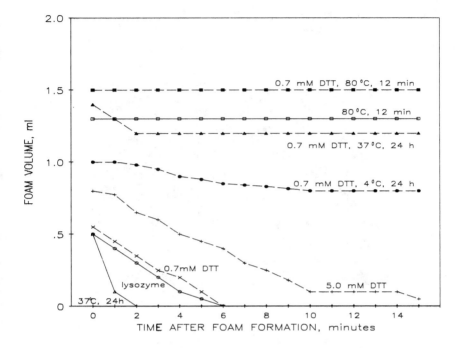

Fig. 14.10. Foaming properties of 0.1% lysozyme solutions after various reducing and/or temperature-time treatments.

ments. Moderate increases in foam volume and foam stability resulted by 15 minute incubation of 0.5% lysozyme solution in the presence of 5 mM DTT, prior to the foaming determination. Greater improvement was seen for lysozyme samples after a longer period of incubation with DTT (24 h, 4°C, 0.7 mM DTT). The greatest improvements in foam volume as well as foam stability were achieved in samples which were heat treated at 80°C in the absence or presence of DTT, or at 37°C in the presence of DTT. Incubation of lysozyme alone at 37°C resulted in poorer foaming properties.

Although the foam volume curves depicted in Figure 14.10 do not indicate any distinction between the stability of foams formed from lysozyme solutions heated at 80°C in the absence or presence of DTT, it should be pointed out that these data were obtained by measuring foam volume over time without any physical disturbance of the foams. When the foams were transferred from the graduated tubes, in which they were formed, onto Petri plates for observation under the stereomicroscope, it was observed that the foam formed from lysozyme heated at 80°C in the absence of DTT quickly collapsed, whereas the foam formed from lysozyme heated in the presence of DTT remained stable. Figure

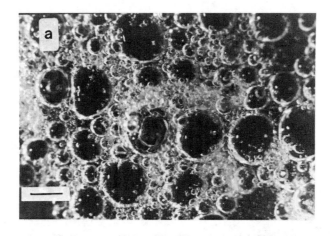

14.11A

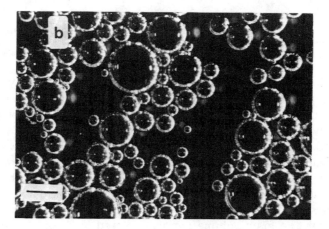

14.11B

Fig. 14.11. Light microscopic structure of foams formed from 0.5% lysozyme solution heated at 80°C (a) in the presence of 0.7 mM DTT and (b) in the absence of DTT, and (c) and (d) of the corresponding foams after standing overnight. (Bar represents 1 mm length.)

Effect of Molecular Changes 249

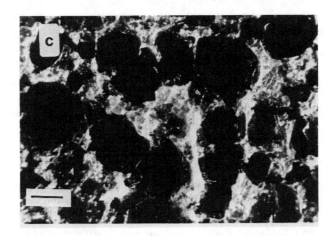

14.11C

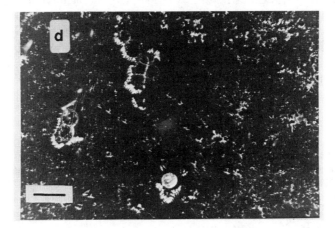

14.11D
Fig. 14.11. Light microscopic structure of foams formed from 0.5% lysozyme solution heated at 80°C (a) in the presence of 0.7 mM DTT and (b) in the absence of DTT, and (c) and (d) of the corresponding foams after standing overnight. (Bar represents 1 mm length.)

14.11 illustrates the differences in the microscopic structures of the protein film stabilizing the air bubbles of these foams. In the case of foam formed from lysozyme heated in the presence of DTT, particles can be seen to be deposited on the surface of the larger bubbles as well as linking the smaller bubbles (Fig. 14.11a). These particles are not evident in the foam formed from lysozyme solution heated without any DTT (Fig. 14.11b). The more extensive intermolecular associations of DTT-heated lysozyme, apparent in its greater turbidity and tendency to insolubilize at higher temperatures, may have been conducive to the formation of structured protein layers at the air-water interface, thus stabilizing the foam. The differences in the foams are also apparent by comparing the film of foam after standing overnight (Fig. 14.11c and 14.11d).

Conclusions

The critical role of disulfide bonds in stabilizing protein structure, restricting the unfolding of the molecule and preventing complete exposure of buried hydrophobic regions, must be considered when evaluating the role of hydrophobic interactions in food protein functionality. Partial reduction of the intramolecular disulfide bridges which maintain the conformational stability of the native lysozyme molecule results in greater molecular flexibility. This increased flexibility is evident in the dramatic increase of exposed hydrophobic regions by subsequent heat treatment at 80°C or at 37°C. The greater flexibility and surface hydrophobicity are manifested in improved gelling, emulsifying and foaming properties.

References

1. Stuper, A.J., W.E. Brugger, and P.C. Jurs, *Computer Assisted Studies of Chemical Structure and Biological Function*, Wiley Inc., New York, NY, 1979.
2. Hansch, C., and J.M. Clayton, *J. Pharm. Sci. 62*:1 (1973).
3. Nakai, S., and E. Li-Chan, *Hydrophobic Interactions in Food Systems*, CRC Press Inc., Boca Raton, FL, 1988.
4. Kato, A., and S. Nakai, *Biochim. Biophys. Acta 624*:13 (1980).
5. Hayakawa, S., and S. Nakai, *J. Food Sci. 50*:486 (1985).
6. Li-Chan, E., S. Nakai, and D.F. Wood, *Ibid. 49*:345 (1984).
7. Bigelow, C.C., *J. Theor. Biol. 16*:187 (1967).
8. Bigelow, C.C., and M. Channon, in *Handbook of Biochemistry and Molecular Biology, Proteins*, edited by G.D. Fasman, Vol. I, CRC Press, Cleveland, OH, 1976, p. 209.
9. Townsend, A.-A., and S. Nakai, *J. Food Sci. 48*:588 (1983).
10. Voutsinas, L.P., S. Nakai, and V.R. Harwalkar, *Can. Inst. Food Sci. Technol. J. 16*:185 (1983).

11. Shimizu, M., T. Kamiya, and K. Yamauchi, *Agric. Biol. Chem.* 45:2491 (1981).
12. Kato, A., N. Tsutsui, N. Matsudomi, K. Kobayashi, and S. Nakai, *Ibid.* 45:2755 (1981).
13. Kato, A., Y. Osako, N. Matudomi, and K. Kobayashi, *Ibid.* 47:33 (1983).
14. Kato, A., K. Komatsu, K. Fujimoto, and K. Kobayashi, *J. Agric. Food Chem.* 33:931 (1985).
15. Kato, A., K. Fujimoto, N. Matsudomi, and K. Kobayashi, *Agric. Biol. Chem.* 50:417 (1986).
16. Kato, A., H. Yamaoka, N. Matsudomi, and K. Kobayashi, *J. Agric. Food Chem.* 34:370 (1986).
17. Li-Chan, E., and S. Nakai, *CRC Crit. Rev. Poultry Biol.* 2:21 (1989).
18. Hirose, M., H. Oe, and E. Doi, *Agric. Biol. Chem.* 50:59 (1986).
19. Oe, H., M. Hirose, and E. Doi, *Ibid.* 50:2469 (1986).
20. Oe, H., E. Doi, and M. Hirose, *Ibid.* 51:2911 (1987).
21. Canfield, R.E., and A.K. Liu, *J. Biol. Chem.* 240:1997 (1965).
22. Donovan, J.W., C.J. Mapes, J.G. Davis, and J.A. Garibaldi, *J. Sci. Food Agric.* 26:73 (1975).
23. Blake, C.C.F., D.E.P. Grace, L.N. Johnson, S.J. Perkins, D.C. Phillips, R. Casse, C.M. Dobson, F.M. Paulson, and R.J.P. Williams, in *Molecular Interactions and Activity in Proteins*, CIBA Foundation Symposium 60, Excerpta Medica, New York, NY, 1978, p. 137.
24. Hayakawa, S., and R. Nakamura, *Agric. Biol. Chem.* 50:2039 (1986).
25. Weber, G., and L.B. Young, *J. Biol. Chem.* 239:1415 (1964).
26. Li-Chan, E., S. Nakai, and D.F. Wood, *J. Food Sci.* 50:1034 (1985).
27. Beveridge, T., S.J. Toma and S. Nakai, *Ibid.* 39:49 (1974).
28. Pearce, K.N., and J.E. Kinsella, *J. Agric. Food Chem.* 26:716 (1978).
29. Graham, D.E., and M.C. Phillips, in *Foams*, edited by R.J. Akers, Academic Press, Inc., London, England, 1976, p. 237.
30. Hatta, H., N. Kitabatake, and E. Doi, *Agric. Biol. Chem.* 50:2083 (1986).
31. Utsumi, S., and J.E. Kinsella, *J. Food Sci.* 50:1278 (1985).
32. *Handbook of Biochemistry*, edited by H.A. Sober, CRC Chemical Rubber Co., Cleveland, OH, 1970, p. C-83.
33. Layne, E., in Methods in Enzymology, Vol. III, edited by S.P. Colowick and N.O. Kaplan, Academic Press, Inc., New York, NY, 1957, p. 448.

Chapter Fifteen

Relationship of SH Groups to Functionality of Ovalbumin

Etsushiro Doi, Naofumi Kitabatake, Hajime Hatta and Taihei Koseki

Research Institute for Food Science
Kyoto University
Uji, Kyoto 611, Japan

Ovalbumin is a main protein component of egg white. Many of the characteristics for functional properties of egg white, such as gel formation and foam stability, are related to the properties of ovalbumin. Large amounts of pure ovalbumin can easily be prepared. Ovalbumin is a phosphoglycoprotein with a single peptide chain (1). One molecule contains one disulfide bond and four free sulfhydryl groups. By the simplicity of the structure, the ease of obtaining the pure protein in large quantity, and the presence of disulfide and sulfhydryls in appropriate numbers, ovalbumin is a suitable protein for studying the relationship of sulfhydryl groups and the functional properties of food proteins.

Reactivity of Sulfhydryl Groups

Amino acid residues in an ovalbumin molecule are arranged according to the hydropathic index of Kyte and Doolittle (2) in Figure 15.1. One disulfide and four free sulfhydryls are contained in the molecule of MW 45,000. The disulfide is in a hydrophilic or intermediate region. All of the sulfhydryls are in hydrophobic regions. Therefore, these sulfhydryls are not very reactive against various sulfhydryl reagents as shown in Table 15.1.

p-Chloromercuribenzoate (pCMB) reacts with three of the sulfhydryls at neutral pH. It takes more than one hour for a complete reaction (5,6). N-ethylmaleimide (NEM) reacts only partly (6,7). 5,5'-Dithiobis(2-nitrobenzoic acid)(DTNB) does not react at all. 2,2'-Dithiodipyridine (2-PDS), a colorimetric reagent similar to DTNB, reacts very slowly with two or three of the sulfhydryls at pH 7.0, taking more than 24 hours (8). At acid pH, one of the sulfhydryls reacts rapidly with 2-PDS, which means that some conformational change occurs at acid pH (8). Four sulfhydryls of ovalbumin react with DTNB after denaturation by either

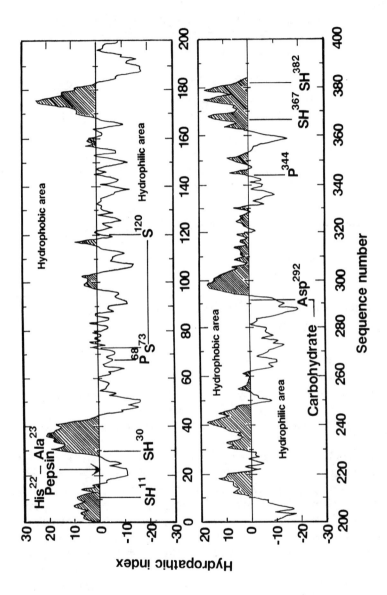

Fig. 15.1. Hydropathic indices of the ovalbumin molecule according to Kyte and Doolittle (2). The amino acid sequence is cited from Nisbet *et al.* (3). The arrow shows the site of limited proteolysis by pepsin (4).

TABLE 15.1
Reactivity of Various Sulfhydryl Reagents Against Native Ovalbumin

Reagent	pH	SH (mol/mol)	Reference
pCMB	7.0	3.0	(5)
	4.6	2.8	(6)
NEM	7.0	0	(7)
	7.0	0.4-0.6	(6)
DTNB	8.0	0	(7)
2-PDS	7.0	2.0-2.5	(8)
	2.2	1.0	(8)

sodium dodecylsulfate (SDS)(7), or SDS and urea (9). Therefore, the change of reactivity of DTNB with ovalbumin can be used as a measure of the conformational change of the ovalbumin molecule.

Conformational Change of Ovalbumin During Foam Formation (9)

Ovalbumin solutions form stable foams (10). Such stable foams might be made by two dimensional networks of proteins, probably an association of denatured proteins. However, no direct evidence of the conformational change of ovalbumin during foam formation, such as surface denaturation, has been reported. We proved that there is a conformational change of ovalbumin during foam formation by the change of reactivity of sulfhydryls with DTNB.

Ovalbumin solutions were foamed as shown in Figure 15.2. First, 15 ml of 4% ovalbumin dissolved in buffer A (20 mM sodium phosphate buffer, pH 7.5, containing 0.1 M EDTA) was stirred for 1 min at 10,000 rpm in a homogenizer. After the stirring, the ovalbumin foam was left tightly sealed for one hour. The drained liquid (Drain 1 in Figure 15.2) was separated from the foam. The foam was collapsed by centrifugation at 3000 × g for 10 min. The supernatant solution (Drain 2) was separated from the precipitate (Coagulum). The coagulum was washed with buffer A and dissolved in buffer B (40 mM sodium phosphate buffer, pH 8.0, containing 0.48% SDS, 8 M urea, and 1 mM EDTA) bubbled with N_2. The sulfhydryl contents of Drains 1 and 2 and the coagulum were then analyzed (Table 15.2).

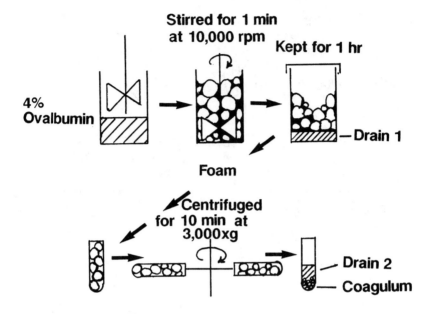

Fig. 15.2. Fractionation of foam. See the text for details.

TABLE 15.2

Sulfhydryl Contents of Ovalbumin in Drains and Coagulum

Fractions	SH (mol/mol)[a]
Original ovalbumin	4.08
Drain 1	4.14
Drain 2	4.12
Coagulum	1.32

[a]Sulfhydryl contents were assayed as follows: 4.9 ml of buffer B was added to 0.5 ml of ovalbumin solution or Drain. Then, 33 μl of 10 mM DTNB was dissolved in 40 mM sodium phosphate buffer, and pH 7.0 was added to the mixture, which was then incubated for 1 hr at 50°C. After the mixture was cooled, the absorbance was measured at 412 nm. Coagulum dissolved in buffer B (5 ml) was added to 33 μl of 10 mM DTNB solution and treated as described above.

Drains 1 and 2 contained four moles of sulfhydryl groups per mole of ovalbumin, the same as the original ovalbumin. However, the coagulum contained 1.32 moles of sulfhydryl groups per mole of ovalbumin. This

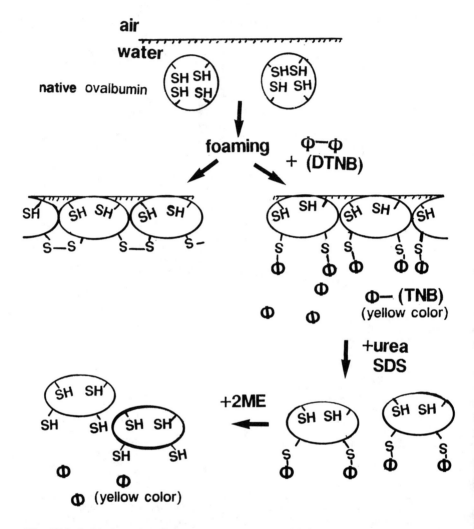

Fig. 15.3. Oxidation of sulfhydryl groups during foam formation, and reaction of sulfhydryl groups with DTNB. Yellow color develops with the release of trinitrobenzoic acid (TNB).

indicates that some of the sulfhydryl residues of ovalbumin molecules in the coagulum are oxidized to form disulfide bonds.

SDS-polyacrylamide gel electrophoresis of the original ovalbumin and Drains 1 and 2 in the presence of iodoacetamide was done without a reducing reagent. The three samples showed a single band, indicating that there were no intramolecular disulfide bonds in these samples. Electrophoresis of the coagulum under the same conditions gave many bands of higher molecular weights. The coagulum treated with 2-mercaptoethanol (2ME) showed a single band corresponding to the monomer, which indicates the presence of intermolecular disulfide linkages. Similar findings were obtained by the gel permeation chromatography of these samples in the presence of 6 M guanidine hydrochloride.

These results suggest that a conformational change of ovalbumin occurred during foaming and some of the sulfhydryl residues became reactive and formed intermolecular disulfide bonds as shown in Figure 15.3A.

To get direct evidence of conformational changes in the molecule during foaming, foam was prepared in the presence of DTNB as in Figure 15.2. The original ovalbumin solution incubated with DTNB for one hour did not have any color (Fig. 15.6A), but Drains 1 and 2 were yellow. That means DTNB reacted with some of the sulfhydryls in the ovalbumin molecule. The sulfhydryls reacted with DTNB and combined with thionitrobenzoic acid (TNB). The addition of a reducing reagent to these molecules yields free TNB molecules (yellow) in the solution as shown in Figure 15.3. Drains 1 and 2, which were separated from free TNB by gel chromatography, did not turn color upon the addition of 2ME, but the coagulum became yellow, corresponding to 2.15 moles of sulfhydryl groups per molecule.

The results indicate that the denaturation of ovalbumin molecules occurred during foam formation, and the interaction between denatured proteins probably stabilized the foam. The presence of DTNB or addition of iodoacetamide to the ovalbumin solution did not affect the stability of the foam. Therefore, the essential factor for the formation of stable foams is not the formation of intermolecular disulfide bonds but rather the network formation by other noncovalent interactions such as hydrophobic interactions.

Heat-induced Gels of Ovalbumin

Ovalbumin solution gave gels upon heating. The rigidity of the gels changed with the pH of the solution (11). The pH and salt concentrations effect the thermal aggregation of ovalbumin (12). We also examined the effects of pH and salt concentrations on the properties of heat-induced gels of ovalbumin (13,14). A 5% ovalbumin solution with 20 mM NaCl was adjusted to various pHs and heated for one hour at 80°C. The

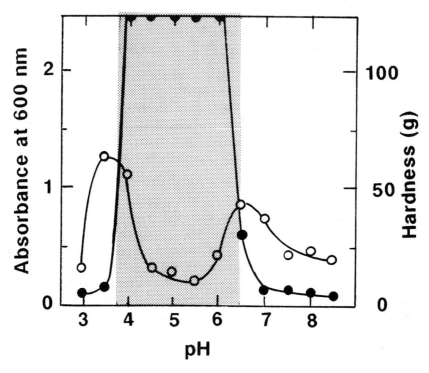

Fig. 15.4. Hardness and turbidity of a heat-induced gel from ovalbumin at various pHs. The 5% (w/v) ovalbumin solution containing 20mM NaCl was heated at 80°C for one hour. Inside the stippled area, turbid gels were obtained (13). O——O, hardness; ●——●, turbidity (absorbance at 600 nm).

gel hardness and turbidity, which were measured by the absorbance at 600 nm, are shown in Figure 15.4. Turbid gels were obtained around the isoelectric point (pH 4.7) of ovalbumin. Transparent gels were obtained at a lower acid pH than 3.5 and a higher alkaline pH than 6.5. Gel hardness had two maxima at pH 3.5 and 6.5. These were the pHs critical for gel turbidity.

The effects of NaCl concentrations on the hardness and turbidity were examined at pH 3.5, 5.5, and 7.5 (Fig. 15.5). At pH 5.5, all the samples were turbid and the gel hardnesses were very low. In the absence of NaCl, a turbid suspension containing a coagulum of ovalbumin was obtained at pH 5.5. At pH 3.5 and 7.5, in the absence of NaCl, transparent solutions were obtained. At fairly low concentrations of NaCl, the gels

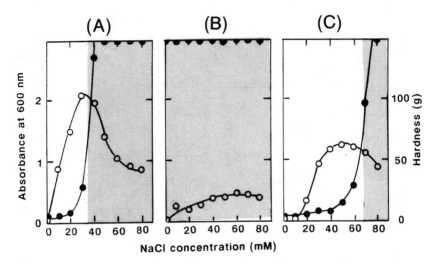

Fig. 15.5. Hardness and turbidity of a heated sample from ovalbumin at various NaCl concentrations. (A) pH 3.5, (B) pH 5.5, (C) pH 7.5. Heating conditions were the same as in Figure 16.4. Inside the stippled area, turbid gels or solutions were obtained (13). ○——○, hardness; ●——●, turbidity (absorbance at 600 nm).

were transparent and their hardness increased with the increase of NaCl concentration. When the NaCl concentrations were increased more, the turbidity of the gel rapidly increased and the hardness gradually decreased. Transparent hard gels were obtained in a narrow range of pH and salt concentrations. For example, at pH 7.5, the ovalbumin solution containing 20-50 mM of NaCl, gave transparent hard gels.

However, we could make transparent gels in a wide salt concentration range using two-step heating. The experiments shown in Figure 15.6A (one-step heating) are almost the same as that in Figure 15.5C, but NaCl concentrations were higher than in Figure 15.5. Turbid gels were obtained at the NaCl concentrations higher than 0.1 M. In the experiments shown in Figure 15.6B (two-step heating) 5% ovalbumin solution, pH 7.5, was heated at 80°C for one hour in the absence of NaCl, which gave a transparent solution. After being cooled, various concentrations of NaCl were added to the solution.

Each sample was heated again at 80°C for one hour. The hardness of gels increased depending on the NaCl concentration, but the turbidity did not increase as much as with one-step heating. The transparent gel prepared by two-step heating with 0.3 M NaCl was harder than the

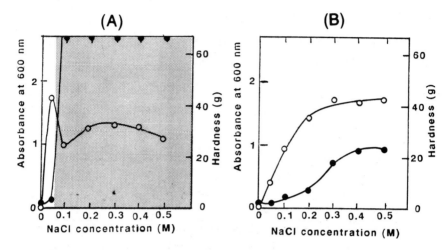

Fig. 15.6. Comparison of one-step heating method (A) and two-step heating method (B). (A) Ovalbumin solution (5% w/v, pH 7.5) was heated at 80°C for one hour at various NaCl concentrations. (B) Ovalbumin solution (5% w/v, pH 7.5) was heated at 80°C for one hour in the absence of NaCl. After being cooled, each sample was heated again with various NaCl concentrations. Inside the stippled area, turbid gels or solutions were obtained (14). O———O, hardness; ●———●, turbidity (absorbance at 600 nm).

turbid gel prepared by the one-step heating with 0.3 M NaCl. The results indicate that during the first heating at 80°C for one hour without NaCl some conformational change in the ovalbumin occurred, which was required for the formation of a transparent gel.

Conformational Change of Ovalbumin by Heat

The conformational change of ovalbumin arising from heat was examined for the transparent solution obtained by heating ovalbumin in 20 mM phosphate buffer and pH 7.0 at 80°C for one hour(15). The CD spectrum of heat denatured ovalbumin is not very different from that of native ovalbumin (15,16), which shows the secondary structure of heat-denatured ovalbumin is not a random coil, but it still has a globular conformation. Sedimentation analysis indicates the presence of polymers of high molecular weight. Transmission electron microscopy of the clear solution obtained by heating ovalbumin shows the presence of fibrous polymers (15). The distribution of molecular weight of the soluble aggregates obtained after heat treatments of ovalbumin has been

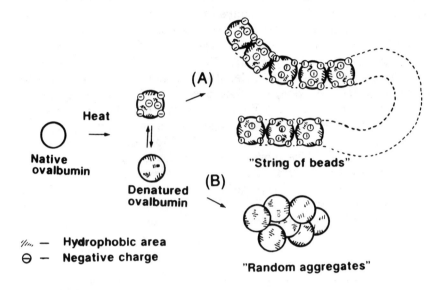

Fig. 15.7. Model for the heat denaturation and formation of linear polymers at low ionic strength (A), or aggregation at high ionic strength (B) of ovalbumin.

analyzed by low-angle laser light scattering by Kato and Takagi (17). The average molecular weight is about 10×10^6. We also examined the molecular properties of the soluble aggregates by the combination of light scattering, intrinsic viscosity and gel-permeation chromatography. The average molecular weights ranged from $1-10 \times 10^6$, depending on the heating time at 80°C. The linear polymer fitted into a worm-like chain model (18,19) whose stiffness was about half of double-helical DNA (data to be published).

We postulate the formation of linear polymers as shown in Figure 15.7. The conformation of the heat-denatured ovalbumin was not very different from that of the native molecule, but some hydrophobic areas that had been buried in the molecule are exposed to the surface of the protein after denaturation. At pH near the isoelectric point and at high ionic strength, denatured proteins randomly aggregate by hydrophobic interaction. At pH far from the isoelectric point and at low ionic strength, electrostatic-repulsive forces hinder the formation of random aggregates, and linear polymers, sometimes called a "string of beads"(20), are formed.

Sulfhydryl Groups in Heat-Denatured Ovalbumin

The reactivity of sulfhydryl groups of heat-denatured ovalbumin was tested with DTNB (Fig. 15.8). DTNB did not react with sulfhydryl groups in native ovalbumin but reacted rapidly with four sulfhydryls after denaturation with urea and SDS (Fig. 15.8A). DTNB reacted rapidly with three sulfhydryls in heat-denatured ovalbumin, which means two or three of the hydrophobic regions are exposed on the outside of the protein molecule (Fig. 15.1). Four sulfhydryl groups in heat-denatured ovalbumin reacted with DTNB in the presence of urea and SDS, which shows the oxidation of sulfhydryls to disulfides did not occur during heat denaturation. Therefore, intermolecular disulfide bonds can be made only by a sulfhydryl-disulfide exchange reaction.

The heated ovalbumin preparations at various pH and salt conditions, 80°C for one hour, were solubilized by 0.1 M sodium phosphate buffer, pH 7.2, containing 1% SDS. The distribution of polymers connected by inter-

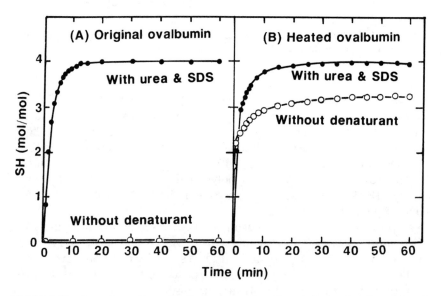

Fig. 15.8. Number of reactive sulfhydryl groups of ovalbumin before heating (A), and after heating at pH 7.0, 80°C for 10 min. (B). Ovalbumin solution (0.1 mg/ml) and 0.1 mM DTNB reacted in 40 mM phosphate buffer, pH 8.0 at 25°C. The change of absorbance was measured at 412 nm. The complete denaturation of ovalbumin was done in the presence of 8 M urea and 0.48% SDS at 50°C for one hour before the reaction with DTNB.

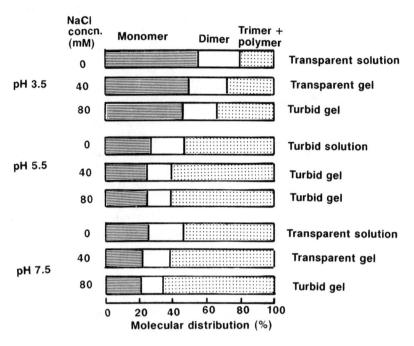

Fig. 15.9. Distribution of molecular species connected by intermolecular disulfide linkages in heated ovalbumin. The heat-treated ovalbumin solutions shown in Figure 16.5 were dissolved in 0.1 M Na phosphate buffer containing 1% SDS, pH 7.2, and analyzed by gel-permeation chromatography in the presence of SDS (14).

molecular disulfide bonds was analyzed by gel-permeation chromatography (Fig. 15.9). At pH 5.5 and 7.5, the proportions of monomers and dimers are smaller than those at pH 3.5. The amounts of trimers and polymers increased in the presence of higher concentrations of salt. However, large amounts of monomers and dimers were always observed in all preparations. The degree of polymerization of the polymers was very small compared to the molecular weights of polymers found in heated solutions (about ten million). When the samples in Figure 15.9 were treated with dithiothreitol and analyzed by gel-permeation chromatography, a single peak having the same retention time as that of native ovalbumin was obtained.

The addition of 2ME to the ovalbumin solution did not inhibit the formation of a heat-induced gel. These results indicate that the disulfide bond is not of primary importance in gel formation. However, the presence of sulfhydryl reagents or reducing reagents during heat treatments effects the functional properties of ovalbumin gels.

Effects of Sulfhydryl Reagents on the Functional Properties of Ovalbumin Gels

The textures of heat-induced ovalbumin gels are effected by various factors, such as pH, salt concentrations, protein concentrations, and the times and temperatures of the heat treatments. Generally speaking, transparent gels obtained at low ionic strength were harder and more adhesive than the turbid gels obtained at high ionic strength. More detailed analyses of the textures of ovalbumin gels are now in progress.

We show an example of the effects of sulfhydryl reagents on the texture of transparent gels containing 6% ovalbumin (Table 15.3). The addition of a sulfhydryl reagent (NEM, iodoacetic acid, or iodoacetamide) decreased the hardness, cohesiveness, and adhesiveness of the heat-induced gels. A reducing reagent, 2ME, also lowered the hardness and adhesiveness of the gels. The results indicate that intermolecular disulfide linkages did not connect all of the molecules in the linear polymers, but affected the physical properties of the gels.

TABLE 15.3
Effects of SH-Reagents on the Textural Properties of Ovalbumin Gel[a]

	Hardness $\times 10^5$ (dyne/cm^2)	Cohesiveness	Adhesiveness $\times 10^3$ (dyne/cm^2)
Control	1.82	0.80	8.80
NEM	1.02	0.54	5.61
Iodoacetic acid	1.00	0.75	5.17
Iodoacetamide	1.27	0.76	5.63
2ME	1.68	0.80	5.05

[a]The gel-forming solution of the control sample contained 6% (w/v) ovalbumin, 0.02 M, pH 7.5, Na phosphate buffer and 0.1 M NaCl. 0.0125 M NEM, 0.05 M iodoacetic acid, 0.1 M iodoacetamide or 0.5% (v/v) 2ME was added to the mixture. The mixture (0.90 ml) was heated at 80°C for 75 min in a cylindrical tube (10 mm i.d. x 10 mm height). Textural properties of gels were measured with a Rheometer RE-3305 (Yamaden Co., Ltd., Tokyo) using a 16 mm diameter plunger. Hardness, cohesiveness, and adhesiveness were measured and calculated by the method described by Szczesniak (20).

Model of Gel Formation by Ovalbumin

The sulfhydryl-disulfide relationship is concerned with the aggregation of proteins, especially the gelling processes (21). The viscosity of urea-denatured ovalbumin solution increases with time, and eventually gelling of the solution occurs (22). This is interpreted as an aggregation process due to the formation of intermolecular disulfide linkages. The urea-induced aggregation and viscosity change are suppressed at the same rates by the presence of pCMB (22). Therefore, the intermolecular disulfide linkages are very important for the gelling process of urea-denatured ovalbumin. In the case of heat-induced ovalbumin gel, the formation of intermolecular disulfide linkage was observed but it was not indispensable for the gelling process.

A model of gel formation by heat-denatured ovalbumin to explain the experimental results is shown in Figure 15.10. Electron micrographs of heated ovalbumin (15,23) and other proteins (24) were consulted to make this model. At very low protein concentrations, at a pH far from the pI of the protein and low ionic strength, the polymer is not formed by heating. Foster and Rhees (25) reported that aggregation of heated ovalbumin is minimal at pH 2.4 and 9–10. We failed to find monomers at pH 2.4, but they should be present at very low protein concentrations. Fibrous polymers are formed at the usual protein concentration and low ionic strength. With increasing ionic strength or decreasing electrostatic repulsive force, a three-dimensional gel network is formed by interpolymer interaction. At high ionic strength and a pH near the pI,

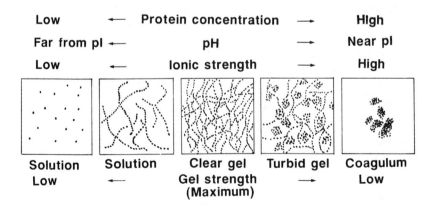

Fig. 15.10. Model for the formation of gel networks by heated ovalbumin.

proteins aggregate to form a coagulum. The presence of such a coagulum interferes with the formation of the gel network and results in the formation of a soft, turbid gel. The presence of intermolecular disulfide bonds affects the texture of gels.

This model was made mainly to fit our data. Ovalbumin gels at high pH may be close to the gel made of random aggregates (23). This model explains fairly well the properties of heat-induced gels of ovalbumin described in this paper.

References

1. Osuga, D.T., and R.E. Feeney, in *Food Proteins*, J.R. Whitaker and S. Tannenbaum (eds), Avi Publishing Co., Inc., Westport, CT, 1977, pp. 209-266.
2. Kyte, J., and R.F. Doolittle, *J. Mol. Biol. 157*:105 (1982).
3. Nisbet, A.D., R.H. Saundry, A.J.G. Moir, L.A. Forthergill and J.E. Forthergill, *Eur. J. Biochem. 115*:335 (1981).
4. Kitabatake N., and E. Doi, *Agric. Biol. Chem. 49*:2457 (1985).
5. Boyer, P.D., *J. Am. Chem. Soc. 76*:4331 (1954).
6. Diez, M.J.F., D.T. Osuga and R.E. Feeney, *Arch. Biochem. Biophys. 107*:449 (1964).
7. Leslie, J., L.G. Butler and G. Gorix, *Ibid. 99*:86 (1962).
8. Koseki, T., N. Kitabatake and E. Doi, *J. Biochem. 103*:425 (1988).
9. Kitabatake, N., and E. Doi, *J. Agric. Food Chem. 35*:953 (1987).
10. Kitabatake, N., H. Sasaki and E. Doi, *Agric. Biol. Chem. 46*:2881 (1982).
11. Egelandsdal, B., *J. Food Sci. 45*:570 (1980).
12. Hegg, P.O., *Ibid. 47*:1241 (1982).
13. Hatta, H., N. Kitabatake and E. Doi, *Agric. Biol. Chem. 50*:2083 (1986).
14. Kitabatake, N., H. Hatta and E. Doi, *Ibid. 51*:771 (1987).
15. Doi, E., N. Kitabatake and T. Koseki, *J. Am. Oil Chem. Soc. 64*:1697 (1987).
16. Egelandsdal, B., *Int. J. Peptide Protein Res. 28*:560 (1986).
17. Kato, A., and T. Takagi, *J. Agric. Food Chem. 35*:633 (1987).
18. Benoit. H., and P. Doty, *J. Phys. Chem. 57*:958 (1953).
19. Yamakawa, H., and H. Fujii, *Macromolecules 7*:128 (1974).
20. Szczesniak, A.S., in *Texture Measurements of Foods*, A. Kramer and A.S. Szczesniak, (eds), D. Reidel Publishing Company, Dordrecht, Holland, 1973, pp. 71-108.
21. Huggins, C., D.F. Taplay and E.V. Jensen, *Nature 167*:592 (1951).
22. Frensdorff, H.K., M.T. Watson and W. Kauzmann, *J. Am. Chem. Soc. 75*:5157 (1953).
23. Van Kleef, F.S.M., *Biopolymers 25*:31 (1986).
24. Clark, A.H., F.J. Judge, J.B. Richards, J.M. Stubbs and A. Suggett, *Int. J. Peptide Protein Res. 17*:380 (1981).
25. Foster, J.F., and R.C. Rhees, *Arch. Biochem. Biophys. 40*:437 (1952).

Chapter Sixteen
Use of Radio-labeled Proteins to Study the Thiol-Disulfide Exchange Reaction in Heated Milk

Bong Soo Noh and Tom Richardson

Department of Food Science and Technology
University of California
Davis, CA 95616

Heat induced complexes between milk proteins are of considerable importance in determining the heat stability and rennin clottability of milk products. Thiol-disulfide interchange reactions have been suggested as the principal reaction mechanism for complex formation. Studies to date have not adequately established the mechanism and stoichiometry of complex formation *in situ* in total milk system.

Tracer amounts of ^{14}C-β-lactoglobulin were added to skim milk, which was heated under various conditions. After clotting with rennet, radioactivity retained in the curd was counted to estimate the extent of the interaction of β-lactoglobulin with casein. Also, milk containing ^3H-κ-casein and ^{14}C-β-lactoglobulin was subjected to various heat treatments and protein complexes containing ^3H and ^{14}C, which were separated on Sephacryl S-300 eluted with 6 M guanidine hydrochloride (pH 6.58). Further separations of protein complexes were obtained using controlled pore glass chromatography in the presence of 6 M guanidine hydrochloride (pH 6.58). Ratios of radioactivities in protein complex peaks suggested various stoichiometries of β-lactoglobulin and κ-casein in complexes were possible as a function of heat treatment.

The proteins of milk can be classified as either caseins (precipitated from milk by acidification to pH 4.6 or by rennet, an enzyme) or whey proteins (1). The caseins, if heated alone at 120°C for several minutes, do not appear to change. By contrast, the whey proteins readily precipitate under the same conditions. However, if milk is heated, there is no change in its appearance, but if the heated milk is acidified to pH 4.6, the precipitate consists of the caseins and the bulk of the whey proteins. The caseins exist in milk as colloidal aggregates (the casein micelle) with calcium and phosphate as integral parts of their structure.

This phenomenon, whereby the whey proteins interact with casein during heat treatments, has practical importance in slowing or preventing rennet action from coagulating milk and slowing the heat gelation of evaporated milk during sterilization. Our present understanding (2) suggests that one of the caseins, κ-casein, is very important in stabilizing the casein micelle against precipitation with calcium. Its special character is caused by an unusual distribution of its amino acids so that one of its two domains is clearly apolar and carries a net positive charge while the other carries a net negative charge and is dispersed and very hydrophilic. The hydrophobic domain carries two cysteine residues which, in the casein micelle, cross-link the κ-casein into a shell or skin with the hydrophilic domains projecting into the solvent surrounding the micelle and the hydrophobic positively charged domains interacting with the α_{s1}-, α_{s2}- and β-caseins that constitute the core of the micelle.

Early studies (3) on the results of heating milk demonstrated that the order of sensitivity of the whey proteins was immunoglobulins > bovine serum albumin > β-lactoglobulin > α-lactalbumin (4). The technique used in that study (free boundary electrophoresis) did not have an especially good resolution, but the conclusions have not been refuted by any later study. The judicious use of thiol blocking agents, careful analysis and studies in model systems of purified proteins demonstrated that κ-casein was a key reactant in the heat induced reactions in milk and that disulfide interchange reactions were involved (3,5,6).

In 1978 Creamer et al. (7) showed that the whey proteins and κ-casein formed small particles up to approximately 10 nm in length and 1 nm wide that adhered to the casein micelles at a pH of about 0.15 less than that of the natural pH of milk (6.70) but not at about 0.15 greater than the natural pH. Creamer (8) later showed that the size (on Sephacryl S-1000 in an SDS buffer system) of these particles decreased with increasing pH of heating and that, by gel electrophoresis of the reduced proteins, the κ-caseins and the major whey proteins appeared to be incorporated into the complex. Singh and Fox (9) showed that κ-casein absorption to the casein micelle was pH-dependent and that (10) β-lactoglobulin influenced this absorption by complexing with the κ-casein. This influence was diminished by the reduction of disulfides of either β-lactoglobulin or κ-casein to thiols followed by alkylation of thiols with N-ethylmaleimide (11).

There is some difficulty in obtaining quantitative results with the techniques used in these earlier studies and it is not possible to unambiguously follow the various protein species through the denaturation and degradation pathways. The use of radioactive tracers can be a very valuable adjunct to other studies. The least intrusive modification method appears to be reductive methylation of protein lysine residues (12,13) with radioactive formaldehyde and sodium cyanoborohydride (14,15). Rowley et al. (13,16) showed that both methylated κ-casein and β-lactoglobulin were similar to the precursor proteins and that they could be used to measure the transfer of β-lactoglobulin into micellar heat-induced complexes in heated milk systems (16).

The present study confirms and extends these two earlier reports.

Results and Discussion

Heating at 95°C. When a small quantity of β-lactoglobulin with a low level of ^{14}C-methylation was added to milk which was then heated for various times at 95°C, the β-lactoglobulin was increasingly associated with the colloidal particles that separated out on centrifugation (Table 16.1). The addition of calcium and rennet to the heated milks enhanced the ease of precipitation of the colloidal casein micelles. After washing the sedimented micelle fraction four times, a small proportion (about 15%) of the radioactivity had been washed out. A small amount of this in the less-heated samples may have been absorbed in the serum β-lactoglobulin, but most is likely to have been smaller complexes of whey protein and κ-casein that were readily dissociated from the initial sediment. It is also possible that in the more extensively heated samples the rennet enzymes did not have access to the chymosin-sensitive bonds of all of the κ-casein so that the sedimentation was not as effective.

TABLE 16.1
Distribution of ^{14}C-β-Lactoglobulin in Milk Heated at 95°C

Heating time (min)	Whey (%)	Washed curd (%)	Other[a] (%)
0.5	38.1 ± 4.8	52.3 ± 6.7	9.6
1.0	28.0 ± 4.1	57.2 ± 6.6	14.8
3.0	17.6 ± 3.3	65.3 ± 6.9	17.1
5.0	15.4 ± 2.8	71.0 ± 7.3	13.6
10.0	9.6 ± 2.6	73.3 ± 7.8	17.1
20.0	7.3 ± 2.3	76.9 ± 5.9	15.8

[a]From difference between expected dpm and sum of those found in whey and washed curd fractions (calculated back to original milk).

TABLE 16.2
Quantity of Protein in Pellets After Heating Milk at 95°C

Heating time (Min)	Protein (g/100 ml milk)	Increase in protein content (g/100 ml milk)	(% change)	Increase due to β-lactoglobulin (g/100 ml milk)[a]	Increase due to other protein (g/100 ml milk)[b]
0	2.14 ± 0.33	—	—	—	—
0.5	2.40 ± 0.32	0.26	12.1	0.16	0.10
1.0	2.46 ± 0.45	0.32	14.9	0.18	0.14
3.0	2.49 ± 0.48	0.35	16.3	0.20	0.19
5.0	2.56 ± 0.32	0.42	19.6	0.22	0.20
10.0	2.58 ± 0.12	0.44	20.5	0.23	0.21
20.0	2.73 ± 0.16	0.49	27.5	0.24	0.25

[a]0.315% of β-lactoglobulin in milk was obtained using SDS gel electrophoresis and a laser densitometer.
[b]By difference from columns 3 and 5.

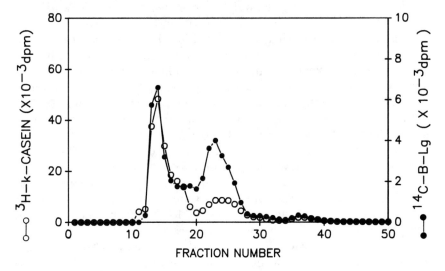

Fig. 16.1. Separation of a sample of milk heated at 95°C for 30 sec on a column of Sephacryl S-300 (35 × 2.5 cm). The milk had small quantities of ^{14}C-β-lactoglobulin and ^{3}H-κ-casein added to it prior to heat treatment. The elution buffer contained 6 M guanidine hydrochloride at pH 6.58. Fraction volume was 5.0 ml.

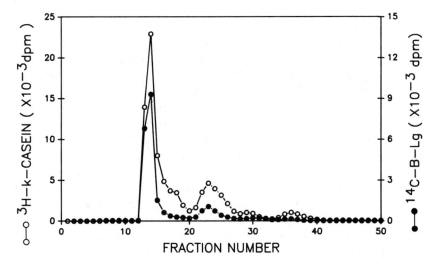

Fig. 16.2. Separation of a sample of milk that had been heated at 95°C for 20 min. Conditions as for Figure 17.1.

The final freeze-dried pellet weights are shown in Table 16.2 and it can be seen that these increase with heating time. The increase in weight was estimated and this was then sub-divided (on the basis of the radioactivity measurements) into β-lactoglobulin and other proteins. Within the error of the measurements, the amount of β-lactoglobulin sedimented is approximately half of the total weight increase due to heat treatment.

Double labeling study. A sample of milk was heated with additions of both ^{3}H-κ-casein and ^{14}C-β-lactoglobulin. It was then analyzed by chromatography in the dissociating buffer of 6 M guanidine hydrochloride at pH 6.58 and selective counting of the two radioisotopes. The results (Figs. 16.1 and 16.2) show that most of the κ-casein is eluted in the front peak (tubes 13–15) while some is eluted near tube 23. There is some dependency on the length of heating time, with more κ-casein being eluted after tube 16 in the sample heated for 30 seconds (Fig. 16.1), as compared with that heated for 20 minutes (Fig. 16.2). In contrast, the proportion of β-lactoglobulin that eluted subsequent to tube 16 diminished dramatically with the increase in heating time. The control sample (no heat) showed no β-lactoglobulin in tubes 13–18, and a large symmetrical peak near tube 22. The κ-casein in the control was essentially the same as that in Figure 16.1.

The reason for the poly dispersity of κ-casein is not clear. Possibilities include the following: The sample of κ-casein used for labeling may have been contaminated with β-casein; methylation of the κ-casein may have involved some S-alkylation; some κ-casein may not have been physically accessible to the β-lactoglobulin for interaction, (e.g., inside the micelle).

Table 16.3 shows the distribution of ^{14}C-β-lactoglobulin between the complex (e.g., tubes 13–15, in Fig. 17.1) and the low molecular weight fraction (e.g., tubes 20–26, in Fig. 17.1).

The small quantity of activity that eluted between the two peaks was not counted in either fraction. The values shown in Table 16.3 are a little greater than the corresponding values shown in Table 16.1 (column 3) and may be related to the material that is lost during the washing to the pellets (column 4 of Table 16.1).

When the material from the first peak (tubes (13–15, Fig. 16.2) was collected and chromatographed in the same buffer (pH 6.58, 6 M guanidine hydrochloride) on controlled pore glass, the pattern shown in Figure 16.3 was obtained, indicating that the ratio of κ-casein to β-lactoglobulin was slightly greater for the smaller complexes. This is shown in more detail (after converting radioactivity to molar ratios of the proteins) in Figure 16.4, along with data from the 30 sec heating time. Clearly the ratio of the protein components is quite different

TABLE 16.3
Distribution of ^{14}C-β-Lactoglobulin in Milk Heated at 95°C and Fractionated on a Column of Sephacryl S-300 in 6 M Guanidine Hydrochloride Buffer

Heating time (Min)	High MW complex[a] (%)
0.0	0
0.5	53.1%
1.0	13.2%
20.0	82.5%

[a]Calculated by 100 (%) x peak I/(peak I + peak II).

between the two heating times for the major peak. These results could be taken to suggest that the initial β-lactoglobulin denatures to interact with any available κ-casein, while the later denaturing β-lactoglobulin interacts with previously denatured β-lactoglobulin in preference to the κ-casein, thus enlarging these complexes at the expense of the higher κ-casein complexes. A similar but more likely explanation is that with the longer heating time there is more opportunity for the heated protein system to come to equilibrium with the smaller complexes containing a higher proportion of κ-casein.

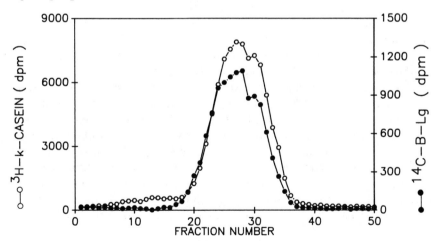

Fig. 16.3. Separation of the major peak (tubes 13–15) from Figure 17.2 on a column of CPG10/3000 (105 × 1.5 cm). The elution buffer contained 6 M guanidine hydrochloride at pH 6.58. Fraction volume of elution was 5 ml.

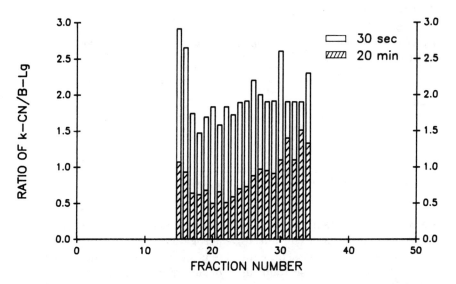

Fig. 16.4. Ratio of κ-casein to β-lactoglobulin for each fraction eluted from the CPG10/3000 column for the complexes derived from the milk samples that had been heated at 95°C for 30 sec and 20 min (Fig. 17.3).

In order to confirm that disulfide interactions were an essential part of the reaction under study, 2-mercaptoethanol was added to the milk mixture prior to heat treatment. After heating at 95°C for 20 min the elution profile from Sephacryl S-300 was indistinguishable from the unheated, but for the S-alkylated control (Fig. 16.5).

Preliminary experiments showed that when milk was heated with ^{14}C-β-lactoglobulin at 70°C and analyzed following rennet treatment, centrifugation, washing etc., that there was a time dependent loss of β-lactoglobulin from the whey fraction, but most of it did not remain with the colloidal fraction during the washing steps. The proportion lost appeared to be nearly constant. This suggests that the complexes may be small and not strongly associated with casein micelles. These results are one of the heat treatments of particular interest to dairy scientists because these times encompass pasteurization conditions where the quantities of denatured protein are small but important in terms of processing.

The amount of ^{14}C-β-lactoglobulin incorporated into the colloidal micelle particles when milk was heated at 140°C for either 2 or 4 sec in

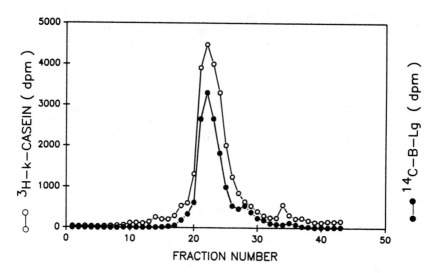

Fig. 16.5. Effect of 2-mercaptoethanol on the aggregation in a sample of milk that contained ^{14}C-β-lactoglobulin and ^{3}H-κ-casein and was heated at 95°C for 20 min. After treatment of 2-mercaptoethanol, iodoacetamide was used for alkylation. The chromatographic conditions are described in Figure 17.1.

further preliminary experiments showed that there was about 50% transfer to the colloidal fraction.

The early results for loss of β-lactoglobulin from milk at 70 and 140°C confirm the data of Dannenberg and Kessler (17), which was obtained by analysis of the acid whey from heated milks by gel electrophoresis. The present results for milk heated at 95°C are lower than would be expected for either genetic variant on the basis of Dannenberg and Kessler's results (17).

The present results are consistent with the following model for protein denaturation and interaction in milk during heating. The proportion of κ-casein, which is partly associated with the outer surface of the casein micelle and partly soluble in the whey, is influenced in each phase by factors such as temperature, pH and calcium level. When a protein such as β-lactoglobulin is heated, it changes its conformation near 70°C and its single thiol group becomes exposed to the environment. Once exposed, it quickly reacts with either the disulfides of β-lactoglobulin or those of κ-casein. At or slightly below the pH of milk, the β-lactoglobulin/κ-casein complex remains bound to the casein micelle and affects its properties during further processing. On the basis that the various whey protein species denature at different rates, it is likely that there are no special catalytic properties associated with β-lactoglobulin as far as other whey proteins are concerned.

Acknowledgments

The authors gratefully acknowledge the support given by the National Dairy Promotion and Research Board, the College of Agricultural and Environmental Sciences, University of California, Davis, the Peter J. Shields' Endowment Fund, and Dr. L.K. Creamer for assistance with the manuscript.

References

1. Eigel, W.N., J.E. Butler, C.A. Ernstrom, H.M. Farrell, Jr., V.R. Harwalker, R. Jenness and R. McL. Whitney, *J. Dairy Sci. 67*:1599 (1984).
2. Walstra, P., and R. Jenness, *Dairy Chemistry and Physics*, John Wiley and Sons, New York, NY, 1984.
3. Sawyer, W.H., *J. Dairy Sci. 52*:1347 (1969).
4. Larson, B.L., and G.D. Rolleri, *Ibid. 38*:351 (1955).
5. Fox, P.F., in *Development in Dairy Chemistry*, 1. *Proteins*, edited by P.F. Fox, Applied Science Publishers Ltd., London, England, 1982, pp. 189-228.
6. Mulvihill, D.M., and M. Donovan, *Irish J. Food Sci. Technol. 11*:43 (1987).
7. Creamer, L.K., G.P. Berry and A.R. Matheson, *New Zealand J. Dairy Sci. Technol. 13*:9 (1978).
8. Creamer, L.K., *J. Chromatogr. 291*:460 (1984).
9. Singh, H., and P.F. Fox, *J. Dairy Res. 52*:529 (1985).
10. Singh, H., and P.F. Fox, *Ibid. 54*:509 (1987).
11. Singh, H., and P.F. Fox, *Ibid. 54*:347 (1987).
12. Koch, G.K., I. Heertje and F. Van Stijn, *Radiochim. Acta 24*:215 (1977).
13. Rowley, B.O., D.B. Lund and T. Richardson, *J. Dairy Sci. 62*:533 (1979).
14. Dottavio-Martin, D., and J.M. Ravel, *Anal. Biochem. 87*:562 (1978).
15. Jentoft, N., and D.G. Dearborn, *J. Biol. Chem. 254*:4359 (1979).
16. Rowley, B.O., W.J. Donnelly, D.B. Lund and T. Richardson, in *Modification of Proteins: Food, Nutritional, and Pharmacological Aspects* (Adv. Chem. Ser. #198), edited by R.E. Feeney and J.R. Whitaker, Am. Chem. Soc., Washington, D.C., 1982, pp. 125-147.
17. Dannenberg, F., and H-G. Kessler, *J. Food Sci. 53*:258 (1988).

Chapter Seventeen

Genetic Modifications of Milk Proteins

Lawrence Creamer, Sang Suk Oh, Robert McKnight, Rafael Jimenez-Flores and Tom Richardson

Department of Food Science and Technology
University of California
Davis, CA 95616

Milk Proteins as a Model to Study Functionality

Studies on the relationship between food protein functionality and structure have been dominated by the effect of chemical and enzymatic modifications on the physicochemical properties of the proteins. Defining changes in the physicochemical properties of food proteins resulting from genetic alterations in their primary sequences offers a useful approach to understanding the structural aspects of functionality. These mutations in the primary sequence of proteins can occur naturally or can be produced in the laboratory using genetic engineering techniques (site-directed mutagenesis).

The milk proteins have been extensively studied, and we know that some natural mutations have a marked effect on the functional properties of the proteins. For example, the B variant of β-lactoglobulin, a protein with a known three-dimensional structure, has Ala_{118} in place of Val_{118} and Gly_{64} substituting for Asp_{64}. It is more soluble than the A variant and its rate of denaturation in heated milk is different.

The A variant of α_{s1}-casein lacks a sequence of 13 hydrophobic amino acids compared to the common B and C variants. The A variant confers lower viscosity on casein solutions (an order of magnitude in 15% solutions) and softer texture in young cheeses. This casein forms casein micelles and associates with the other casein components (in dilute solutions) in the normal manner. These two examples show that for the globular protein with its well-defined, stable core structure a minor change can cause an observable effect, and in the more randomly structured α_{s1}-casein the deletion of 6.5% of its amino acid chain causes technologically significant but minor changes in the physical behavior of the protein.

Available knowledge of milk protein chemistry provides the basis for constructing an ideal model in which to observe the following: (a) a system that could be reconstituted from purified, well-characterized components; (b) structural information of these components in the milk as well as their interactions during processing; and (c) chemical or physical tests that measure properties directly associated with the functionality of any given component in the mixture. When coupled with systematic alterations in the primary sequences of the milk proteins using genetic engineering techniques, the foregoing model can serve as a basis for eventually improving the functional characteristics of milk and other food proteins. For example, it should be possible to engineer enzyme-sensitive bonds or enzyme-binding sites and/or region or domains that could enhance the foaming or emulsifying properties of the proteins.

Methodology to Understand Structure/Function of Food Proteins

In food systems there are some major generalizations that may prove useful even in the absence of an ideal model, or when results of limited macroscopic tests seem contradictory. For example, Kato and Nakai (1) determined the relative surface hydrophobicity of some proteins and its correlation with their surface properties, using hydrophobic probes that fluoresce in a hydrophobic environment. These surface properties, in turn, influence the interfacial activities of the proteins that govern emulsification and foaming properties of the proteins. An examination of a series of proteins has shown direct correlation between the ability of a protein to reduce interfacial tension and its effectiveness as an emulsifying agent (2).

Improvements in the prediction and determination of the secondary structure of various proteins using a variety of physicochemical methods such as nuclear Overhauser enhancement N.M.R. and laser Raman spectroscopy provide valuable information on the three-dimensional structure of proteins in the absence of X-ray crystallography data (3,4). In addition, traditional methods such as infrared and circular dichroism spectroscopy have been refined. Thus, the combination of genetic modifications of specific amino acids in the primary sequence of a protein, its secondary structure (protein folding and topology) prediction and determination, and its macroscopic functionality or biochemical activity represent tools which can elucidate the relationship between structure and function of a food protein which lacks enzymatic activity and detailed X-ray crystallographic data.

In the following sections we present a possible strategy for modifying milk proteins genetically and present some examples to illustrate procedures.

Strategy for Engineering Milk Proteins

Possible strategies for the eventual engineering of the milk proteins are shown in Figure 17.1. The general strategy includes cloning and characterizing the cDNAs coding for the milk proteins (cDNAs are complementary to the messenger RNA that translates into the protein; cDNAs are simpler in structure than the structural genes found in the nucleus which contain extraneous DNA sequences). The cDNAs could be inserted into microbial expression vectors to produce quantities of desired food proteins. Selective mutation of the cDNA sequences using techniques that will be described later makes it possible to produce variants of the natural proteins for structure-function studies of the protein's behavior in foods.

These cDNAs can also serve as hybridization probes for isolating and characterizing the structural genes and controlling elements from a genomic library (prepared from an enzymatic digest of nuclear DNA) of bovine genes. Regulatory sequences of DNA govern the biosynthesis of the milk proteins *in vivo* and could be used to regulate expression of the altered genes. Once the DNA-controlling regions (probably preceding the milk protein genes) have been characterized, it should be possible to make various DNA constructions to enhance selected protein biosynthesis, to produce novel proteins, and to ensure that production of new and novel milk proteins in the mammary gland is under hormonal control.

DNA constructs containing a gene and a regulatory region of interest that is expressed in the mammary gland can be microinjected into embryos of animals such as mice or cows. Thus, microinjected embryos are implanted into pseudopregnant mothers and the resulting transgenic females may then express the foreign gene during lactation.

Recently, Simons *et al.* (5) reported mammary gland-specific expression of the sheep milk protein gene, β-lactoglobulin, in the mouse. Lee and coworkers (6) also showed that β-casein from rat can be expressed tissue-specifically in mice. It may be possible to increase the success over embryo microinjection with the use of cell culture techniques and to generate transgenic animals. One approach is to collect the undifferentiated internal cell mass from an embryonic blastocyst and culture these cells *in vitro*. The cells can be transformed and reintroduced into

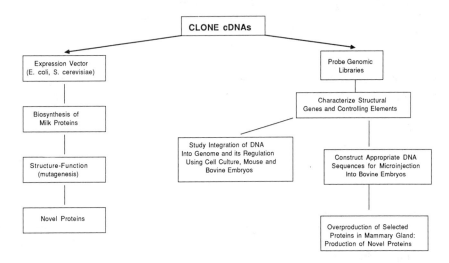

Fig. 17.1. Engineering of milk proteins by employing a mirobial expression system for *in vitro* structure-function studies and a long-term *in vivo* production strategy.

another blastocyst. The resulting chimera can be mated (males) or used as an embryo donor (females) to determine whether chimerism extends to germ cells.

Directed Mutagenesis of Proteins (Protein Engineering)

Directed mutagenesis (7,8) of proteins can be performed in several ways, from the change of a single base in the DNA to alter a codon coding for a single amino acid to the deletion or the insertion of a domain or polypeptide region. All of these possibilities could be useful for tailoring food proteins. However, currently all the examples that exist in protein engineering relate to the modification of enzymes rather than food commodity proteins. An iterative cycle for the directed mutagenesis and assessment of the functionality of a protein is depicted in Figure 17.2. To successfully engineer a food protein, one must develop a structure-function hypothesis. Since X-ray crystallographic data are often lacking for food proteins, a suggested new protein structure is evolved from examination of available structural information derived from various sources and the effects of chemical, physical and enzymatic modifications on protein functionality. Directed mutagenesis can then be

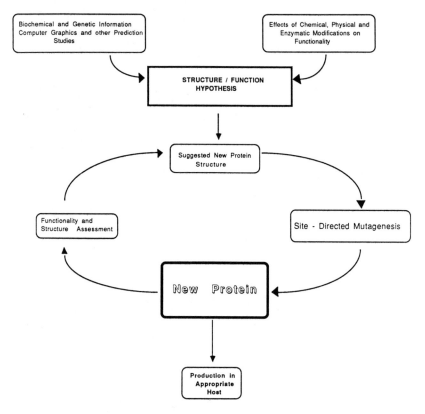

Fig. 17.2. Iterative cycle for engineering food proteins using recombinant DNA technology coupled with structure-function information.

used to alter primary sequences in the protein. The new protein can then be evaluated for functionality and, if found suitable, can be produced in an appropriate host. The cycle can be repeated to eventually obtain a desirable protein. Synthetic oligonucleotides can be used to construct mutations in the genetic code of that protein. These mutations can be single nucleotide base substitutions, insertions of bases or deletions of bases, either singly or in any combination (7,8). The enormous versatility of this mutagenesis technique derives from the fact that the mutating agent is a short synthetic oligonucleotide that differs from the original gene sequence by planned mismatches in nucleotide bases that determines the original type of mutation that occurs in the protein.

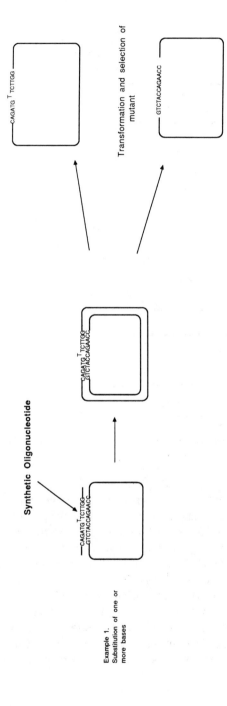

Fig. 17.3. Some examples of different types of site-directed mutagenesis.

1. The complementary synthetic oligonucleotide containing base mismatch is annealed to a single-stranded form of the wild-type gene. The short heteroduplex serves as a primer for extension by DNA polymerase using the wild-type sequence as a template. The DNA polymerase extends the newly synthesized strand around the circular template until it reaches the 5′ end of the oligonucleotide to form a nicked, double stranded DNA. Covalent closure of the nick by DNA ligase generates a duplex circular molecule that has wild-type sequence in the template strand and a mutant sequence in the newly synthesized strand. On transformation into a cell in which the DNA molecule undergoes replication, the two genotypes segregate and subsequently can be isolated as pure clones. The pure mutant oligonucleotide, labeled with ^{32}P, is used as a hybridization probe to physically screen for mutant alleles among the progeny molecules generated on replication of the mutagenized DNA.

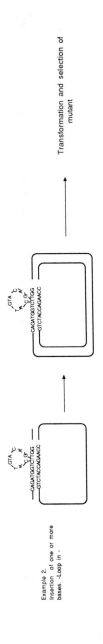

Example 2.
Insertion of one or more
bases. -Loop in -

Transformation and selection of mutant

2. After annealing the probe to the single-stranded template, the homologous complementary parts of the probe will hybridize precisely with their complement and the added bases will form a loop of nonhybridized bases. Once this step is completed, the synthesis of the double strand, transformation and selection of the mutant is performed in the same way as in example 1.

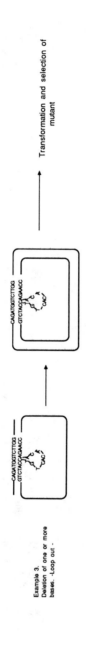

Example 3.
Deletion of one or more bases: -Loop out -

3. The "loop out" method is essentially the same as example 2 except that the single-stranded template forces formation of a loop of bases to be deleted.

Genetic Modifications of Milk Proteins

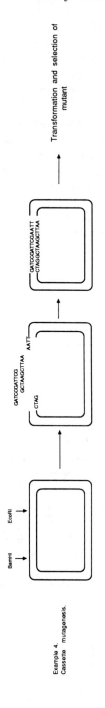

4. In this replacement strategy, the restriction nucleaes (*EcoRI* and *HindIII* in the example) act on the plasmid. The two fragments are separated and the synthetic duplex is annealed into the longer fragment.

Basically, there are seven general conditions required for this oligodeoxynucleotide site-directed mutagenesis (9): cloning the relevant gene into an appropriate vector that is capable of autonomous replication in a host cell such as *Escherichia coli* (e.g., plasmids and viral genomes), obtaining replication of the cloned gene, determining the primary sequence of the DNA insert, making the site (to be mutagenized) available in a suitable form, using the synthetic oligodeoxynucleotide containing base mismatches for altering the cloned sequence, performing site-directed mutagenesis wherein the sequences that are to remain wild-type must be regenerated with fidelity, and identifying mutant colonies with provisions for isolating and characterizing the mutant DNA.

Several strategies are available to achieve oligodeoxynucleotide site-directed mutagenesis. Schemes for some examples of different strategies to direct mutations in a protein are shown in Figure 17.3.

Milk Proteins

There are two major classes of milk proteins: the caseins which comprise the colloidal micellar particles in milk, and the globular whey proteins. Each group of these proteins is of interest, either because of biological functions or because of interactions among themselves or with various constituents in milk and food systems. Predominate in bovine milk are α_{s1}- and β-casein. Thus, in foods, their interactions with each other and with other components are likely to dominate any food system that might contain them. They both have hydrophobic domains which are important in protein-protein and protein-lipid interactions. Biologically they are of interest because neither human nor goat milk has appreciable quantities of α_{s1}-casein and yet the micelle system in goat milk seems very similar to that of cows' milk.

A colloid-stabilizing protein that predominates on the surface of casein micelles to prevent their aggregation is κ-casein. It also contains the principal chymosin-sensitive bond hydrolyzed during the formation of milk clots in the stomach of the suckling infant and the development of a coagulum in cheese-making.

Important in the reactions that occur in heated milk is β-lactoglobulin, and there are biochemical interests because of its possible function as a retinol-carrying species.

The modulator of galactosyl transferase is α-lactalbumin that controls the lactose synthesizing function of this enzyme in the mammary gland.

TABLE 17.1
Some Structural and Genetic Characteristics of Milk Proteins[a]

Protein	Structural Information	cDNA	Structural Gene	Expression
α_{s1}-casein	secondary structure prediction by CD and ORD[b] 50–70% random coil (10,11)	—variant B (12–14) —variant A (unpublished results)	—variant B, partial bovine sequence (15)	—expressed in *E. coli* (16)
α_{s2}-casein		—(17)		
β-casein	secondary structure prediction by CD and ORD[b] 77–80% random coil (10,11,18)	—variant B (17,19,20)	—(21)	—expressed in *S. cerevisiae* (unpublished results) —rat β-casein gene expressed in transgenic mice (6)
κ-casein	secondary structure prediction by CD. 23% α-helix 24% β-turns (10)	(13,22) —(17,21)		—expressed in *E. coli* (22)
β-lactoglobulin	3-dimensional structure by X-ray crystallography, low resolution (23). Structure and similarity to retinol-binding protein (24–26).	—variant A (27)		—sheep β-Lg expressed in transgenic mice (5). —bovine β-Lg expressed in *E. coli* (C. Batt, personal communication)
α-lactalbumin	crystallographic analysis of baboon α-lactalbumin at low resolution (28)	—(29,30)	—(29)	

[a]Bovine proteins unless otherwise stated.
[b]Abbreviations: CD - circular dichroism, ORD - optical rotary dispersion. The references are in parentheses.

A summary of some of the relevant information about the foregoing proteins and research on cDNAs and structural genes that code for them is shown in Table 17.1. Available techniques allow us to outline the preparation of two casein mutants that may contain altered primary sequences with enhanced enzyme susceptibility. The incorporation of enzyme-susceptible bonds into milk proteins or bonds that may be cleaved at an enhanced rate could provide the basis for modified or faster aging cheese or the use of lower levels of enzyme in the cheese-making process. Moreover, one reason for selecting the foregoing system for detailed examination is that low concentrations of mutant proteins can be studied as substrates for proteases. This is an important consideration since a major limitation in much of the protein mutant research is the quantity of desired protein that can be obtained readily.

One of the biochemical processes important in cheddar cheesemaking is the enzymic coagulation of milk by chymosin to form a curd and, later, the action of this same enzyme in the proteolysis pathway in the maturation of cheddar cheese. Of particular interest are the reasons κ-casein is so susceptible to hydrolysis at pH 6.5 by chymosin.

Characteristics of Acid Proteases

Acid proteases are found in stomachs of animals, in microorganisms, in lysosomes of many cell types, and in many tissues (31). Acid proteases are inhibited by pepstatin and by the active site-directed affinity labels, diazoacetyl norleucine methyl ester EPNP-epoxy (*p*-nitrophenoxy) propane (31). They have molecular weights between 35 and 40,000 and have their maximal activity in an acidic pH environment (32). Acid proteases of interest to the dairy industry include porcine pepsin, calf chymosin, chicken pepsin, penicillopepsin and Aspergillus pepsin. Extensive homology exists among these enzymes with particular conservation of residues around the catalytic aspartic acid residues. This together with the X-ray structures indicate that the acid proteases have similar three-dimensional conformations. These enzymes appear to have an extended active site cleft (33) which can accommodate at least seven amino acids of a substrate in the S4-S3' subsites (34) with a preference for cleavage of the bond between two hydrophobic residues occupying the S1 and S1' sites (35). The differences in specificity and activity may be explained by discrete alterations in some or all of the subsites in the various enzymes.

Pearl (36) suggested that a water molecule is tightly bound to the two active-site aspartates. The geometry of the water-carboxyl interaction, in which the proton is involved in the 32-215 (pepsin numbering) hydrogen bond, is equally likely to be close to either carboxyl group. The carboxyl group of the peptide bond to be cleaved becomes polarized thus increasing the susceptibility of the carboxyl carbon atom to nucleophilic attack and leading to the formation of a tetrahedral intermediate. This resultant tetrahedral intermediate may be stabilized, in a manner analogous to the "oxyanion hole" of serine proteases, by interaction with the peptide. Productive decay of this transition state produces the free products, and the activity of the enzyme is restored by binding of a water molecule to replace the original bound water which is incorporated into the free carbonyl product. By means of substrates of the general structure Z-His-X-Y-OMe, the primary specificity of pepsin was defined as a preference for aromatic L-amino acid residues in both the X- and Y-positions; the best substrates are those in which X=Phe and Y=Trp or Phe (37) exist. Powers (35) investigated the cleavage sites of 177 proteins or peptides to determine subsite specificity or porcine pepsin. Analysis of the specificity by subsite showed that P1 was the primary determinant of pepsin specificity with hydrophobic animo acids Phe (probability = 0.51), Met (0.43) and Leu (0.41), and P1 was Try (0.34), Phe (0.29) and Ile (0.26) was preferred. The acid proteases are therefore considerably less specific in their action than other proteases such as the serine proteases, trypsin and chymotrypsin. When insulin beta chain was used as an extended substrate for several acid proteases, cleavage of the bond joining hydrophobic residues was favored, although hydrolysis was not confined to such bonds (Fig. 17.4). These studies showed that chymosin appeared to be more specific than pepsin. The other fungal proteases also showed different specificity patterns (38). Note (Fig. 17.4) that the common cleavage sites of all acid proteases are Phe-Phe and Val-Tyr which are all hydrophobic residues.

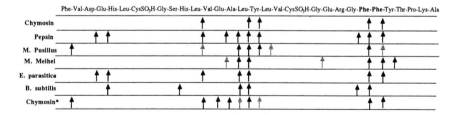

Hydrolysis of κ-Casein by Chymosin

Cheese curd formation involves the initial hydrolysis of κ-casein located on the surface of casein micelles. Cleavage of the Phe_{105}-Met_{106} bond releases a soluble segment from κ-casein, commonly referred to as the glycomacropeptide. Release of the glycomacropeptide reduces the electrostatic repulsion between the resulting p-casein micelles allowing for attractive forces, such as ionic bonding and hydrophobic interactions that lead to aggregation. The ability to enhance the rate of the cleavage reaction may be advantageous in terms of decreasing chymosin requirements in the manufacture of cheese products. The specificity of chymosin for κ-casein is related in part to the presence of the cluster of cationic residues in the κ-casein structure (39), while the particular amino acids in the peptide between residues 97 and 112, which is well conserved in bovine, ovine and caprine κ-casein show their functional importance (40,41). The dipeptide, Phe-Met, and the tri- and tetra peptides, Phe-Met-Ala and Leu-Phe-Met-Ala, were not hydrolyzed by rennin (41–43).

Visser (44,45) examined the influence of short peptides containing the Phe-Met bond in series, and a selection of his results is shown in Fig. 17.5. These and other results demonstrated that not only the length but the composition and sequence of the peptide substrate are important in determining the chymosin and κ-casein interaction. Ser_{104} of κ-casein is believed to form a hydrogen bond to a specific site on chymosin. The strong involvement of Ser_{104} is shown in Fig. 17.5. Substitution of a Gly

His-Pro-His-Pro-His-Leu-Ser-Phe=Met-Ala-Ile-Pro-Pro-Lys-Lys														
98	99	100	101	102	103	104	105	106	107	108	109	110	111	112	Kcat/Km

Peptide	Kcat/Km
Phe=Met	0.00
Ser-Phe=Met-Ala-Ile	0.04
Ser-Phe=Met-Ala-Ile-Pro	0.11
Leu-Ser-Phe=Met-Ala-Ile	21.6
Leu-Ser-Phe=Met-Ala-Ile-Pro	55
His-Leu-Ser-Phe=Met-Ala-Ile	31
Leu-Ser-Phe=Met-Ala-Ile-Pro-Pro	100
Pro-His-Leu-Ser-Phe=Met-Ala-Ile	100
His-Pro-His-Pro-His-Leu-Ser-Phe=Met-Ala-Ile-Pro-Pro-Lys-Lys	2500

for Ser_{104} decreased not only K_{cat} but K_m. Another important feature of the κ-casein and chymosin interaction is involvement of hydrophobic residues (Leu_{103}, Phe_{105}, Met_{106}, Ala_{107} and Ile_{108}) which are known to contribute to the effectiveness of the enzyme-substrate interaction (45). Because of the hydrophobic nature of the active site cleft in the tertiary structure of acid protease such as chymosin, the hydrophobic residues close to the κ-casein susceptible bond play an important role in its hydrolysis.

Based on the studies of the subsite specificity of pepsin and κ-casein and chymosin interaction using various sizes of synthetic peptides, it is clear that Phe-Phe may well be a better cleavage site for chymosin and, if it is not better, then there are factors involved in κ-casein site specificity other than S_1 and S_1' interaction alone.

Site-directed Mutagenesis of κ-Casein

Recent advancements in recombinant DNA technology allowed the change of Phe_{105}-Met_{106} in κ-casein to Phe_{105}-Phe_{106} (Fig. 17.3). The κ-casein cDNA (22) was cloned into the replicative form of bacteriophage M13mp18. Single-stranded DNA was prepared by denaturing the double-stranded replicative form of bacteriophage M13 mp18 containing the cDNA of interest (46). An oligonucleotide was synthesized to induce the change of Met_{106} of κ-casein to Phe_{106}. Mutagenesis was done by forming a gapped duplex (Fig. 17.3). After polymerization and ligation, the mixture was transformed into competent cells of *E. coli* JM107. The ^{32}P-radiolabelled oligonucleotide was used to screen for the putative mutant which was then screened to isolate the mutant DNA which was subjected to restriction enzyme analysis to confirm mutagenesis. One of the positive mutants from restriction enzyme analysis was subcloned into M13 mp19 for DNA sequencing which confirmed the change of Phe_{105}-Met_{106} to Phe_{105}-Phe_{106}. The next step is to isolate the expressed protein in sufficient quantity and to develop a test that measures the rate of chymosin cleavage of mutant and wild-type κ-casein that has been expressed through *E. coli*.

Preparation of an α_{s1}-Casein Mutant

The work in the foregoing section is designed to increase the rate of enzymatic hydrolysis of a bond which is susceptible to chymosin. The intention of the present study on α_{s1}-casein A is to clone a new chymosin-sensitive site into the protein. We know α_{s1}-casein A does not contain the chymosin-sensitive site Phe_{23}-Phe_{24} present in α_{s1}-casein B that has been implicated in cheddar cheese ripening. In selecting the location to introduce a new chymosin cleavage site in the cDNA for bovine α_{s1}-casein A, a number of factors were considered. If the site was introduced into the middle of the protein this would result in the greatest reduction in size of the protein with a single cut. However, since α_{s1}-casein is actually hydrolyzed at a number of sites, the decision included the effects of multiple cuts as well as the effect of flanking residues on the accessibility of the new site. The most susceptible bond in α_{s1}-casein B, which is absent in the A variant due to a deletion, is Phe_{23}-Phe_{24} within the domain Leu-Leu-Arg-Phe-Phe-Val-Ala-Pro between residues 20–27. Given this information it seemed reasonable to assume that the best site would be flanked by a high number of hydrophobic residues. In order to evaluate the results easily it is also desirable that a significant change in the peptide fragment size occur when the new site is hydrolyzed. Finally, consideration had to be given as to how extensive the base change(s) would be. Therefore, we elected to change Ile_{98}-Val_{99} to Phe-Phe by changing the DNA sequence ATT GTT to TTC TTT because a Phe-Phe bond, as already indicated, is highly susceptible to chymosin hydrolysis. Due to degeneracy of the genetic code, either of two Phe codons—TTC or TTT—could be used. Based on the usage of these codons in α_{s1} (both occur 4 times within the gene), there is no apparent codon bias for expression in the bovine. However, if TTC were used for the first Phe and the preceding residue was Glu, it would result in introducing a new *EcoRI* restriction site which could be used to screen for the mutation. *EcoRI* recognizes the sequence GAATTC which codes for Glu-Phe. If no consideration was given for flanking regions, there are eleven possible sites that could be used. However, many of these sites are already near existing chymosin cleavage sites, near serine residues that are phosphorylated and are likely to inhibit chymosin action, require significant base changes, or include flanking residues with minimal hydrophobicity.

The Ile_{98}-Val_{99} bond occurs within a 17 amino acid peptide, assuming complete hydrolysis of all existing sites, and could be converted to Phe-Phe using both codons. If cleaved, the 17 residue peptide would be reduced to peptides of 7 and 10 amino acids. This mutation required changing only three bases and generated a convenient *EcoRI* restriction site for screening purposes. This mutant has been obtained and expression of the wild type and mutant clones in *E. coli* is necessary in order to assess the effects of the alteration on chymosin hydrolysis. As with the κ-casein mutants, sufficient protein is needed to carry out a functional chymosin sensitivity test on the expressed proteins.

Additional Goals

The long-term future of using engineered mutants rather than those that arise by chance in food products will depend on their acceptability. Their place in research into protein interactions will be far-reaching, particularly when cassette mutagenesis becomes a commonplace laboratory technique. In structure-function studies of milk proteins, the globular proteins, α-lactalbumin and β-lactoglobulin, have the benefit of having defined three-dimensional structures so that molecular modeling can be used to predict the likely effects of making changes to the protein sequences. The role of the calcium and zinc binding sites of α-lactalbumin and the hydrophobic binding site of β-lactoglobulin can be explored, elucidated and exploited. Although the three-dimensional structures of the caseins are not well understood, genetic modifications will give a greater understanding of their functionality in food systems. An important step will be the development of functionality tests that can adequately assess milligram samples of protein since these are quantities of engineered protein that can be conveniently isolated from recombinant microbial systems.

Acknowledgments

This review was made possible by support from the Peter J. Shields Endowment Fund, Dairy Research Foundation, the National Dairy Promotion and Research Board, the New Zealand Dairy Research Institute, the Fulbright Foundation and the College of Agricultural and Environmental Sciences, University of California, Davis, CA 95616.

References

1. Kato, A., and S. Nakai, *Biochim. Biophys. Acta 642*:13 (1980).
2. Nakai, S., *J. Agric. Food Chem 31*:676 (1983).
3. Yada, Y.R., R.L. Jackman, and S. Nakai, *Int. J. Peptide Protein Res. 31*:98 (1988).
4. Hammer, R.E., V.G. Pursel, C.E. Rexroad, Jr., R.J. Wall, D.J. Bolt, K.M. Ebert, R.D. Palmiter, and R.L. Brinster, *Nature 315*:680 (1985).
5. Simons, J.P., M. McLenaghan, and A.J. Clark, *Ibid. 328*:530 (1987).
6. Lee, K.-F., F.J. DeMayo, S.H. Atiee, and J.M. Rosen, *Nucleic Acids Res. 16*:1027 (1988).
7. Bolstein, D., and D. Shortle, *Science 229*:1193 (1985).
8. Zoller, M.J., and M. Smith, *Methods Enzymol. 100*:468 (1983).
9. Craik, C.S., *Bio-Techniques 3*:12 (1985).

10. Swaisgood, H., *Developments of Dairy Science*, edited by P.F. Fox, Applied Sciences, New York, NY, 1982, pp. 1-55.
11. Creamer, L.K., T. Richardson, and D.A.D. Parry, *Arch. Biochem. Biophys. 211*:689 (1981).
12. Nagao, M., M. Maki, R. Sasaki, and H. Chiba, *Agric. Biol. Chem. 48*:1663 (1984).
13. Stewart, A.F., I.M. Willis, and A.G. Mackinlay, *Nucleic Acids Res. 12*:3895 (1984).
14. Gorodetskii, S.I., V.M. Zakar'ev, D.R. Kyarshulite, T.V. Kapelinskay, and K.G. Skyrabin, *Biochemistry (USSR) 31*:1641 (1986). Engl. translation *31*:1402 (1987).
15. Yu-Lee, L., L. Richter-Mann, C.H. Couch, A.F. Stewart, A.G. Makinlay, and J.M. Rosen, *Nucleic Acids Res. 14*:1883 (1986).
16. Nagao, M., Y. Nakagawa, A. Ishii, R. Sasaki, H. Tanaka, and H. Chiba, *Agric. Biol. Chem. 52*:191 (1988).
17. Stewart, A.F., J. Bonsing, C.W. Beattie, F. Shah, I.M. Willis, and A.G. Mackinlay, *Mol. Biol. Evol. 4*:231 (1987).
18. C. Carles, J.-C. Huef, and B. Ribadeua-Dumas, *FEBS Lett. 229*:265 (1988).
19. Jimenez-Flores, R., Y.C. Kang, and T. Richardson, *Biochem. Biophys. Res. Commun. 142*:617 (1987).
20. Ivanov, V.N., D.R. Kershulite, A.A. Bayer, A.A. Akhundova, G.E. Sulimova, E.S. Judinkova, and S.I. Gorodetski, *Gene 32*:381 (1984).
21. Gorodelsky, S.I., T.M. Tkach, and T.V. Kapelinskaya, *Ibid. 66*:87 (1988).
22. Kang, Y.C., and T. Richardson, *J. Dairy Sci. 71*:29 (1988).
23. Sawyer, L., M.Z. Papiz, A.C.T. North, and E.E. Eliopoulos, *Biochem. Soc. Trans. 13*:265 (1985).
24. Papiz, M.Z., L. Sawyer, E.E. Eliopoulos, A.C.T. North, J.B.C. Findlay, R. Sivaprasadarao, T.A. Jones, M.E. Newcomer, and P.J. Kraulis, *Nature 324*:383 (1986).
25. Pervaiz, S., and K. Brew, *Science 228*:335 (1985).
26. Godovac-Zimmerman, J., *TIBS 13*:64 (1988).
27. Jamieson, A.C., M.A. Vandiyar, Y.C. Kang, J.E. Kinsella, and C.A. Batt, *Gene 61*:85 (1987).
28. Smith, S.G., M. Lewis, R. Aschaffenburg, R.E. Fenna, I.A. Wilson, M. Sundaralingam, D.I. Stuart, and D.C. Phillips, *Biochem. J. 242*:353 (1987).
29. Vilotte, J.L., S. Soulier, J.-C. Mercier, P. Gaye, D. Hue-Delahaie, and J.P. Furet, *Biochemie 69*:609 (1987).
30. Gaye, P., D. Hue-Delahaye, J.C. Mercier, S. Soulier, J.-L. Vilotte, and J.P. Furet, *Ibid. 68*:1097 (1986).
31. Kay, J., in *Aspartic Proteinases and Their Inhibitors*, edited by V. Kostka, de Gruyter, New York, NY, 1985, pp. 1-17.
32. Hofmann, T., in *Food Related Enzymes*, edited by J.R. Whitaker, Adv. in Chem. Ser. No. 136, Am. Chem. Soc., Washington, DC, 1974, pp. 146-188.

33. Tang, J., M.N.G. James, I-N. Hsu, I.A. Jenkins, and T.L. Blundell, *Nature* 271:618 (1978).
34. Schechter, I., and A. Berger, *Biochem. Biophys. Res. Commun.* 27:157 (1967).
35. Powers, J.C., A.D. Harley, and D.V. Myers, in *Acid Proteinases—Structure, Function and Biology*, edited by J. Tang, Plenum Press, New York, NY, 1977, pp. 141–157.
36. Pearl, L.H., and T.L. Blundell, *FEBS. Lett.* 174:96 (1984).
37. Fruton, J.S., in *Acid Proteases; Structure, Function, and Biology*, edited by J. Tang, Plenum Press, London, England, 1976, p. 131.
38. Dalgleish, D.G., in *Developments in Dairy Chemistry*, Vol. 1, edited by P.F. Fox, Applied Sci., New York, NY, 1982, pp. 157–188.
39. Bringe, N.A., and J.E. Kinsella, in *Developments in Food Proteins -5*, edited by J.F.B. Hudson, Elsevier Applied Sci., New York, NY, 1986, pp. 159–194.
40. Jolles, P., and A. Henschne, *Trends Biochem. Sci.* 7:325 (1982).
41. Brignon, G., A. Chtourou, and R. Ribadeau-Dumas, *FEBS Lett. 188*:48 (1985).
42. Hill, R.D., *Biochem. Biophys. Res. Commun. 33*:659 (1968).
43. Vonick, I.M., and J.S. Fruton, *Proc. Natl. Acad. Sci. USA* 68:257 (1971).
44. Visser, S., P.J. Van Rooijen, C. Schettenkerk, and K.E.T. Kerling, *Biochem. Biophys. Acta 438*:265 (1976).
45. Visser, S., P.J. Van Rooijen and K.J. Slangen, *Eur. J. Biochem. 108*:415 (1980).
46. Messing, J., *Methods Enzymol. 101*:20 (1983).

Chapter Eighteen

Inactivation and Analysis of Soybean Inhibitors of Digestive Enzymes

Mendel Friedman, Michael R. Gumbmann, David L. Brandon and Anne H. Bates

Western Regional Research Center
Agricultural Research Service
U.S. Department of Agriculture
Albany, CA 94710

Soybean proteins are widely used in human foods in a variety of forms, including infant formulas, flour, soy protein concentrates, soy protein isolates, soy sauces, textured soy fibers and tofu (1-116). The presence of inhibitors of digestive enzymes in soy proteins impairs the nutritional quality and possibly the safety of this important legume. Normal processing, based on the use of heat or fractionation of protein isolates, does not completely inactivate these inhibitors [Table 18.1 (1, 3, 5, 6, 9-12, 17, 18, 23-27, 32, 38, 46, 48, 50, 53, 55, 56, 59-63, 69, 81-85, 90-93, 98, 105, 106, 108, 111, 114)], so that residual amounts of inhibitors are consumed by animals and humans. Efforts to develop new soybean

TABLE 18.1

Effect of Autoclaving on the Protease Inhibitor Content of a Standard Soybean Variety and of a Variety Lacking the Kunitz Inhibitor

Variety and treatment	Trypsin inhibited (units/gram flour)		Chymotrypsin inhibited (units/gram flour)	
Williams 82		*% Left*		*% Left*
Unheated	4800	100	150	100
10 min, 121°C	3067	64	125	83
20 min, 121°C	1225	26	56.7	38
30 min, 121°C	797	17	26.7	18
Low-TI				
Untreated	2500	100	143	100
10 min, 121°C	2083	83	56.7	38
20 min, 121°C	232	9	15.0	10
30 min, 121°C	56.7	2	10.0	6

TABLE 18.2
Protease Inhibitor and Lectin Content of Different Soybean Cultivars

Cultivar	mg chymotrypsin inhibited per gram flour	mg trypsin inhibited per gram flour
Williams 82	4.50	37.1
Low-TI	3.73	16.2

TABLE 18.3
Protease inhibitors of soybeans

Kunitz trypsin inhibitor
    isoforms
    germination forms
Bowman-Birk inhibitors
    isoforms
    germination forms
Glycine-rich trypsin inhibitor
Non-protein inhibitors
    fat
    phytate

cultivars with low inhibitor content have been only partially successful. A soybean variety lacking the Kunitz inhibitor is now available [Table 18.2 (70)], whereas varieties lacking Bowman-Birk inhibitor (BBI) are not.

Inhibitors of digestive enzymes are present not only in legumes, such as soybeans, lima beans and kidney beans, but also in nearly all plant foods, including cereals and potatoes, albeit in much smaller amounts. The antinutritional effects of inhibitors of proteolytic enzymes have been widely studied and can be ameliorated by food processing and also by sulfur amino acid fortification. Of more urgent concern are reports that rats fed diets containing even low levels of soybean-derived inhibitors (which are found in foods such as soy-based infant formulas) may develop pancreatic lesions leading eventually to neoplasia or tumor formation (49,54). On the other hand, recent studies suggest that certain enzyme inhibitors from plant foods may prevent cancer formation in other tissues (115A, 115B). A key question, therefore, is whether inhibitors from plant foods constitute a human health hazard.

If inhibitors of proteolytic enzymes present in plant foods are indeed a health hazard, a need exists to devise processing conditions that inactivate or remove these hazards or, in the case of soybeans, to develop new varieties that lack the inhibitors. Processing conditions to inactivate inhibitors are based largely on heat treatments. However, heat does not inactivate all of the inhibitors present. The possible adverse effects of residual inhibitors in food are largely unknown. In addition, conditions used to inactivate inhibitors may destroy important amino acids including sulfur amino acids and lysine. A related unsolved problem is that conventional enzyme assays of inhibitors often give inaccurate measures of activity and do not specifically differentiate among the major types of inhibitors, Kunitz and Bowman-Birk and their respective isoforms [Table 18.3 (21,64,71-73,77,98,103,107,109,109A,109B)].

In this paper we discuss our studies on the following aspects of soybean inhibitors: (a) Effects of heat on disulfide-interchange in the inactivation of inhibitors; (b) nutritional benefits of the inactivation; and (c) monoclonal antibody-based immunoassays (ELISA) for specific inhibitors. The cited references (1-116) offer a more comprehensive coverage of agronomic, analytical, biochemical, chemical, food processing, medical, nutritional and toxicological aspects of soybean inhibitors of trypsin and chymotrypsin.

Experimental Section
Materials.

Soy flour was a gift from Central Soya, Ft. Wayne, IN. Soybean cultivars were a gift from Prof. T. Hymowitz, Department of Agronomy, University of Illinois, Urbana. The cultivars were milled in a Wiley mill and the flours were then defatted by extraction with hexane (4). Raw soy protein isolate was obtained from Kraft, Glenview, IL. Soybean Bowman-Birk inhibitor (BBI), prepared by affinity chromatography was obtained from Prof. Y. Birk, Faculty of Agriculture, Hebrew University, Rehovot, Israel (8,9). Soybean Kunitz trypsin inhibitor (KTI), L-cysteine, N-acetyl-L-cysteine (NAC), and all other reagents including N-benzoyl-DL-arginine-Ω-p-nitroanilide (BAPNA), benzoyl-L-tyrosine ethyl ester (BTEE), 5,5'-dithiobis(2-nitrobenzoic acid) (DTBA), ethylenediaminetetraacetic acid (EDTA), 3-(N-morpholino propanesulfonic acid (MOPS), N,N-bis(2-hydroxyethyl)glycine (BICINE) were obtained from Sigma Chemical Company, St. Louis, MO.

Trypsin inhibitor inactivation.

Soy flour (700 g) and 0.128 moles of cysteine were suspended in 2.5 liters of prewarmed 0.5 M Tris, pH 8.5 buffer. The mixture was then heated in a water bath at 45°C, 65°C, or 75°C for one hour with occasional stirring. The contents were then dialyzed in a cold room for three days with frequent changes of deionized water and lyophilized. Control samples were prepared without cysteine or *N*-acetylcysteine.

Trypsin inhibitor assay.

Many units are used to define the inhibitor activity of protease inhibitors (21, 64, 71-73,77, 98, 103, 107). The following method, based on the defined stoichiometry of enzyme, inhibitor and substrate was adapted from Mikola and Suolinna (87). It is based on titrating the residual trypsin after complexing the inhibitor with a known excess of trypsin.

Inhibition of trypsin activity was measured at pH 8.2 and 37°C, with BAPNA as substrate (44-46, 87). Specifically, 100 mg of soy flour was shaken for one hour at room temperature in 15 ml 0.5 M Tris-HCl buffer, pH 8.5. The material was then centrifuged and 0.5 ml of the supernatant was diluted with 4.5 ml 0.05 ml M Tris - 0.02 M $CaCl_2$ (1:10), pH 8.2. A trypsin solution was prepared by dissolving 2.5 mg of twice-crystallized bovine trypsin in 25 ml 1 mM HCl. A substrate solution was prepared by dissolving 30 mg of BAPNA in 1 ml dimethyl sulfoxide (DMSO) and diluting to 100 ml with 0.05 M Tris-HCl buffer, pH 9.2, containing 0.02 M $CaCl_2$. Next, 0.25 ml of the enzyme solution was added to 0.5 ml of the soy extract solution, the control and a blank. The three solutions were then incubated for 5 min. Then 3 ml of the substrate solution was added to the extraction solution, the control and the blank, to start the reaction. The reaction was stopped after either 5 or 10 min by adding 0.5 ml 30% acetic acid.

Determination of active trypsin.

The concentration of active trypsin was determined by active site titration with *p*-nitrophenyl guanidinobenzoate (NPGB), essentially according to Friedman *et al.* (46). Absorbance of free *p*-nitrophenol was followed in a Cary 15 spectrophotometer at 410 nm with the 0.1 absorbance slidewire. NPGB (0.01 M) was dissolved in dimethylformamide-acetonitrile (1:4). Trypsin was dissolved in 2 mM HCl/0.02 M $CaCl_2$ at 5-6 mg/ml, and then diluted (1:10) in the same solvent. Each cuvette contained 0.05-0.09 M Na barbital buffer (pH 8.35 or 8.39 at

23°C), 0.02 M $CaCl_2$, 0.1 M KCl. Trypsin (3-5 nmol of active enzyme) was added to the sample cuvette as 0.20 ml of the 1:10 dilution of 25 μL of the stock solution. After the baseline was determined, 5 μL of NPGB was added to each cuvette. Total volume was 1.0 ml. In our instrument, 1 A_{410}=61.1 mol of p-nitrophenolate. The trypsin used was 61% active enzyme, based on its protein content as determined by absorbance at 280 nm and an active site titration.

Calculations.

The following sample calculation was used to estimate the trypsin inhibitor activity per milligram of active enzyme, where

A_{410} = absorbance measured at 410 nm
ϵ_M = molar extinction coefficient
V_o = rate of BAPNA hydrolysis by trypsin in the absence of inhibitor.
V_i = rate of BAPNA hydrolysis by trypsin in the presence of inhibitor.
TU = trypsin unit, defined as the amount of trypsin that catalyzes the hydrolysis of 1 μmol of BAPNA per min at 37°C and pH 8.2.
TIU = trypsin inhibitor unit, defined as reduction in activity of trypsin by one trypsin unit (TU).

Using an ϵ_M for p-nitroaniline of 8800 L/mol-cm (R), and a total volume of 4.25 ml, 1 TU corresponds to an increase in A_{410} of 2.07 min^{-1}.

We found trypsin activity of 2.51 TU per mg of active enzyme (see sample calculation). Using molecular weights of 23,400 for trypsin and 21,500 for soybean trypsin inhibitor (STI), we calculated 2.92 TIU per milligram of STI.

For trypsin only, 0.358 A_{410}/5 min; Rate, V_o = 0.716/min; for trypsin and inhibitor, 0.389 A_{410}/10 min.

The difference in the rate of BAPNA hydrolysis by trypsin in the absence and presence of inhibitor is given by:

$$V_o - V_i = \frac{A_{410} \text{ (no inhibitor)}}{t_o} - \frac{A_{410} \text{ (inhibitor)}}{t_i} = \frac{0.358}{5} - \frac{0.389}{10}$$

$$= 0.0716 - 0.0389 = 0.0327 \; A_{410} \; min^{-1}$$

Since 2.07 $A_{410} min^{-1}$ is equivalent to 1 trypsin unit (TU), the amount of inhibitor present is:

$$\frac{0.0327}{2.07} = 0.0158 \text{ TIU}$$

For pure KTI, the inhibitor content of the samples used is:

$$\frac{0.0158 \text{ TIU}}{2.92 \text{ TIU/mg KTI}} = 5.41 \times 10^{-3} \text{ mg KTI}$$

Since 1 mg of active trypsin = 2.51 TU, 0.0158 TIU represent

$$\frac{0.0158 \text{ TU}}{2.51 \text{ TU/mg trypsin}} = 6.29 \times 10^{-3} \text{ mg}$$

active trypsin inhibited in the pure soybean trypsin inhibitor sample.

If inhibitors other than KTI are also present (as is apparently the case in most plant foods), a more precise estimate is:

$$\frac{\text{TIU in sample}}{\text{wt. of sample}} = \text{TIU/mg (or g)}$$

The following average values demonstrate the reproducibility of the assay. Pure commercial KTI inhibited an average of 1.23 ± 0.03 mg of active trypsin per milligram KTI (eight replicates). This value is higher than greater than one, the expected value of 1.08 for the trypsin—KTI complex, because of possible minor contributions of lower molecular weight compounds such as phytate to the total enzyme inhibitory process. Untreated soy flour (the average for four determinations) contained 109.51 ± 1.52 TIU and inhibited 43.63 ± 0.605 mg trypsin per gram.

Chymotrypsin inhibition assay.

Inhibition of chymotrypsin was determined titrimetrically, with 0.01 M BTEE as substrate. The procedure was adapted from Friedman *et al.* (46). A chymotrypsin solution was prepared by dissolving 1.4 mg of chymotrypsin in 2 ml 1 mM HCl. A substrate solution was prepared by dissolving 25 mg of BTEE in 0.5 ml 100% ethyl alcohol. The soy solution (0.5 ml) was incubated in a Radiometer pH Stat with chymotrypsin (10 µL of 0.7 mg/ml) solution and then equilibrated for 10 min at 25°C in 1.1 ml of H_2O. After the baseline uptake of titrant (0.1 N NaOH) was recorded, 100 µL of 0.2 M BTEE was injected into the solution. Initial reaction rates were determined from the linear portion of the plot of base uptake against time. The chart speed was 3 cm/min. The nonenzymatic breakdown of BTEE and the effect of order of adding BTEE and chymotrypsin (no inhibitor) were also measured and used to correct assays of inhibition.

Immunoassays.

Immunoassays for KTI were conducted as previously described (13-15) using polystyrene plates (Costar, Inc., Cambridge, MA). The enzyme-linked immunosorbent assay (ELISA) was performed using plates coated (either 4 hr at room temperature or overnight at 4°C) with 100 µl/well of purified IgG anti-KTI at 10 µg/ml. Plates were then washed with PBS-Tween, rinsed with distilled water, and blocked by incubation with PBS-Tween + 1% BSA (200 µl/well, 1 hr at room temperature) then washed and rinsed again. Unknowns or standard samples containing KTI were premixed with equal volumes of KTI-HRP and incubated in the assay wells (100 µl/well, 1-2 hr at room temperature) with shaking. Pilot experiments indicated that the binding reaction was complete within one hour, with no change in binding between one and two hours. Plates were washed and rinsed again, and then bound KTI-HRP was assayed using the chromogenic substrate 2,2' - azinobis (3 - ethylbenzthiazoline sulfonic acid).

Monoclonal antibodies were derived from Balb/c mice as previously described (ALB), using the "NS-1" cell line or a variant, P3-X63-Ag8-653, which does not produce a light chain. Antibodies were characterized with regard to specificity and isotype. All of the monoclonal antibodies described in this report are $IgG_1(K)$. Antibodies were purified by ammonium sulfate fractionation and ion exchange chromatography from ascites fluids or sera. The recent development of an immunoassay for BBI is described in reference 16.

Proteolytic digestion assays.

The extent of peptide cleavage by trypsin was estimated on a pH-Stat (46,67,102). Samples were dissolved or dispersed (5 mg/ml) in 0.05 M KCl with stirring and adjusted to pH 8.5. To determine digestibility by trypsin, aliquots (5 mg of protein) were pipetted into plastic titration vessels (Radiometer V 524) and diluted to 4.0 ml with 0.05 M KCl. Samples were incubated under N_2 to minimize CO_2 absorption. In a typical run, the substrate was brought to pH 8.5 at 37°C and the baseline rate of uptake of titrant (0.02 N NaOH) was determined. Then, 20 µL of bovine trypsin (5 mg/ml in 2 mM HCl-0.02 M $CaCl_2$) was added. More was added when the uptake of alkali became very slow. The net uptake was determined when the reaction was essentially complete.

Since liberation of hydrogen ions by proteolysis is incomplete at pH values where some new α-amino groups are ionized, calculation of the number of peptide bonds cleaved requires a value for the average pK of the newly formed peptide α-amino groups. To calculate the actual average number of peptide bonds cleaved per chain corrected for incomplete liberation of hydrogen ions, we also assumed an average molecular weight of 23,000 for the soy protein.

The theoretical rationale for measuring the extent of peptide bond cleavages is based on potentiometric titration of hydrogen ions liberated when peptide bonds are hydrolyzed (Equation A).

$$\text{P-CO-NH-P'} \xrightarrow[H_2O]{\text{Trypsin}} \text{P-COO}^- + (H_3N^+\text{-P} \rightleftharpoons H^+ + H_2N\text{-P'}) \qquad (A)$$

P, P' = Protein chains.

Above pH 6.5, the number of hydrogen ions liberated is equal to the number of dissociated α-amino groups produced. Since the average pK of these new amino groups is near 7.5, about 90% of the hydrolyzed peptide bonds will produce titratable hydrogen ions at the pH (8.5) used in the present study. Since pH is kept constant during the titration, the degree of ionization of lysine ε-amino groups (pK near 10) does not change. Similarly, the slight contribution of hydrogen ions from the single terminal α-amino group can also be neglected.

The relation between the number of peptide bonds cleaved and hydrogen ions produced is given by Equation B.

$$pH - pK = \log \frac{NH_2}{NH_3^+} = \log \frac{H^+}{NH_3^+} \qquad (B)$$

The number of enzyme-catalyzed peptide bonds cleaved per unit (grams or moles) of protein is given by Equation C.

$$N = H^+ \left[1 + 10^{(pK - pH)}\right] \qquad (C)$$

where N = number of peptide bonds cleaved per unit of protein.
H^+ = number of hydrogen ions liberated and titrated during the enzymatic hydrolysis (Equation A).
pH = pH value at which the enzymatic digestion was carried out.
Thus, at pH 7.5, cleavage of two peptide bonds liberates one hydrogen ion since N = $H^+(1 + 1)$. At pH 8.0 and 8.5, the corresponding values are 1.32 and 1.1, respectively.

Half-cystine and methionine content.

The half-cystine (cysteine plus cystine) and methionine content of treated and untreated materials was determined on an amino acid analyzer as cysteic acid and methionine sulfone, respectively, after performic acid oxidation of dialyzed and lyophilized samples (41).

SH content.

The cysteine (SH) content was determined by a modified Ellman procedure (33,104) as follows: One gram soy flour in 9 ml 0.1M phosphate buffer (pH 8.0) and 1 ml (0.02 M) solution ethylenediamine tetraacetic acid (EDTA), were shaken together in a 37°C water bath for 10 min. The mixtures were then centrifuged for 30 min at 8,000 rpm. Next, 0.5 ml samples of the supernatant were mixed with 1 ml 0.02M EDTA, 1 ml 0.1M phosphate buffer (pH 8), and 2.5 ml deionized H_2O. Next, 25 µL 0.01M 5,5'-dithiobis (2-nitrobenzoic acid) dissolved in 0.1M phosphate buffer (pH 7.0) was added to each supernatant. The solution was left standing for 20 min for maximum color development. The absorption maximum of the solution at 412 nm was then determined with a Cary Model 14 spectrophotometer against a reagent blank. A cysteine standard curve in the range 0.01 to 0.1 µmole was prepared and the SH content was calculated by means of the equation:

$$\text{SH content (mg/g soy flour)} = \frac{\text{Absorbance of sample} \times \text{Dilution factor} \times 120}{13{,}600}$$

where 120 represents the molecular weight of L-cysteine, and 13,600 the molar extinction coefficient of DTNB.

Protein analysis.

The percent protein (% nitrogen × 6.25) was calculated from the nitrogen values determined by an automated micro-Kjeldahl procedure (4).

Protein efficiency ratios (PER).

Protein efficiency ratios were determined by the official AOAC method (4) using male, weanling Sprague-Dawley rats from Simonsen Laboratories, Inc., Gilroy, CA. Each assay ran 28 days. The composition of diets was: protein, 10%; corn oil, 8%; water, 5%; cellulose (Alphacel), 3%; complete vitamin mixture, 2%; mineral mixture, 5%; glucose, 20%; and corn starch to make 100%. Nitrogen digestibility was based on feed intake and fecal collections from day 14-21. The following nutritional parameters were evaluated:

PER: weight gain/protein intake.
Apparent Nitrogen Digestibility: (N intake - fecal N)/N intake × 100.

Mean body weight, feed consumption and PER were analyzed statistically by Duncan's Multiple Range Test (30).

Results and Discussion

Inactivation of enzyme inhibitors by thiol amino acids.

The effect of pH on inactivation in the presence and absence of NAC at 45°C is shown in Table 18.4. Inactivation by heat alone is greatest in acid (pH 3.5) and alkali (pH 9.5 and 11.0), and somewhat less in the intermediate range. Adding thiol greatly increases inactivation.

Additional studies designed to assess the practical potential of the method show that the NAC-facilitated inactivation is even more effective when carried out in concentrated suspensions (Table 18.5). The higher the concentration of thiol or inhibitor, the greater the probability for reaction between added thiol and inhibitor disulfide bonds. Shaking further increases the extent of inactivation by thiols (results not shown).

Effect on disulfide content.

Forming mixed disulfides with cysteine should increase the cystine content of the proteins. This is a desirable objective because methionine is a limiting amino acid in legumes and cystine has a sparing effect on methionine. This expectation was realized, since our studies show that

TABLE 18.4
Effect of pH and N-Acetyl-L-Cysteine (NAC) on Trypsin-Inhibitory Activity of Soy Flour at 45°C[a]

pH	Percent inhibiting activity remaining	
	— NAC	+ NAC
3.5	45.1	25.4
4.4	68.2	17.2
5.3	78.6	29.2
6.3	80.1	9.2
7.4	66.5	13.6
8.5	68.8	8.1
9.5	38.7	7.8
10.0	23.6	0.0
11.0	9.8	1.7

[a](44, 46).

TABLE 18.5
Enhancement of Inactivation of Trypsin Inhibitors in Legumes with Increasing Concentration of N-Acetyl-L-Cysteine (NAC)[a]

NAC (g)	% Inhibitory activity remaining		
	Lima bean	Soybean	Great Northern bean
0	75.6	44.2	50.0
1	25.7	11.4	15.1
2	19.6	4.6	8.5
3	13.0	0.0	2.5
4	13.7	0.0	0.0
5	10.3	0.0	0.0

[a]Conditions: 100 g flour in 500 ml (slurry); Tris pH 8.5 buffer; 65°C; 1 hr.

the half-cystine contents of treated commercial soy flours are significantly higher than the corresponding values for the starting materials (Table 18.6).

The action of thiols such as cysteine in inhibiting the activity of disulfide-containing enzyme inhibitors is postulated to involve formation of mixed disulfides among added thiols, enzyme inhibitors and structural proteins. Since oxygen was not excluded, formation of new disulfide bonds probably proceeded by sulfhydryl-disulfide interchange

TABLE 18.6
Half-Cystine Content of Treated and Untreated Materials

Material	Treatment	Half-cystine (nmoles/mg)[a]
Soy flour	dialyzed in H_2O	67.0
Soy flour	700 g + 21 g NAC; pH 8.5	139.0

[a]Determined on an amino acid analyzer as cysteic acid after performic acid oxidation and acid hydrolysis.

(equations a-e in Scheme A) and oxygen-catalyzed oxidation of two SH groups to an S-S bond. One of the sulfurs of the mixed disulfide originated from the protein and the other from added sulfhydryl compound. The added compound, therefore, disappears and becomes part of the protein structure. Because of structural alterations due to formation of mixed disulfide, the modified inhibitors lose their ability to complex with the active sites of trypsin and other proteolytic enzymes.

SCHEME A

Sulfhydryl-Disulfide and Oxidation Interchange Pathways. Net Effects: Network of New Disulfide Bonds and Altered Protein Configuration

	R-SH + In-S-S-In	$\rightleftarrows$	R-S-S-In + HS-In	(a)
	R-SH + Pr-S-S-Pr	$\rightleftarrows$	R-S-S-Pr + HS-Pr	(b)
	In-SH + Pr-S-S-Pr	$\rightleftarrows$	In-S-S-Pr + HS-Pr	(c)
	In-SH + HS-Pr + ½ O_2	$\rightarrow$	In-S-S-Pr + H_2O	(d)
	In-SH + HS-In + ½ O_2	$\rightarrow$	In-S-S-In + H_2O	(e)
R-SH:	Added thiol (Cysteine, N-acetyl-L-cysteine etc)			
In-S-S-In:	Inhibitor (In) disulfide bonds			
Pr-S-S-Pr:	Protein (Pr) disulfide bonds			

Although equations a-c in Scheme A are, in principle, reversible, statistical considerations (35) suggest a low probability for the regeneration of the native inhibitor configuration from the complex new network of mixed disulfide bonds. However, formation of mixed disulfide bonds may not occur with inhibitors with few or no disulfide bonds.

Additional evidence for the postulated reaction scheme was obtained by measuring the cysteine (SH) content of soybean flours treated by the Ellman procedure (33,104). The results in Table 18.7 show that cysteine

TABLE 18.7
Trypsin Inhibitor, SH, and Half-Cystine Content of Treated Soy Flours

Protein source	mg trypsin inhibited/g soy flour	SH content (nanomoles/ mg)	Half-cystine (nanomoles/ mg)
Native soy flour	37.5	3.85	71.6
Soy flour, heated at 45°C	15.83	1.29	73.1
Soy flour, plus cysteine, heated at 45°C	6.24	1.67	108.9
Soy flour plus acetyl-L-cysteine, heated at 45°C	4.76	1.32	147.5
Soy flour, heated at 65°C	8.18	1.12	59.1
Soy flour plus cysteine, heated at 65°C	2.87	0.89	86.2
Soy flour plus acetyl-cysteine, heated at 65°C	1.48	0.85	86.2
Soy flour, heated at 75°C	5.84	0.48	63.5
Soy flour plus cysteine, heated at 75°C	2.23	0.52	133.9

treatment did not increase the SH content, as would be expected if the added thiols had not participated in the indicated sulfhydryl-disulfide interchanges, but that the cysteine (SH) content of treated materials was also less than the value found for native soybean flour. This result, together with the observed increase in the total half-cystine content of the products determined by amino acid analysis, implies that the added cysteine or acetylcysteine (a) participated in the indicated sulfhydryl-disulfide interchange and oxidation reactions and (b) became covalently attached to soybean protein chains in the form of mixed disulfides produced by SH and S-S groups from added thiol, enzyme inhibitors and structural protein chains, as illustrated in Scheme A.

Extent of reduction of disulfide bonds.

From both chemical and nutritional viewpoints, inactivation of a protease inhibitor by an added thiol via a sulfhydryl-disulfide interchange should be distinguished from the related reductive cleavage of disulfide bonds followed by alkylation of the generated SH groups. For example, 2-vinylpyridine reacts to produce pyridylethylcysteine derivatives which can be determined by amino acid analysis or by ultraviolet spectroscopy [Table 18.8 (42)]. The former treatment produces new disulfide bonds. The latter eliminates the original ones from the protein (29,42,116). As mentioned earlier, possibilities for sulfhydryl-disulfide interchange include interaction between inhibitor SH or S-S bonds with soybean protein SH and S-S groups, respectively, to form a complex network that may make it difficult for the original disulfide bonds to be reformed. The relative concentration of SH and S-S bonds in the thiol and protein may dictate whether the thiol is considered a catalyst or stoichiometric reagent in the inactivation. For these reasons, optimum conditions for the inactivation of disulfide-containing protease inhibitors by thiols have to be established in each case. Moreover, as noted above the introduction of new half-cystine residues into legume proteins is an additional, desirable benefit because cystine has a sparing effect on methionine (52).

Effect on digestibility and nutritional quality.

Besides inactivating trypsin inhibitors, modifying disulfide bonds of inhibitors and associated structural proteins should also improve the digestibility and nutritional quality of inhibitors since disulfide cleavage is expected to decrease protein crosslinking. The mixed-disulfide proteins should be more accessible to proteolytic enzymes, thus permitting more rapid digestion.

Measurement of *in vitro* digestibilities of NAC-treated soy flour by trypsin (Table 18.9) showed that although NAC had no effect at 25°C, its beneficial influence at higher temperatures was striking. Treatment with NAC at 45°C increases the bonds cleaved by trypsin per mole of soy protein from about 6-10; treatment at 75°C, from about 6-18. This last value is nearly equal to all the available trypsin-susceptible arginine and lysine residues in soy proteins.

Table 18.10 shows the PER and *in vivo* digestibility results for soy flour heated at 45°C, 65°C, and 75°C with and without cysteine or *N*-acetyl-L-cysteine. At each temperature, apparent attachment of cysteine or *N*-acetyl-L-cysteine resulted in a higher PER than in the corresponding untreated control. As would be expected, increasing temperature by

TABLE 18.8
2-PEC (Half-Cystine) Contents and Inhibitory Activities of Reduced Alkylated Soybean, Lima Bean and Ovomucoid Trypsin Inhibitors

Protein	Time of reduction (min)	Moles half-cystine/ mole protein	S-S bonds cleaved (%)	Percentage of initial specific inhibitory activity against	
				Trypsin	Chymotrypsin
2-PEC-STI	30	4.0	100	0	
	60	4.6	100	0	
	120	4.4	100	0	
	240		100	0	
2-PE-LBI	30, No urea		14	32	19
	30, 8 M urea		40	10	15
	240, 8 M urea	7.4	51	0	0
2-PE-OMI	240, 8 M urea	15.3	81	0	

TABLE 18.9
Effect of Temperature and N-Acetyl-L-Cysteine (NAC) on Trypsin Inhibitor Activity and Trypsin Digestion of Dialyzed Soy Flour

Temperature (°C)	Percent inhibitory activity remaining		Number of peptide bonds cleaved by trypsin/mole of protein[a]	
	− NAC	+ NAC	− NAC	+ NAC
25	98.7	84.4	5.6	4.9
45	—	47.2	—	10.6
75	53.2	0.0	6.4	18.2

[a]Based on calculated protein content (N x 6.25) and an average molecular weight per protein chain of 23,000.

TABLE 18.10
Effect of Soy Flour Treated with L-Cysteine and N-Acetyl-L-Cysteine on PER and Nitrogen Digestibility

Soy flour treatment		Protein efficiency ratio (PER)	Nitrogen digestibility (%)
Reaction temperature	Amino acid		
Experiment 1			
45°C	none	0.95	73.7
	Cysteine	2.01	81.7
	Acetyl-cysteine	2.20	82.7
65°C	none	1.61	79.9
	Cysteine	2.43	82.9
	Acetyl-cysteine	2.02	77.0
Casein control		3.05	93.3
Experiment 2			
75°C	none	2.14	79.0
	Cysteine	2.53	82.7
Casein control		3.19	94.6
Pooled standard error		±.08	±0.7

itself also tended to improve PER, but much less than in protein treated by thiols.

TABLE 18.11
Trypsin Inhibitor and Sulfite Content of Raw, Heated and Sulfite-Treated Soy Flour

Material	TIU/g	Sulfite (%)
Soy flour, raw	50.43	0.0
Soy flour, heated 75°C	9.72	0.0
Soy flour, sulfite-treated[a]	0.0	0.0

[a]700 g soy flour plus 0.03 moles (3.78 g) sodium sulfite, 75°C.

TABLE 18.12
PER and Digestibility of Raw, Heated and Sulfite-Treated Soy Flour

| Diet | PER | Digestibility | | Pancreas weight |
| | | diet | nitrogen | |
		(%)	(%)	(% body weight)
Raw soy flour	1.55	90.9	78.3	0.51
Raw soy flour (dialyzed)	1.65	90.9	79.2	0.56
Heated soy flour	2.11	91.3	81.9	0.43
Heated soy flour (+0.03M sulfite)	2.49	91.7	84.0	0.40
ANRC casein	3.44	94.4	93.0	0.40
S.E.	±0.06	±0.4	±0.7	±0.02

Nitrogen digestibility generally paralleled PER (with the possible exception of N-acetyl-L-cysteine treatment at 65°C). Thus, incorporation of cysteine residues into soy flour protein improves both PER and nitrogen digestibility.

Sulfite ions can, in principle, also cleave protein disulfide bonds to form a thiol anion (P—S⁻) and a S-sulfocysteine derivative (P-S-SO$_3^-$) (20), as illustrated by the following equations:

$$\text{P-S-S-P'} + \text{SO}_3^= \rightleftharpoons \text{P-S-} + \text{P'-S-SO}_3^=$$
$$\text{P'-S-SO}_3^= + \text{P-S-} \rightleftharpoons \text{P'-S-S-P} + \text{SO}_3^= \text{ (regenerated)}$$

The S-sulfocysteine can interact further with the generated P—S⁻ to form a new disulfide bond and SO$_3^=$. The net effect is a rearrangement of protein disulfide bonds catalyzed by $SO_3^=$.

Thus, exposure of disulfide-containing trypsin inhibitors to sulfite ions should alter their structure and their inhibitory properties in analogy with the cysteine treatments described earlier. This hypothesis was tested in a preliminary experiment by exposing soy flour to 0.03 M sodium sulfite and feeding the dialyzed and lyophilized materials to rats.

Table 18.11 shows that the sulfite treatment at 75°C lowers the trypsin inhibitor level in soy flour to zero. This value is usually difficult to achieve even at high temperatures without marked deleterious effects on protein quality (101). These findings suggest that the sulfite treatment is highly effective in inactivating trypsin inhibitors and improving the nutritional quality of soy flour.

Table 18.11 also shows that after dialysis, the sodium-sulfite-treated soy flour had no measurable amounts of sodium sulfite. This finding may alleviate concern about the use of sodium sulfite to improve the nutritional quality and safety of foods containing enzyme inhibitors. Precipitating soy proteins at their isoelectric point near pH 4.5 may be a more practical way of separating the proteins from sodium sulfite.

Table 18.12 shows that heat significantly improved protein quality of raw soy flour. Treatment with sodium sulfite further increased the protein efficiency ratio (PER). This improvement in protein quality is also reflected in an increase in *in vivo* nitrogen digestibilities. Table 18.12 shows that pancreas weights were elevated in rats fed raw soy flour, but not in those fed heated soy flour, with or without sodium sulfite.

Sulfur amino acids also may play an important role in the feedback mechanism that has been proposed to explain the adverse effects of trypsin inhibitors on the pancreas (39,49). According to this hypothesis, complexation in the intestinal tract between proteolytic enzymes (trypsin, chymotrypsin) and enzyme inhibitors (Kunitz- and Bowman-Birk-type) creates a deficiency of proteolytic enzymes. This deficiency triggers an endocrine (cholecystokinin, gastrin) sensing mechanism which in turn induces increased protein synthesis in the pancreas. In this chain of events, undenatured trypsin inhibitors and proteolytic enzymes secreted by the pancreas, which are both rich in sulfur amino acids, are lost as tightly bound enzyme-inhibitor complex. This loss of sulfur amino acids is predicted to be an important modulating factor for the pancreas which is now under hormonal stimulation to increase synthesis of protein especially rich in sulfur amino acids.

Gumbmann and Friedman (52) found that diets containing raw soy protein isolate fortified with L-methionine or L-cystine resulted in significantly heavier rat pancreata than control diets without added sulfur amino acids. This result may reflect a preferential ability of the pancreas to mobilize needed sulfur amino acids for pancreatic growth (and synthesis of sulfur amino acid-containing enzymes such as trypsin and chymotrypsin) over other body organs. These observations suggest the need for additional studies to clarify the role of sulfur amino acids in pancreatic physiology and histology as related to trypsin inhibitors in the diet and their mechanism of action.

Immunoassays

The protease inhibitory activity of soy protein is largely inactivated by conventionally applied heat treatments of soy flour, but, as mentioned earlier, 10-15% of the residual activity remains. Since soybeans contain two major classes of protease inhibitors, Kunitz trypsin inhibitor and the double-headed Bowman-Birk trypsin and chymotrypsin inhibitor, and since additional inhibitors and isoforms may be present which are derived from different genes or produced by proteolysis, it is impossible to establish the exact protease inhibitor composition of a sample through enzymatic assays. Moreover, enzyme assays often give inaccurate results with processed samples having low residual activity.

Although immunochemical methods for estimating KTI were reported nearly two decades ago (19) they have not been widely utilized for food analysis. These methods relied on precipitin analysis, which is insensitive, or binding of complement, which is an unstable mixture of serum proteins. In contrast, ELISA seemed an appropriate choice for assaying inhibitors in processed food samples because of the potential sensitivity and specificity of the assay, combined with the ease of sample preparation and stability of reagents. Our objective was the development of immunoassays for research and quality control applications, which would permit measurements of residual levels of inhibitors in complex samples such as infant formulas and other prepared foods.

In one ELISA method which we have developed (13-16) solid-phase KTI-specific antibody was used as immunosorbent to coat the plastic assay plate, and KTI-HRP conjugate was used as labeled ligand. The addition of analyte (sample containing KTI) established a competitive equilibrium, with the amount of enzyme-labeled KTI bound to the assay plate inversely proportional to the analyte concentration. The advantage of this ELISA format is that the titration of analyte and enzyme labeling steps are combined, resulting in a very rapid assay. Figure 18.1 illustrates the titration of KTI—peroxidase conjugate on three assay plates. Polyclonal antibody (R276) and monoclonal antibodies 134 and

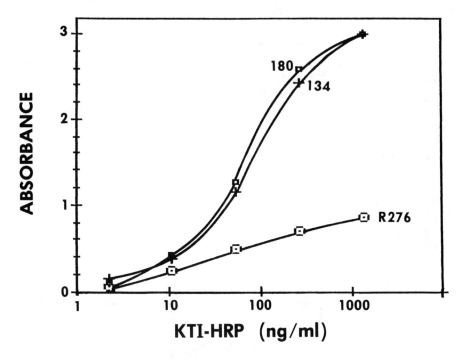

Fig. 18.1. Titration of KTI-peroxidase conjugate on three antibody-coated solid phases: monoclonal antibodies 134 and 180 and polyclonal antibody R276.

180 all produced effective solid phases, but the monoclonals had greater binding capacity and antibody 180 was therefore used in subsequent studies. Antibodies 134 and 180 are representative of antibodies, which are most specific for the native conformation of KTI.

A standard curve for assay of KTI by ELISA is illustrated in Figure 18.2. The working range of the assay is 30 t0 3000 ng/ml. The assay was insensitive to BBI, up to a concentration of 1 mg/ml. The assay plates appear to be critical to the success of this assay. We have noted consistent variation in the evenness of solid phases prepared with mouse monoclonal IgG. We recommend that each lot of assay plates be tested for consistent binding activity prior to use. Standards and controls must be included on each plate. When these precautions are taken, day-to-day variability in the assay is less than 15%, expressed as a coefficient of

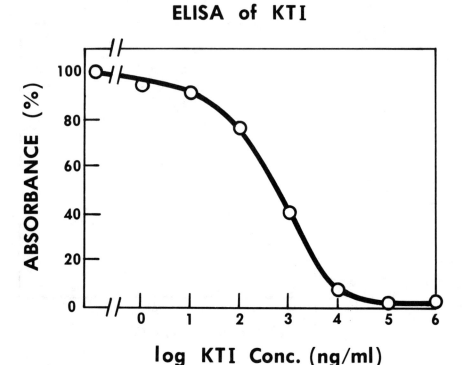

Fig. 18.2. Standard curve for ELISA of KTI with antibody 180.

variation. Certified and/or high-binding capacity plates available from some manufacturers have proven to be worthwhile in further reducing the variabilities of assays. Most important is the sensitivity of the assay, which permits analysis of processed samples having low residual TI content or extracts as little as 30 ng KTI/ml.

With the present assay format, 1 mg of monoclonal antibody can be used for approximately 2000 assays. The concentration of antibody coating the assay plate well could probably be reduced 10-to 100-fold if the monoclonal antibody is attached indirectly through a first layer of polyclonal rabbit anti-mouse IgG. We routinely prepare ascitic fluids which can be used without purification at 10^5-fold dilution for ELISA assays, with a primary antibody coating.

Since soybean products are generally subjected to thermal processing, it was of primary importance to test the validity of the immunoassay

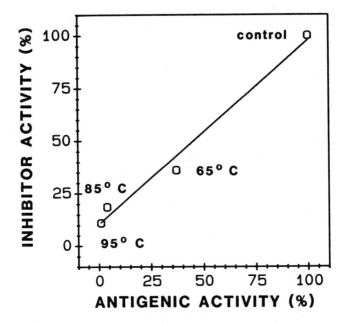

Fig. 18.3. Correlation of antigenicity with the inhibitory activity of heat-treated KTI (r = 0.99). Relative antigenic activity was determined by titration of each sample in the ELISA, and calculated of fitted exponential curves. The concentration of KTI resulting in half maximal (50%) binding of KTI—HRP was determined as a percent relative to the dialyzed control, which produced 50% binding at 201 ng/ml.

for heat-treated samples. For KTI which had been subjected to heat treatment, the activity determined by ELISA correlated strongly with activity determined enzymatically (Fig. 18.3).

The ELISA analysis of raw and processed soy flours is illustrated in Figures 18.4 and 18.5. The analysis of raw flour indicated 7.4 mg KTI/g flour with reference to a standard curve generated with purified KTI. The content was also estimated by the shift in the assay curve caused by the addition of a "spike" of native KTI. In this method, purified KTI is added to achieve approximately a 2- to 3-fold increase in the total KTI of the sample. Figure 18.4 illustrates that analysis. Figure 18.5 yielded an estimate of 6 mg KTI/g flour. Figure 18.5 shows that the processed flours with low activity (82-98% reduction of trypsin inhibitory activity) also had low immunologic activity (98-99% loss in activity in ELISA). The correlation of the antigenic and inhibitory activities is complicated by the presence of BBI's which have greater thermal stability than KTI. The

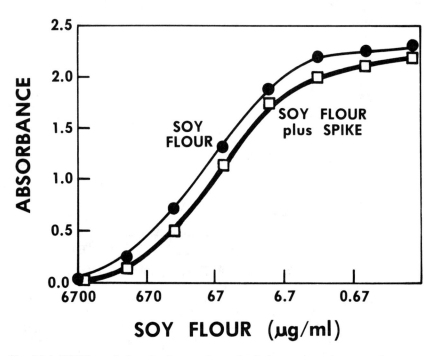

Fig. 18.4. ELISA analysis using internal standard. An extract of raw soy flour was analyzed with and without the addition of purified KTI (29.1 µg/ml). The "spike" increased the KTI content of the sample by 75% resulting in an estimate of KTI as 6 mg/g soy flour.

contribution of other inhibitors to the trypsin inhibitory activity can be estimated by combining the results of enzymatic and ELISA analyses. Using the value of 2.9 TIU/mg for purified KTI and the results of ELISA analysis, the KTI content of soy flours can be computed as a percentage of total TI activity measured enzymatically (Table 18.13). Analyses of other commercial soy-based food products by ELISA are summarized in Table 18.14.

To assess the isoform specificity of the ELISA using antibody 180, extracts were prepared from soybean seeds representing three genotypes expressing different TI genes. The results (Fig. 18.6) indicate that the assay is most sensitive to isoform Ti^b. Extracts of seeds expressing Ti^c had only about 1% crossreactivity in this assay.

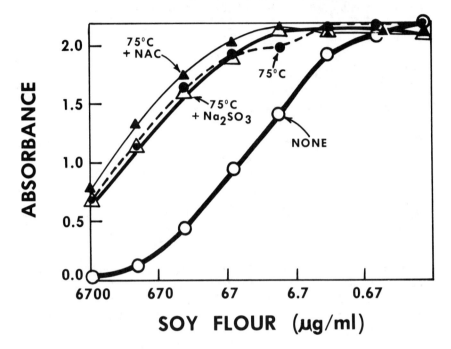

Fig. 18.5. ELISA analysis of soy flours.

TABLE 18.13
Analysis of Soy Flours

Treatment	Total trypsin inhibitor content by enzyme assay (TIU/g)	KTI content by ELISA[a]		
		mg/g	TIU/g	% of total TIU
None	71	7.5	22	31
Heat (75°C)	13	0.15	0.44	3.5
Sodium sulfite (75°C)	1.6	0.15	0.44	27
N-acetyl cysteine (75°C)	1.5	0.094	0.27	18

[a]ELISA data are converted to TIU using the value of 2.9 TIU/mg KTI and expressed as a percent of the total TIU determined enzymatically.

TABLE 18.14
Kunitz Trypsin Inhibitor Content of Commercial Foods Assayed by ELISA

Product	Concentration[a]	mg/g protein	mg/serving[b]
Infant formulas (concentrated)			
Prosobee	12.7	0.31	1.2
Soyalac	5.0	0.12	0.5
Isomil	7.5	<0.003	<0.01
Similac	<0.1	<0.003	<0.01
Tofu	4.8	0.06	0.54
Soy sauce	1.3	0.013	0.02
Soy flour (raw)	7750	19	78

[a]The KTI concentration is expressed as µg/g/ml for liquid samples, µg/g for flour and tofu.
[b]Serving size was 200 ml of reconstituted formula, 112 g of tofu, 18 g of soy sauce and 10 g of flour.

Since soy protein is often processed commercially under alkaline conditions, we also studied the influence of alkalinity on the antigenic activity of KTI. A set of titration curves for samples treated at 65° at various pH's is illustrated in Figure 18.7. It is apparent that both heat and alkalinity cause antigenic and functional changes in KTI.

The results demonstrate that a monoclonal antibody-based enzyme immunoassay can be effective in measuring levels of active KTI in a variety of processed soy food products. The parallel titration curves for spiked and unspiked samples and purified KTI suggest that the same antigen is being measured in each of the samples and that nonspecific interferences with the assay are minimal.

For purified KTI samples, there is an excellent correlation between measurements of antigenic and inhibitory activities. For the more complex flour samples, the correlation is not as close. A likely explanation is the presence of both Kunitz and Bowman-Birk inhibitors in the samples. The enzymatic assay measures total trypsin inhibitory activity of the sample, but the immunoassay measures the Kunitz trypsin inhibitor specifically. The results summarized in Table 18.13 support this interpretation. Heat treatment in the presence of NAC or sodium sulfite eliminates about 98% of total trypsin inhibitor activity, with a relatively small change in the ratio of KTI to total trypsin inhibitory activity. However, flour treated under milder conditions (heating at 75°C) retains 18% of trypsin inhibitor activity, but only 2% of KTI as compared to control.

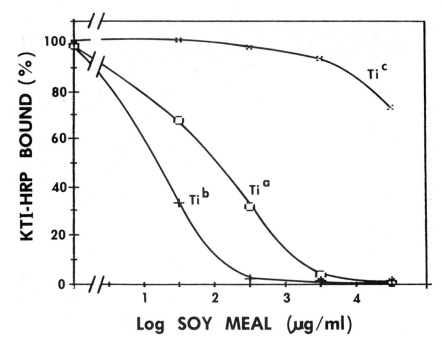

Fig. 18.6. ELISA of soy meal from seeds of different cultivars expressing the Ti^a, Ti^b, or Ti^c gene product. Extracts were compared by titration in the ELISA.

The more severe conditions apparently inactivate more of the relatively stable, highly-disulfide-crosslinked BBI's [Obara and Watanabe (93)]. Similarly, our ELISA results and enzymatic analysis of soy infant formulas suggest that less than 10% of the trypsin inhibitor activity is contributed by KTI.

There is no theoretical reason for antigenicity and activity of a protein to correlate closely, unless the antigenic region includes the enzymatic active site. Indirect evidence suggests that the antigenic determinant recognized by antibody 180 may include the active site, residues 63 (arginine) and 64 (isoleucine) (74). These investigators reported that isoform Ti^c differs from Ti^a only in a change from glycine to glutamic acid at position 55. Isoform Ti^b retains glycine (55), but differs at eight other positions from isoform Ti^a and Ti^c. The known specificity of antibody 180 and other Group 3 antibodies, such as 134, for the native conformation of Ti^a (14,15) together with the antigenic differences among the isoforms, suggests that the binding site of these antibodies could include

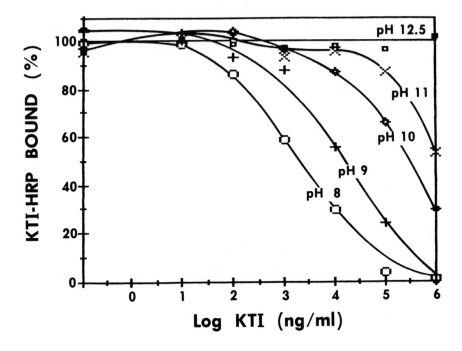

Fig. 18.7. Effect of temperature and pH on KTI antigenicity. KTI was treated at 65°C at the indicated pH. Samples were then titrated in the ELISA.

residues at or near the active site. These conclusions concerning the antigenic structure of the isoforms were supported by subsequent binding studies with purified Ti^b and Ti^c (12B).

The ELISA method documented here can readily measure KTI isoforms Ti^a and Ti^b, permitting analysis of all commercially grown soybeans. Ti^c was found in only 0.3% of tested cultivars from the USDA soybean germplasm collection, and these accessions originated in Japan (70). When complemented by assay which detects BBI (16), the trypsin inhibitor content can accurately be determined. Even in its present form, the ELISA could be used for quality control monitoring.

Active trypsin inhibitors can be removed from foodstuffs in two ways. One is through conventional heat treatment which eliminates 85-90% of trypsin inhibition activity. Problems with this approach include: (a) Only partial inactivation is achieved, (b) The nature of residual activity is difficult to characterize, and (c) The process is costly in energy usage.

A second method of reducing the trypsin inhibitor content of soy products is through genetic modification of the soybean plant. This has been accomplished by Hymowitz (70) with the development of soybean cultivars lacking KTI. Nevertheless, there are inherent limitations to this approach:

(a) Desirable nutritional value (cysteine and methionine) and potential anticarcinogenic activity (112) and pest resistance may be lost concomitant with the reduction of toxicant.

(b) Cross-pollination of the genetic variant by a wild-type cultivar could result in reexpression of trypsin inhibitor gene.

(c) Bowman-Birk inhibitors remain in the low-KTI cultivars.

With either approach to the problem, levels of trypsin inhibitor must be monitored.

The antigenic changes which accompany thermal processing of KTI, especially under alkaline conditions, are dramatically evident in the ELISA results. These results suggest that the processing of proteins could impact positively or negatively on the allergenicity and potential immunotoxicity of food proteins. The immune response to dietary soy protein has been implicated in digestive disorders of cattle (2,13,12A), in the failure of some celiac patients to respond to treatment (57), and in food allergies (89).

Conclusions

Treatment of raw soy flour with cysteine or N-acetyl-L-cysteine results in the introduction of new half-cystine residues into native proteins, with a corresponding improvement of nutritional quality. The proteins are modified through formation of mixed disulfide bonds among added thiols, protease inhibitors and structural proteins. This leads to decreased inhibitory activity and increased protein digestibility and nutritive value. Exposure of soy flour to sodium sulfite was also nutritionally beneficial.

Additional studies are desirable to demonstrate whether enzyme-catalyzed disulfide interchange and isomerization may lead to commercially viable food processes for inactivating soybean and other inhibitors of digestive enzymes. (34,66).

New monoclonal antibody-based immunoassays were developed which can measure low levels of the soybean Kunitz trypsin inhibitor and Bowman-Birk inhibitor in processed foods and could also be used to study the expression and regulation of the gene which encodes these protease inhibitors.

Finally, it is noteworthy that enzyme inhibitors can prevent the development of oxygen radical-induced lesions of the mucosa of the feline intestine (95) and of colon cancer in mice (112). Although the molecular basis for these beneficial effects has not been elucidated, one possibility is that inhibitors or inhibitor-protease complexes act as free radical traps, whereby the free electrons on oxygen radicals are transferred or dissipated to the sulfur atoms of the sulfur-rich inhibitors or complexes. These considerations suggest the need for further studies to enrich our knowledge about possible beneficial effects of plant protease inhibitors.

References

1. Albrecht, W.J., G.C. Mustakas and J.E. McGhee, *Cereal Chem. 43*:401 (1966).
2. *An Evaluation of the Potential of Dietary Protein to Contribute to Systemic Disease*, edited by R. Allison, FASEB, Rockville, MD, 1982.
3. Anderson, R.L., J.J. Rackis and W.H. Tallent, in *Soy Protein in Human Nutrition*, edited by H.L. Wilcke, D.T. Hopkins and D.H. Waggle, Academic Press, New York, NY, 1979, p. 209.
4. AOAC, *Official Methods of Analysis*, 13th edn., Assoc. Official Analytical Chemists, Washington, D.C., 1980.
5. Baker, E.C., and G.C. Mustakas, *J. Am. Oil. Chem. Soc. 50*:137 (1973).
6. Baker, E.C., and J.J. Rackis, in *Nutritional and Toxicological Significance of Enzyme Inhibitors in Food*, edited by M. Friedman, Plenum Press, New York, NY, 1986, p. 349.
7. Begbie, R., and A. Pusztai, in *Absorption and Utilization of Amino Acids*, Vol. 3, edited by M. Friedman, CRC, Boca Raton, FL, 1989, p. 243.
8. Birk, Y., *Int. J. Peptide Protein Res. 24*:113 (1985).
9. Birk, Y., and A. Gertler, *J. Nutr. 75*:379 (1961).
10. Boonvisut, S., and J.R. Whitaker, *J. Agric. Food Chem. 24*:1130 (1976).
11. Borchers, R., and C.W. Ackerson, *J. Nutr. 31*:339 (1950).
12. Borchers, R., L.D. Manage, S.O. Nelson and W. Stetson, *J. Food Sci. 37*:333 (1972).
12A. Brandon, D.L., in *Nutritional and Toxicological Aspects of Food Safety*, edited by M. Friedman, Plenum Press, New York, NY, 1984, p. 65.
12B. Brandon, D.L., and A.H. Bates, *J. Agric. Food Chem. 36*:1336 (1988).
13. Brandon, D.L., S. Haque and M. Friedman, in *Nutritional and Toxicological Significance of Enzyme Inhibitors in Foods*, edited by M. Friedman, Plenum Press, New York, NY, 1986, p. 449.
14. Brandon, D.L., S. Haque and M. Friedman, *J. Agric. Food Chem. 34*:195 (1987).
15. Brandon, D.L., A.H. Bates and M. Friedman, *J. Food Sci. 53*:102 (1988).

16. Brandon, D.L., A.H. Bates and M. Friedman, *J. Agric. Food chem.* 37:1192 (1989).
17. Buera, M.P., A.M.R. Pilosof and G.B. Bartholomai, *J. Food Sci.* 49:124 (1984).
18. Burn, R., *J. Food Prot.* 50:161 (1987).
19. Catsimpoolas, N., and E. Leuthner, *Anal Biochem.* 31:437 (1969).
20. Chan, W.W.C., *Biochemistry* 7:4247 (1968).
21. Charpentier, B.A., and D.E. Lemmel, *J. Agric. Food Chem.* 32:908 (1984).
22. Chen, X.J., H.M. Bau, F. Giannageli and G. Derby, *Sci. des Aliments* 6:257 (1986).
23. Chompreeda, P.T., and M.L. Fields, *J. Food Sci.* 49:563 (1984).
24. Churella, H.R., B.C. Yao and W.A.B. Thomson, *J. Agric. Food Chem.* 24:393 (1976).
25. Collins, J.L., and B.F. Beaty, *J. Food Sci.* 45:542 (1980).
26. Collins, J.L., and G.G. Sanders, *Ibid.* 41:168 (1976).
27. Colvin, B.M. and H.A. Ramsey, *J. Dairy Sci.* 52:270 (1969).
28. Cravioto, R.O.B., G.J. Guzman and M.H. Guillermo, *Ciencia (Mexico)* 11:81 (1951).
29. DiBella, F.P., and I.E. Liener, *J. Biol. Chem* 244:2824 (1969).
30. Duncan, D.B., *Biometrics* 11:1 (1959).
31. Eldridge, A.C., and J. Wolf, *Cereal Chem.* 46:471 (1969).
32. Ellenrieder, G., H. Geronazzo and A.B. De Bojarski, *Cereal Chem* 57:25 (1980).
33. Ellman, G.L., *Arch. Biochem. Biophys.* 82:70 (1959).
34. Freedman, R.B., *Trends Biochem. Sci.* 9:438 (1984).
35. Friedman, M., *The Chemistry and Biochemistry of the Sulfhydryl Group in Amino Acids, Peptides, and Proteins*, Pergamon Press, Oxford, England, 1973, p. 199.
36. Friedman, M., in *Nutritional and Toxicological Aspects of Food Safety*, edited by M. Friedman, Plenum Press, New York, NY, 1984, p. 31.
37. Friedman, M., and M.R. Gumbmann, *J. Food Sci.* 51:1239 (1986).
38. Freidman, M., and M.R. Gumbmann, in *Nutritional and Toxicological Significance of Trypsin Inhibitors in Foods*, edited by M. Friedman, Plenum Press, New York, NY, 1986, p. 357.
39. Friedman, M., and M.R. Gumbmann, *J. Nutr.* 118:388 (1988).
40. Friedman, M., and R. Liardon, *J. Agric. Food Chem* 33:666 (1985).
41. Friedman, M., and A.T. Noma and J.R. Wagner, *Anal. Biochem.* 98:293 (1979).
42. Friedman, M., J.C. Zahnley and J.R. Wagner, *Ibid.* 106:27 (1980).
43. Friedman, M., J.C. Zahnley and P.M. Masters, *J. Food Sci.* 46:127 (1981).
44. Friedman, M., O.K. Grosjean and J.C. Zahnley, *Nutr. Rep. Int.* 25:743 (1982).

45. Friedman, M., O.K. Grosjean and J.C. Zahnley, in *Mechanisms of Food Protein Deterioration*, edited by J.P. Cherry, ACS Symposium Series, No. 206 1982, p. 359.
46. Friedman, M., O.K. Grosjean and J.C. Zahnley, *J. Sci. Food Agric. 33*:165 (1982).
47. Friedman, M., M.R. Gumbmann and O.K. Grojean, *J. Nutr. 114*:2241 (1984).
48. Fuller, J.C., Jr., W.J. Owings and G.E. Fanslow, *Nutr. Rep. Int., 37*:104 (1988).
49. Gallaher, D., and B.O. Schneeman, in *Nutritional and Toxicological Aspects of Food Safety*, edited by M. Friedman, Plenum Press, New York, NY, 1984, p. 299.
50. Gandhi, A.P., M.M. Nenwani and N. Ali, *Food Chem. 15*:215 (1984).
51. Gatehouse, J.A., and A.M.R. Gatehouse, *Anal. Biochem. 98*:438 (1979).
52. Gumbmann, M.R., and M. Friedman, *J. Nutr. 117*:1018 (1987).
53. Gumbmann, M.R., M. Friedman and G.A. Smith, *Nutr. Rep. Int. 28*:355 (1983).
54. Gumbmann, M.R., W.L. Spangler, G.M. Dugan and J.J. Rackis, in *Nutritional and Toxicological Significance of Enzyme Inhibitors in Foods*, edited by M. Friedman, Plenum Press, New York, NY, 1986, p. 33.
55. Gupta, Y.P., *Plant Foods Human Nutr. 37*:201 (1987).
56. Hackler, L.R., J.P. Van Buren, K.H. Steinkraus, I. El Rawi and D.B. Hand, *J. Food Sci. 30*:723 (1965).
57. Haeney, M.R., B.J. Goodwin, M.E.J. Barratt, N. Mike, and P. Asquith, *J. Clin. Pathol. 35*:319 (1982).
58. Hafez, Y.S., and A.I. Mohamed, *J. Food Sci. 48*:75 (1983).
59. Hafez, Y.S., and A.I. Mohamed, *Ibid. 48*:1265 (1983).
60. Hafez, Y.S., A.I. Mohamed, G. Singh and F.M. Hewedy, *Ibid. 50*:1271 (1985).
61. Hafez, Y.S., A.I. Mohamed, F.M. Hewedy and G. Singh, *Ibid 50*:415 (1985).
62. Haider, F.G., *Nutr. Rep. Int. 23*:1167 (1981).
63. Ham, W.E., and R.M. Sanstedt, *J. Biol. Chem. 154*:505 (1944).
64. Hammerstrand, G.E., L.T. Black and J.D. Glover, *Cereal Chem. 58*:42 (1981).
65. Hansen, W.E., *Int. J. Pancreatology 1*:341 (1986).
66. Hawkins, H.C., and R.B. Freedman, *FEBS Lett. 58*:7 (1975).
67. Hill, B.S., H.E. Snyder and K.L. Wiese, *J. Food Sci. 47*:2018 (1982).
68. Hogle, J.M., and I.E. Liener, *Can. J. Biochem. 51*:1014 (1973).
69. Honig, D.H., J.J. Rackis and W.J. Wolf, *J. Agric. Food Chem. 35*:967 (1987).
70. Hymowitz, W., in *Nutritional and Toxicological Significance of Enzyme Inhibitors in Foods*, edited by M. Friedman, Plenum Press, New York, NY, 1986, p. 291.
71. Jung, J., and C.Y. Lee, *J. Korean Agric. Chem. Soc. 25*:182 (1982).
72. Kakade, M.L., N. Simons and I.E. Liener, *Cereal Chem. 46*:518 (1969).

73. Kakade, M.L., J.J. Rackis, J.E. McGhee and G. Puski, *Ibid.* *51*:376 (1974).
74. Kim, S., S. Hara, T. Ikenaka, H. Toda, K. Kitamura and N. Kaizuma, *J. Biochem.* *98*:435 (1985).
75. Kunitz, M., *J. Gen. Physiol.* *30*:291 (1947).
76. Kunitz, M., *Ibid.* *32*: 241 (1948).
77. Lehnhardt, W.L., and H.G. Dills, *J. Am. Oil Chem. Soc.* *61*:691 (1984).
78. Lei, M.G., R. Bassette and G.R. Reeck, *J. Agric. Food Chem.* *29*:1196 (1981).
79. Liardon, L., and M. Friedman, *Ibid.* *35*:661 (1987).
80. Liener, I.E., in *Protein Nutritional Quality of Foods and Feeds*, edited by M. Friedman, Marcel Dekker, New York, NY, 1975, p. 523.
81. Liener, I.E., and E.G. Hill, *J. Nutr.* *49*:609 (1953).
82. Liener, I.E., and M. Kakade, in *Toxic Constituents of Plant Foodstuffs*, 2nd. edn., edited by I.E. Liener, Academic Press, New York, NY, 1980, p. 7.
83. Liener, I.E., and S. Tomlinson, *J. Food Sci.* *46*:1354 (1981).
84. Liu, K., and P. Markakis, *Ibid.* *52*:222 (1987).
85. Longnecker, M.L., W.H. Martin and H.P. Sarrett, *J. Agric. Food Chem.* *12*:411 (1964).
86. McNaughton, J.L., *J. Am. Oil Chem. Soc.* *58*:321 (1981).
87. Mikola, J., and E.M. Suolinna, *Eur. J. Biochem.* *9*:555 (1969).
88. Mitchell, G.V., and E. Grundel, *J. Agric. Food Chem.* *34*:650 (1986).
89. Moroz, L.A., and W.H. Yang, *New England J. Med.* *302*:1126 (1980).
90. Mostafa, M.M., A.H. Rady and E.H. Rahma, *Egyptian J. Food Sci.* *14*:409 (1986).
91. Nielsen, E., *J. Am. Oil Chem. Soc.* *53*:305 (1976).
92. Nordal, J., and K. Fossum, *Z. Lebensm. Unters.-Forsch.* *154*:144 (1974).
93. Obara, T., and Y. Watanabe, *Cereal Chem.* *48*:523 (1971).
94. Osborne, T.B., and L.B. Mendel, *J. Biol. Chem.* *32*:369 (1917).
95. Parks, D.A., D.N. Granger, G.B. Bulkley and A.K. Shah, *Gastroenterology* *89*:7 (1985).
96. Rackis, J.J., *J. Am. Oil Chem. Soc.* *58*:495 (1981).
97. Rackis, J.J., J.E. McGhee, I.E. Liener, M.L. Kakade and G. Puski, *Cereal Science Today* *19*:513 (1974).
98. Rackis, J.W., W.J. Wolf and E.C. Baker, in *Nutritional and Toxicological Significance of Enzyme Inhibitors in Foods*, edited by M. Friedman, Plenum Press, New York, NY, 1986, p. 299.
99. Rady, A.H., M.M. Mostafa and E.H. Rahma, *Egyptian J. Food Sci.* *15*:37 (1987).
100. Richardson, M., *Food Chem.* *6*:235 (1980).
101. Rios-Iriarte, B.J., and R.H. Barnes, *Food Technol.* *20*:836 (1966).
102. Rothenbuhler, E., and J.E. Kinsella, *J. Agric. Food Chem.* *33*:433 (1985).
103. Sandholm, M., R. Smith, J. Shih and M. Scott, *J. Nutr.* *106*:761 (1976).
104. Sedlak, J., and R.H. Lindsay, *Anal. Biochem.* *25*:192 (1968).
105. Sessa, D.J., and J.A. Bietz, *J. Am. Oil Chem. Soc.* *63*:784 (1986).

106. Sessa, D.J., and P.E. Ghantous, *Ibid. 64*:1682 (1987).
107. Smith, C., W. Van Megen, L. Twaalfhoven and C. Hitchcock, *J. Soc. Food Agric. 31*:341 (1980).
108. Soetrisno, U., Z. Holmes and L.T. Miller, *J. Food Sci. 47*:530 (1982).
109. Tan-Wilson, A.L., and K.A. Wilson, in *Nutritional and Toxicological Significance of Enzyme Inhibitors in Foods*, edited by M. Friedman, Plenum Press, New York, NY, 1986, p. 391.
109A. Tan-Wilson, A.L., S.E. Cosgriff, M.C. Duggan, R.S. Obach, and K.A. Wilson, *J. Agric. Food Chem. 33*:389 (1985).
109B. Tan-Wilson, A.L., J.C. Chen, M.C. Duggan, C. Chapman, R.S. Obach, and K.A. Wilson, *J. Agric. Food Chem. 35*:974 (1987).
110. Taylor, M.J., and T. Richardson, *J. Food Sci. 45*:1223 (1980).
111. Tidemann, L.J., and J. Schingoethe, *J. Agric. Food Chem. 22*:1059 (1974).
112. Troll, W., K. Frenkel and R. Wiesner, in *Nutritional and Toxicological Significance of Enzyme Inhibitors in Foods*, edited by M. Friedman, Plenum Press, New York, NY, 1986, p. 153.
113. Wallace, J.N., and M. Friedman, *Nutr. Rep. Int. 32*:748 (1985).
114. Weder, J.K.P., in *Nutritional and Toxicological Significance of Enzyme Inhibitors in Foods*, edited by M. Friedman, Plenum Press, New York, NY, 1986, p. 239.
115. Whitaker, J.R., in *Impact of Toxicology on Food Processing*, edited by J.C. Ayres and J.C. Kerschman, AVI, Westport, CT, 1981, p. 57.
115A. Yavelow, J., T.H. Finley, A.R. Kennedy, and Troll, W., *Cancer Res. 43*:2454s (1983).
115B. Yavelow, J., M. Collins, Y. Birk, W. Troll, and A.R. Kennedy, *Proc. Natl. Acad, Sci. USA 82*:5395 (1985).
116. Zahnley, J.C., and M. Friedman, *J. Prot. Chem. 1*:335 (1982).

Chapter Nineteen

The Nutritional Significance of Lectins

Irvin E. Liener

Department of Biochemistry
College of Biological Sciences
University of Minnesota
St. Paul, MN 55108

That a toxic substance must be present in the seeds of certain plants was recognized as early as 1888 when Stillmark (1) observed that the extreme toxicity of the castor bean (*Ricinus communis*) could be attributed to a protein fraction which was capable of agglutinating red blood cells. He coined the name *ricin* for this particular protein. We now know that similar substances are very widely distributed in Nature, not only among plants but also in animal tissue as well. Of primary concern to the nutritionist is the fact that legumes and certain cereal grains are particularly rich in what have sometimes been referred to as "phytohemagglutinins." The latter term has now been largely replaced by the term "lectin" which is taken from Latin word *legere*, which means to choose (2). This name was dictated by the observation that these substances exhibited a high degree of specificity with respect to the type of red blood cells that are agglutinated depending on the origin of these so-called lectins. It is now known that the specificity which lectins exhibit towards red blood cells can be explained by the fact that lectins differ in their ability to combine or form complexes, similar to antibody-antigen reactions, with various kinds of sugars that are present on the cell membranes. Table 19.1 shows a partial list of various lectins derived from edible legumes and their corresponding sugar specificity. A comprehensive coverage of lectins and their properties may be found in reference (3).

Just as variable as the specificity of the lectins towards sugars are their toxicological properties as evaluated by injection or oral ingestion (Table 19.1). It is obvious that only the toxicity associated with the oral ingestion of the plants which contain these lectins is of concern to those who are interested in utilizing these plants for food in the human diet or as a feed ingredient for animals. It is significant to note from Table 2.1 that the oral toxicity of most of the lectins isolated from legumes is

TABLE 19.1
Sugar Specificity of Lectins Present in Edible Legumes and Their Toxicological and Nutritional Effects[a]

Botanical name	Common name	Sugar specificity[b]	Toxicity Peritoneal[c]	Toxicity Oral[d]	Effect of heat on nutritive value[e]
Arachis hypogaea	peanut, groundnut	α-D-Gal	NA	–	+/–
Canavalia ensiformis	jack bean	α-D-Man	+	+	+
Doeclea grandiflora	mucuna	α-D-Man	NA	+	+
Dolichos biflorus	horse gram	α-D-GalNAc	–	+	+
Glycine max	soybean	α-D-Gal, α-D-GalNAc	+	–	+
Lathyrus odoratus	sweet pea	α-D-Man	NA	NA	+
Lathyrus sativus	chickling vetch	α-D-Man	NA	NA	+
Lens esculenta[f]	lentil	α-D-Man, α-D-Glc	–	NA	+
Phaseolus acutifolius	tipary bean	NA	+	NA	+
Phaseolus angularis	Adzuki bean	NA	NA	NA	+
Phaseolus aureus	mung bean	NA	–	NA	–
Phaseolus coccineus	runner bean	α-D-GalNAc	NA	NA	+
Phaseolus lunatus	lima bean	α-D-GalNAc	–	+	+
Phaseolus multiflorus	scarlet runner bean	NA	+	+	+
Phaseolus vulgaris	kidney bean, navy bean	α-D-GalNAc	+	+	+
Psophocarpus tetragonolobus	winged bean	α-D-GalNAc	NA	+	+

TABLE 19.1 — continued
Sugar Specificity of Lectins Present in Edible Legumes and Their Toxicological and Nutritional Effects[a]

Botanical name	Common name	Sugar specificity[b]	Toxicity Peritoneal[c]	Oral[d]	Effect of heat on nutritive value[e]
Ricinus communis[g]	castor bean	α-D-Gal	+	+	+
Solanum tuberosum	potato	α-D-GlcNAc	NA	NA	NA
Vicia faba	broad bean	α-D-Man	NA	NA	+
Vicia sativa	common vetch	α-D-Man	NA	NA	+
Vigna unguiculata[h]	cow peas, black-eyed peas	NA	+	NA	+

[a] Based on material taken from reference (3).
[b] Abbreviations used: Glc, glucose; GlcNAc, N-acetylglucosamine; Gal, galactose; GalNAc, N-acetylgalactosamine; Man, mannose; NA, no information available.
[c] Death produced by intraperitoneal injection of crude seed extract or the purified lectin. + and − denote toxic and non-toxic effects, respectively.
[d] Growth inhibition or death produced by the incorporation of the purified lectin into diets fed to animals.
[e] Growth promoting property of the parent plant material when incorporated into the diet as the main source of protein. + and − denote improvement or no effect, respectively, due to heat treatment. +/− denotes conflicting data from different investigators.
[f] Also known as *Lens culinaris*.
[g] Non-hemagglutinating toxin commonly referred to as *ricin*.
[h] Also known as *Vigna sinensis*.

accompanied by the observation that heat treatment exerts a beneficial effect on the nutritive value of the protein. Despite this apparent association one cannot assume that the beneficial effect of heat on the nutritive value of the protein of a particular plant can be attributed solely to the destruction of the lectin since other antinutritional factors such as trypsin inhibitors may also be responsible for the overall improvement in nutritive value effected by heat treatment. Thus, no broad generalizations concerning the nutritional significance of lectins can be drawn, but each plant must be considered on an individual basis, and any conclusions that are made are dependent on the experimental evidence that is available in each case.

Soybeans

Although it is generally assumed that the trypsin inhibitors are largely responsible for the poor nutritive value of raw soybeans, other lines of evidence suggest that the trypsin inhibitors do not account for all of the growth inhibition observed with rats on a raw soybean diet (4). The search for other growth inhibitors in soybeans ultimately led to the isolation of what subsequently proved to be a lectin. Although this protein proved to be toxic when injected into rats (5,6), feeding experiments involving the feeding of diets to which the purified lectin had been added to heated soybean meal revealed that this protein was capable of inhibiting growth under conditions of *ad libitum* feeding (7). However, when the food intake of a control group of animals not receiving the lectin was restricted to that of animals fed diets containing the lectin, no significant inhibition of growth was observed. It thus appeared that the growth inhibitor effect of the soybean lectin may have been due to a depressing effect on food intake.

A more definitive answer to this problem came from experiments in which the crude extract from raw soybeans which had been "delectinized" by passage through a column of Sepharose-concanavalin A was fed to rats (8). This approach was based on a previous report by Bessler and Goldstein (9) that concanavalin A binds to mannose-containing glycoproteins of which the soybean lectin is an example (10,11). This delectinized soybean extract supported no better growth than the original lectin-containing extract. This would indicate that the soybean lectin probably does not affect the nutritional properties of soybean protein to any significant extent. Jindal *et al* (12) have reported that the feeding of a diet containing 0.27% purified lectin to rats had little effect on growth. However, Grant *et al* (13) found that increasing the level of the soybean lectin to 0.75 % of the diet did cause an inhibition of growth and pancreatic enlargement. This level of lectin, however, is much higher than would be provided by a diet containing 10% protein derived from soybeans which is generally employed for evaluating the PER's in rats. The relative innocuous effect of the oral ingestion of the soybean lectin may perhaps be explained by the fact that it is readily digested by pepsin and by the enzymes of the brush border membrane (14,15).

Phaseolus vulgaris

Many varieties of beans belonging to *Phaseolus vulgaris* provide a significant source of protein for large segments of the world's population (16). Nevertheless, numerous reports may be found in the literature concerning the toxic effect in man and animals that frequently accompany the ingestion of raw or inadequately cooked beans (17,18). Although the presence of an hemagglutinating factor in edible beans was recognized as early as 1908 (19), direct evidence that they might be responsible for the toxicity of the raw bean was first provided by Jaffé and his coworkers (20,21). They isolated a protein fraction from the black bean, which they called "phaseolotoxin A"; this preparation displayed hemagglutinating activity towards rabbit erythrocytes, was toxic when injected into mice, and inhibited the growth of rats when incorporated into the diet. Honavar et al (22) subsequently confirmed the growth-inhibitory effect of lectins isolated from kidney beans and black beans

TABLE 19.2
Effect of Purified Hemagglutinin Fractions from the Black Bean and Kidney Bean on Growth of Rats[a]

Source of hemagglutinin	Purified hemagglutinin in diet, %	Gain in weight, g/day	Mortality[b] days
Black bean	0	+2.51	
	0.5	+1.04	
	0.5[c]	+2.37	
	0.75	+0.20	
	1.2	-0.91	15-19
	2.3	-1.61	12-17
	4.6	-1.72	5-7
Kidney bean	0	+2.31	
	0.5	-0.60	13-16
	0.5[c]	+2.29	
	1.0	-0.87	11-13
	1.5	-1.22	4-7

[a]Taken from Honavar et al. (22).
[b]100% mortality observed during the period recorded. Blank space indicates no deaths observed.
[c]Solution of hemagglutinin boiled for 30 min and dried coagulum was fed at level indicated. Hemagglutinating activity was completely destroyed by this treatment.

(see Table 19.2). At levels in excess of 1 % of the diet not only was the growth of rats markedly inhibited but death followed within 7-10 days, similar to what was observed with the raw bean itself. Similar results were obtained with the chick (23) and the Japanese quail (24,25). Pusztai and his group (26-30) have conducted a series of comprehensive studies that definitively identifies the lectin as the main factor responsible for the toxicity associated with the consumption of *P. vulgaris* in its raw form.

Preliminary soaking prior to cooking or autoclaving sems to be required for the complete elimination of the toxicity although autoclaving alone for 5 min serves to eliminate the toxicity of finely ground navy bean meal (31). However, autoclaving for 30 min was necessary to destroy the hemagglutinating activity of certain African varieties of *P. vulgaris* (32). Of particular significance was the observation that hemagglutinating activity was still detectable after 18 hours of exposure to dry heat. Since beans which have been allowed to germinate lose much of their hemagglutinating activity (33), bean sprouts may be regarded as essentially devoid of any adverse effects that can be attributed to lectins.

One of the complicating factors involved in relating toxicity to hemagglutinating activity is the fact that there are literally hundreds of different strains and cultivars of *P. vulgaris* that may differ in their specificity toward the erythrocytes of various species of animals and blood types. Moreover, their hemagglutinating activity may vary markedly depending on the manner in which the cells have been pretreated with proteases. Jaffé and his colleagues (34-36) have classified the hemagglutinating action of extracts from the seeds of various cultivars of *P. vulgaris* into four main groups: type A agglutinates rabbit cells and trypsinated cow cells; type B agglutinates only rabbit cells; type C agglutinates only trypsinated cow cells; and type D agglutinates only pronase-treated hamster cells. A systematic study of 20 varieties and cultivars of *P. vulgaris*, with respect to the above classification and the toxicity of crude extracts injected into mice (36,37), revealed that only types A and C (active toward trypsinated cow cells) proved to be toxic. Feeding experiments confirmed the fact that those cultivars that could be classified as type A and C likewise supported very poor growth of rats (38). A recent survey of 85 samples of 15 different lines of legumes generally available in the United Kingdom confirmed the fact that those legumes that displayed hemagglutinating activity toward trypsinated cow cells were likewise those that supported very poor growth when fed to animals (39). These results serve to emphasize the importance of testing the hemagglutinating activity of seed extracts against several species of blood cells before one is justified in concluding whether a particular bean is toxic or not.

Other species of *Phaseolus*

Among the other species of *Phaseolus* that exhibit hemagglutinating activity are the lima bean (*P. lunatus*), mung bean (*P. aureus*), white tipary bean (*P. acutifolius*), scarlet runner bean (*P. multiflorus*), Adzuki bean (*P. angularis*), and runner bean (*P. coccineus*) (see Table 19.1). In only a few instances, however, has the purified lectin been tested for injected or oral toxicity. The lectins from the lima bean and mung bean have been reported to be nontoxic when administered by injection (40,41). Despite its lack of toxicity by injection, the oral administration of the lima bean lectin severely retarded the growth of rats (41). No reports on the oral toxicity of mung bean are available, but the failure of heat to improve its nutritional quality (see Table 19.1) would suggest that it is probably not toxic when ingested. The white tipary bean and runner bean are highly toxic when fed to rats in their raw form (39,42), and extracts of these beans are toxic when injected (40). Since the effect of the oral administration of the lectins from the white tipary bean and runner bean has not been reported, one cannot conclude for certain that the lectins are responsible for the poor nutritive value of the protein in the raw beans.

Castor bean *(Ricinus communis)*

Ricin, the lectin of the castor bean, was one of the first lectins to attract the attention of investigators, presumably because of its extreme toxicity; its minimum lethal dose is about 0.001 $\mu g/g$ (mice), which makes it about 1000 times more toxic than any of the other bean lectins (43). This toxicity persists after oral ingestion, and, for this reason, detoxification of castor pomace is essential for its safe handling and for its utilization for animal feeding. Steam heating, as used for the recovery of the solvents employed for the extraction of castor oil, has been found to produce a thousand-fold reduction in toxicity and to render the pomace harmless for sheep, rabbits and rats when incorporated into the diet at a level not in excess of 10% (44). The steaming of castor bean meal for 1 hr at 15 lb of pressure reduced the toxicity of the meal to 1/2000 of its original value (45). Rats fed 23.9% of the autoclaved meal were in good health after four weeks, although growth and food conversion were lower than in the casein controls. Effective detoxification can also be achieved by extraction with hot water (46) or treatment with dilute alkali or formaldehyde (47,48).

In most of the earlier work dealing with toxicity of ricin it was assumed that toxicity and hemagglutinating were associated with the same protein. It is now clear that these are two distinct components: a toxic protein, for which the name ricin has been reserved and which is devoid of hemagglutinating activity, and the so-called castor bean agglutinin which exhibits agglutinating activity, but which is nontoxic (49). It may be reasonably assumed that most of the toxic effects observed with castor bean preparations in the past have been due to the toxic component rather than the agglutinin which may have accompanied it.

Jack bean *(Canavalia ensiformis)*

Although concanavalin A, the lectin of the jack bean, has been the object of considerable study with respect to its physicochemical properties and its biological effects, very little is known regarding the role it plays in the nutritional properties of the seeds from which it is derived. The direct injection of concanavalin A into an animal causes agglutination of red blood cells, followed by hemolysis, and finally death (50,51). Jack bean meal is of poor nutritive value unless heated (52), and the consumption of the raw bean has been reported to cause a wide variety of pathological lesions in rats (53) and cattle (54).

Despite evidence for the *in vivo* toxicity of concanavalin A there appears to be some question of whether the harmful effects that accompany the ingestion of raw jack bean meal are entirely due to this lectin. Dennison et al. (55) removed the agglutinating activity of a crude extract of jack bean meal by selective absorption onto Sephadex and observed that the unabsorbed fraction still retained some toxicity when injected into rats. The isolation of a nonhemagglutinating toxic protein from the jack bean also has been reported by Carlini and Guimaraes (56).

Other edible legumes

The lectin isolated from the pea *(Pium sativum)* does not produce any toxic effects when incorporated into the diet of rats at a level of 1% (57). Presumably this explains why the nutritive value of peas is not enhanced by heat treatment (see Table 19.1). The lectins from the hyacinth bean *(Dolichos lablab)* and the horse gram *(Dolichos biflorus)* inhibit the growth of rats when fed as part of the diet, but the growth-depressing effect was much less than that observed with the raw bean (41,58), indicating that other heat-labile antinutritional factors must be present in these beans. The poor nutritive value of the raw winged bean (*Psophocarpus tetragonolobus*) and the improvement effected by autoclaving have been attributed to a lectin with galactose-binding specificity (15,59).

The broad bean (*Vicia faba*) is known to contain hemagglutinins, but this activity would appear to have little relationship to the disease in humans known as "favism" which sometimes accompanies the ingestion of this bean (60). Although lectins have been isolated from the lentil, its influence on the nutritive value of this legume is not known. Raw peanuts contain a level of hemagglutinating activity comparable to soybeans, but, since heat did not serve to improve the nutritive value of the peanut protein, Sitren *et al* (61) concluded that the peanut lectin is probably of little nutritional importance.

It should be pointed out that legumes eaten in their wet form such as fresh garden peas (*Pisum sativum*) or beans comsumed fresh in the pod such as string or French beans (*Phaseolus vulgaris*) have relatively low levels of lectins and may be considered to be devoid of any toxic effects (62). Other legumes which have been examined for hemagglutinating activity and have been found not to exhibit such activity include the chick pea (*Cicer arietinum*), cowpea (*Vigna sinensis*) and lupines (62,63). Of course, failure to detect hemagglutinating activity may be the consequence of using red blood cells from an inappropriate species of animal which would not react with the lectin even if it were present.

Other Edible Plants

Table 19.3 provides an additional compilation of edible plants that have been demonstrated to exhibit agglutinating activity toward erythrocytes from various species of animals. The size of this list serves to emphasize the degree to which the human population is exposed to dietary sources of lectins other than legumes. However, in the absence of any studies involving the oral intake of lectins purified from these sources, their possible physiological significance cannot be evaluated at this time.

Mode of Action

Interaction with intestinal mucosa.

A number of years ago, Jaffé and his coworkers (20,21) had proposed that the toxicity of bean lectins could be attributed to their ability to

TABLE 19.3
Other Edible Plants that Exhibit Lectin Activity[a]

CEREAL GRAINS

Avena sativa	Oats	*Secale cereale*	Rye
Hordeum vulgare	Barley	*Triticale spp.*	Triticale
Oryza sativum	Rice	*Triticum vulgare*	Wheat
	Zea mays Corn		

VEGETABLES

Abelmoschus esculentus	Okra	*Cucurbita maxima*	Pumpkin
Apium graveolens	Celery	*Cucurbita sativus*	Cucumber
Asparagus officinalis	Asparagus	*Ipomea batatas*	Sweet potato
Beta vulgaris	Swiss chard	*Lactuca scariole*	Prickly lettuce
Brassica camapestris rapa	Turnip/beet	*Lycopersicon esculentum*	Tomato
Brassica napobrassica	Rutabaga	*Medicago sativum*	Alfalfa
Capsicum annum	Sweet Pepper	*Petrosecinum hortense*	Parsley
Cucurbita pepo	Zucchini	*Rheum rhapontium*	Rhubarb
Solanum tuberosum	Potato		

FRUITS

Carica papaya	Papaya	*Malus* species	Apple
Citrus aurantium	Orange/lemon	*Musa paradisiac*	Banana
Citrullus vulgaris	Watermelon	*Prunus americana*	Plum
Citrus medica	Grapefruit	*Prunus avium bigarreaus*	Cherry
Cucumis melo cantalupensis	Cantaloupe	*Ribes rubrum*	Currant
Cydonia oblonja	Quince	*Rubus idaeus*	Raspberry
Fragaria vesca	Strawberry		
	Rubus fructicosus Blackberry		

SPICES

Allium sativum	Garlic	*Myristica fragrans*	Nutmeg
Labiacae origanum	Marjoram	*Menta piperita*	Peppermint
Pimenta officialis	Allspice		

OTHER

Agaricus bisporus	Mushroom	*Juglans regia*	Walnut
Carum carvi	Caraway seeds	*Helianthus annus*	Sunflower seeds
Cocos nucifera	Coconut	*Sesamum indicum*	Sesame seeds
Coffee arabica	Coffee	*Theobroma cacao*	Cocoa
Corylus avellania	Hazelnut		

[a] Taken from reference (3) pp. 535-536.

bind to specific receptor sites on the surface of the epithelial cells lining the intestine. Subsequent studies by numerous other investigators have since served to establish the validity of this concept by demonstrating that lectins of diverse sugar specificity bind to various regions and to different types of cells of the intestinal mucosa (64-71). The *in vivo* binding of the kidney bean lectin (72,73) and the lectin from the Kintoki bean (74) to the intestinal mucosa has been demonstrated by immunofluorescence. Figure 19.1 shows that the lectin ingested by rats in the form of raw kidney beans binds to the luminal surface of microvilli in the proximal region of the small intestine. Accompanying the binding of the kidney bean lectin to the intestines is the appearance of lesions, severe disruption and abnormal development of the microvilli (75-80). Figure 19.2 provides an example of the structural damage inflicted on the microvilli of rats fed raw kidney beans. Morphological changes in the intestinal mucosa have also been reported to accompany the ingestion of lectins present in the tipary bean (*P. acutifolius*) (42,81), the winged bean (*Psophocarpus tetragonolobus*) (15,59,82), wheat germ (83,84) and soybeans (12). The interaction of lectins with the intestines have been the subject of a number of reviews (85-87).

Interference with the absorption of nutrients.

One of the major consequences of lectin damage to the intestinal mucosa appears to be a serious impairment in the absorption of nutrients across the intestinal wall. This was first demonstrated *in vitro* by Jaffé and Comejo (88), who showed that isolated intestinal loops taken from animals that had been fed raw black beans, or the lectin purified therefrom, displayed a significant decrease in the rate at which glucose was transported across the intestinal wall. This phenomenon can also be demonstrated by the technique of vascular intestinal perfusion employing rats that had been fed the lectin purified from the navy bean (89). In the latter experiments [^{14}C]glucose was injected directly into the lumen of the isolated intestine of a rat. The absorption of glucose was monitored by the appearance of radioactivity in the perfusate of the cannulated portal vein. As shown in Figure 19.3, the rate of absorption of glucose across the intestines of rats fed the lectin-containing diet was less than one-half that of controls fed a diet without lectin. An impairment in the absorption of glucose in chicks on diets of raw navy beans (90), of glucose and galactose in rats and chicks on diets of raw field beans (*Vicia faba*) (91,92), and of glucose by the oral administration of ricin to rats (93,94) has also been reported. The ability of lectins to

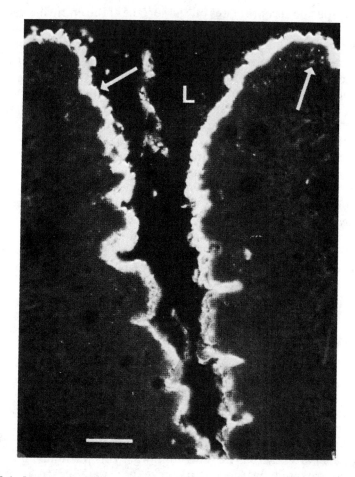

FIG. 19.1. Immunofluorescence micrograph of part of a transverse section through the duodenum of a rat fed on a diet containing kidney beans ("Processor"). Incubation with rabbit anti-lectin immunoglobulins and fluorescein isothiocyanate-conjugated anti-rabbit IgG, showing immunofluorescence in brush border region and within apical cytoplasm of mature enterocytes (arrows). L, Intestinal lumen. Bar: 50 μm. [From King et al. (72)].

decrease the absorptive capacity of the small intestine is not confined to sugars. For example, the lectin purified from the field pea (*Pisum arvense*) decreased the duodenal absorption of L-histidine in mice (95), and the lectins from the pea and lentil (*Lens esculenta*) caused a significant decrease in the absorption of L-phenylalanine in rats (96). The

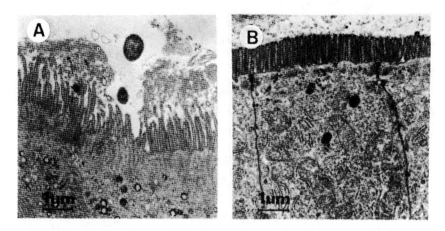

FIG. 19.2. Electron micrographs of sections through the apical regions from rats fed diets containing (A) 5% raw kidney beans ("Processor") and 5% casein compared to (B) 10% casein. [From Pusztai et al. (77)].

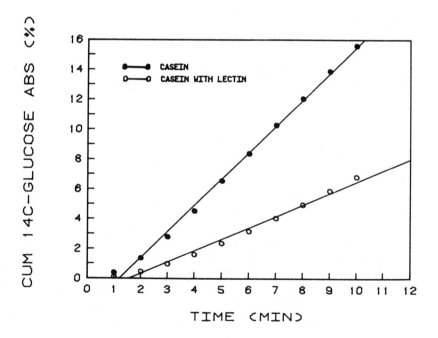

FIG. 19.3. Rate of absorption of glucose by intestines of rats fed navy bean lectin as measured by intestinal vascular perfusion. Animals had been fed diets containing 10.5% casein (●) or 10% casein plus 0.5% lectin (○). [From Donatucci et al. (89)].

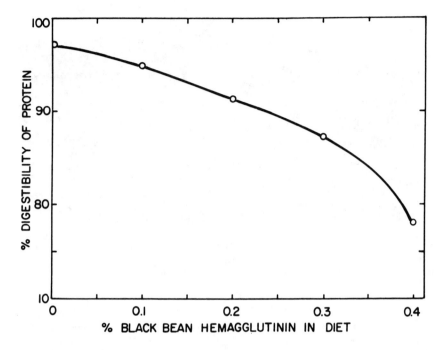

FIG. 19.4. Effect of increasing levels of the black bean lectin on the apparent digestibility of casein fed to rats. [From data in Jaffé and Camejo (88)].

addition of the kidney bean lectin to a casein diet fed to rats resulted in the malabsorption of lipid, nitrogen, and vitamin B_{12}, as well as an interference with ion transport (97,98).

This nonspecific interference with the absorption of nutrients is no doubt one of the reasons that the protein of raw legumes is so poorly utilized. The effect of the black bean lectin (*P. vulgaris*) in causing a decrease in what appears to be the *in vivo* digestibility of casein fed to rats is shown in Figure 19.4. The use of the term "digestibility" may be somewhat misleading since the major effect here is most likely an interference with the absorption of protein-derived nitrogen rather than the digestion of the protein. In rats fed diets containing red kidney beans or its purified lectin, this failure to absorb protein is manifested by extreme diarrhea (99). Not to be overlooked, however, is the fact that lectins have been shown to inhibit a wide array of enzymes which are involved in the direct digestion of proteins and carbohydrates. These include the *in vitro* inhibition of brush border peptidases and disaccharidases (100–

102) as well as the observation that animals fed raw beans, or the lectin derived therefrom, show decreased levels of these enzymes (12,59,82,92,96,97,103-106). Bean lectins have also been shown to inhibit enterokinase (107), the proteolysis of casein and bovine serum albumin by pepsin and/or pancreatin (108), and the degradation of various starches by pancreatic and salivary amylases (109,110). The reason for the inhibition of such diverse enzymes by lectins is not fully understood, nor is it clear to what extent this inhibitory effect may be responsible for the poor utilization of protein and carbohydrate by animals fed raw beans.

Superimposed on the non-specific effect of lectins on the absorption of nutrients and on the activity of membrane-bound enzymes is the observation that the kidney bean lectin caused a marked increase in the weight and number of cells of the small intestine and an increase in the secretion of mucin (111). All of these effects would be expected to lead to an endogenous loss of nitrogen and thus further exacerbate the toxic effects of the lectin with respect to protein utilization.

In dealing with the feeding of crude food sources that contain lectins to animals, it is important to bear in mind that other food components may also interfere with digestion. The most significant of these are the protease inhibitors (112), tannins or polyphenols (113), phytates (114), and even native, undenatured protein (4). Indeed, in the case of polyphenols, there may be a interference with the absorption of nutrients (115,116) accompanied by gross morphological changes of the brush border (117).

Bacterial colonization.

An alternative explanation for the toxic effects of lectins is suggested by the observation that germfree animals are better able to tolerate raw legumes in their diet as compared to conventional animals (97,118-122). An overgrowth or colonization of coliform bacteria has been observed in the small intestines of rats (73,97,123) and chicks (124,125) fed diets containing raw beans or the purified lectins. Banwell *et al* (73) have attempted to identify the various kinds of organisms that populate the different regions of the small intestines of rats fed the purified lectin of raw red kidney beans. The most characteristic change from control rats was a thinning of the mucous layer of the jejunal surface, which was extensively populated by bacterial cells of two distinct morphotypes—a gram-negative rod and a gram-positive coccobacillus. King *et al* (78) have shown as accumulation of bacteria in the region of the disrupted

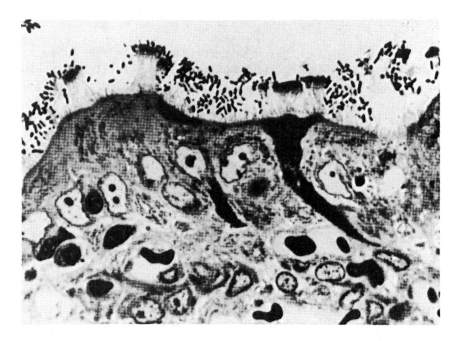

FIG. 19.5. Section through the duodenum of a rat fed for three days on a diet containing 10% kidney bean protein, showing bacteria associated with disrupted microvilli. Stained with toluidine blue. Magnification: X1000. [From King et al. (78)].

microvilli of the duodenum induced by the feeding of raw kidney beans (see Fig. 19.5). Banwell et al (73) are of the opinion that this damage to the microvilli is a direct consequence of bacterial growth. Basteriological examination of the internal organs revealed that the death of animals fed raw beans was accompanied by bacterial infiltration (126). Jayne-Williams (118) has theorized that the lectin-induced damage to the intestinal mucosa alters their permeability so that normally innocuous intestinal bacteria, or the endotoxins that they produce, gain entrance into the bloodstream and produce toxic systemic effects.

The precise mechanism whereby lectins induce the colonization of the small intestine is not known. Wilson et al (123) have suggested the following possibilities: (a) an impairment of the immunological suppression of the growth of certain bacteria, (b) lectin-induced aggregation of bacterial cells, (c) elimination of competitive bacterial strains, and (d) presence of a more suitable substrate for bacterial growth due to the malabsorption of nutrients. A more likely explanation is that the lectins, because of their polyvalency, bind to receptor sites on the brush border as well as to bacteria and thus serve to "glue" bacteria to the surface of the intestines. This lectin-mediated attachment of bacteria to the intestines would be somewhat analogous to the role that lectins are believed to play in the symbiotic relationship between leguminous plants and nitrogen-fixing bacteria (127) or in the lectin-induced attachment of bacteria to oral epithelial cells (128).

Internalization of lectins.
Pusztai and coworkers (87,129-131) have shown that the kidney bean lectin can pass through the intestinal wall and thus enter the circulatory system where they may have a direct toxic effect on certain target tissues. Although the precise mechanism is not known, those systemic effects which have been observed include an increase in protein catabolism, an enhanced breakdown of stored fat and glycogen, a disturbance in mineral metabolism, and an elevation in blood insulin levels.

Significance in the Human Diet

The fact that lectins are so widely distributed in food items commonly consumed by humans (see Tables 19.1 and 19.3) raises the important question of whether they pose any significant risk to human health. Although the lectins of most food items are inactivated by proper heat treatment (31,32,132-136), the possibility of toxic effects produced by the consumption of these same food items in their raw or inadequately cooked state is a very real one. Lectins have even been detected in food items that have been subjected to some form of processing, such as dry cereals and peanuts (137), wheat germ (138) and dry roasted beans (32). One should also consider the fact that the lectins found in most legumes (15,74,76,139-141), wheat germ (138) and tomatoes (142,143) are resistant to the action of digestive enzymes.

There are several reports in the literature of cases of human intoxication in which lectins appear to have been the causative agents. Faschingbauer and Kofler (144) described two cases of poisoning in children who had consumed raw runner beans (*P. coccineus*). Barker (145) described a number of cases of children who had become extremely ill following the ingestion of poke berries from the pokeweed (*Phytolacca americana*) or by systemic exposure to cuts in the hand to berry juice. Examination of the peripheral blood revealed a marked increase in the numbers of plasmacytoid lymphocytes, and it was in fact this observation that led to the discovery of the now well-known mitogenicity of the pokeweed lectin.

In 1948 a severe outbreak of gastroenteritic occurred among the population of West Berlin due to the consumption of partially cooked beans that had been air-lifted into the city during its blockade in 1950 (146). Korte (147) has reported that a mixture of beans and maize prepared by mothers in Tanzania as a porridge for infant food had received an insufficient amount of cooking to destroy its lectin content, and he regarded such preparations as potentially harmful food for infants. This problem presumably arises from the fact that primitive cooking is often done in earthenware pots on a wood fire, so that, with a thick viscous mass like beans and maize, heat transfer may be inefficient. Thus, in the absence of constant and vigorous stirring, significant amounts of the lectin may escape destruction. A reduction in the boiling point of water, such as would be encountered in certain mountainous regions of the world, might also result in the incomplete destruction of the lectin.

A more recent outbreak of intoxication in England has been described that serves to emphasize the risk associated with the consumption of raw or inadequately cooked beans (63,134,148). In 1976 a party of schoolboys on holiday ate kidney beans that had been soaked in water but had not been cooked. All nine who ate the beans became acutely nauseated within 1-1 1/2 hr and began to vomit, followed by diarrhea. As few as four to five beans were sufficient to produce these reactions. Two of the boys were admitted to the hospital and one needed intravenous infusion. However, recovery was rapid in all cases. Several subsequent reports of individuals who had become ill after eating raw beans prompted a television program in which this problem was brought to the attention of the public, who were requested to report similar experiences. This resulted in a response that revealed that 870 individuals had become ill following the consumption of raw beans, usually as part of a salad, or in a stew, casserole, or chili con carne prepared in a slow cooker.

Prompted by these reports, warning labels may now be found on most labels of dried kidney beans sold in retail markets in England (Fig. 19.6). This label recommends that the beans be boiled for at least 10 min. Slow-cooking, without prior boiling of the beans, does not always completely eliminate lectin activity. Detectable levels of hemagglutinating activity still remain after the slow-cooking of kidney beans under household conditions (134-136, 149). For example, cooking kidney beans for 11 hr at 82°C or for 5 hr at 91°C failed to eliminate all of the hemagglutinating activity (150). It is disturbing to note that beans cooked under these

FIG. 19.6. Warning label that has been placed on packets of dry red kidney beans sold in the retail market in England.

conditions were still considered acceptable in terms of texture and palatability by a trained test panel. This would support the documented toxicity of kidney beans cooked in crock pots at relatively low temperatures for long periods of time (148).

Two factors appear to be responsible for these recent outbreaks: (a) the spread of the "back to nature" movement, which extols the virtues of raw foods, and (b) the use of low-temperature cooking vessels. Another problem is created by the use of uncooked or partially cooked bean powders which the consumer is directed to simply mix with hot water. In terms of practical human dietary exposure, only the raw mature beans belonging to the genus *Phaseolus* have sufficiently high levels of lectins to cause untoward reactions. Reference has already been made to the fact that some beans, in their wet immature state, such as green beans in their pods, contain low levels of lectin activity. Other legumes, such as lentils, garden peas, split peas, mung beans and black-eyed peas, have relatively low nontoxic levels of lectins.

Conclusions

For the future we can anticipate an increased world-wide consumption of legumes and oil-seeds as a cheap and nutritious source of protein in the human diet as well as in animal feed. It is evident that the presence of lectins in such plant materials dictates that careful attention be paid to the elimination of this toxic compound by suitable processing techniques. Although lectins can be destroyed by heat in most instances, nevertheless numerous instances can be cited where this has not always been the case. With the advent of modern technology we are apt to overlook the fact that we may not be effectively eliminating the lectins, as exemplified by the slow-cooking process and the introduction of products which can be conveniently eaten after adding water. Even the use of protein isolates offers no assurance that the product is devoid of lectins. Attention should also be given to the possibility that newly introduced plant cultures produced by the plant breeder or by genetic engineering may contain toxic levels of the lectin. Moreover, we are faced with the ever present problem that the absence of toxic manifestations in one animal species does not preclude possible toxic effects in others including humans.

References

1. Stillmark, H., Thesis, University of Dorpat, Tartu, USSR, formerly Estonia, 1888.
2. Boyd, W.C., and E. Shapleigh, *Science 119*:419 (1954).
3. *The Lectins: Properties, Functions and Applications in Biology and Medicine*, edited by I.E. Liener, N. Sharon and I.J. Goldstein, Academic Press, New York, NY, 1986.
4. Kakade, M.L., D.E. Hoffa and I.E. Liener, *J. Nutr. 103*:1772 (1973).
5. Liener, I.E., *J. Biol. Chem. 193*:183 (1951).
6. Liener, I.E., and M.J. Pallansch, *J. Biol. Chem. 197*:29 (1952).
7. Liener, I.E., *J. Nutr. 49*:527 (1953).
8. Turner, R.H., and I.E. Liener, *J. Agric. Food Chem. 23*:481 (1975).
9. Bessler, W., and I.J., Goldstein, *FEBS Lett. 34*:58 (1973).
10. Lis, H., and N. Sharon, *J. Biol. Chem. 253*:3468 (1978).
11. Dorland, L., H. van Halbeck, J.F.G. Vliegenthart, H. Lis, and N. Sharon, *Ibid. 256*:7708 (1981).
12. Jindal, S., G.L. Soni, and R. Singh, *Nutr. Rep. Int. 29*:95 (1984).
13. Grant, G., W.B. Watt, J.C. Stewart, and A. Pusztai, *Med. Sci. Res. 15*:1197 (1987).
14. Liener, I.E., *J. Biol. Chem. 233*:401 (1958).
15. Higuchi, M., M. Suga, and K. Iwai, *Agric. Biol. Chem. (Tokyo) 47*:1879 (1983).

16. Tobin, G., and K.J. Carpenter, *Nutr. Abstr. Rev., Ser. A: Human Exp.* 48:919 (1978).
17. Liener, I.E., *J. Agric. Food Chem.* 22:17 (1974).
18. Jaffé, W.G., *Toxic Constituents of Plant Foodstuffs*, 2nd edn., edited by I.E. Liener, Academic Press, New York, NY, 1980, pp. 69–101.
19. Landsteiner, K., and H. Raubitschek, *Zentralbl. Bakteriol. Parasitenkd., Infektionskr., Abt. 2:Orgin* 45:660 (1908).
20. Jaffé, W.G., A. Planchert, J.I. Paez-Pumar, R. Torrealba and D.N. Francheschi, *Arch. Venez. Nutr.* 6:195 (1955).
21. Jaffé, W.G., *Arzneim.-Forsch.* 10:1012 (1960).
22. Honavar, P.M., C.-V. Shih and I.E. Liener, *J. Nutr.* 77:109 (1962).
23. Wagh, P.V., D.F. Klaustermeier, P.E. Waibel, and I.E. Liener, *Ibid.* 80:191 (1963).
24. Andrews, A.T., and D.J. Jayne-Williams, *Br. J. Nutr.* 32:181 (1974).
25. Jayne-Williams, D.J., and C.D. Burgess, *J. Appl. Bacteriol.* 37:149 (1974).
26. Evans, R.J., A. Pusztai, W.B. Watt, and D.H. Bauer, *Biochim. Biophys. Acta.* 303:175 (1973).
27. Pusztai, A., G. Grant, and R. Palmer, *J. Sci. Food Agric.* 26:149 (1975).
28. Pusztai, A., and R. Palmer, *Ibid.* 28:620 (1977).
29. Pusztai, A., E.M.W. Clarke, T.P. King, and J.C. Stewart, *Ibid.* 30:843 (1979).
30. Pusztai, A., E.M.W. Clarke, G. Grant, and T.P. King, *Ibid.* 32:1037 (1981).
31. Kakade, M.L., and R.L. Evans, *Br. J. Nutr.* 19:269 (1965).
32. DeMuelenaere, H.J.H., *Nature* 201:1029 (1964).
33. Nielsen, S.S., and I.E. Liener, *J. Food Sci.* 53:298 (1988).
34. Brucher, O., M. Wecksler, A. Levy, A. Pallozzo, and W.G. Jaffé, *Phytochemistry* 8:1739 (1969).
35. Jaffé, W.G., O. Brucher and A. Pallozzo, *Z. Immunitaetforsch. Exp. Klin. Immunol.* 142:439 (1972).
36. Jaffé, W.G., and O. Brucher, *Arch. Latinoam. Nutr.* 22:267 (1972).
37. Jaffé, W.G.,, and M.J. Gomez, *Qual. Plant. Plant Foods Hum. Nutr.* 24:359 (1975).
38. Jaffé, W., and C.L. Vega Lette, *J. Nutr.* 94:203 (1968).
39. Grant, G., L.J. More, N.H. McKenzie, J.C. Stewart, and A. Pusztai, *Br. J. Nutr.* 50:207 (1983).
40. DeMuelenaere, H.J.H., *Nature* 206:827 (1965).
41. Manage, L., A. Joshi and K. Sohonie, *Toxicon* 10:89 (1972).
42. Sotelo, A., A.C. Licea, M.T. Gonzalez-Garza, E. Vilasco, and A. Feria-Vilasco, *Nutr. Rep. Int.* 27:329 (1983).
43. Jaffé, W.G., *Qual. Plant Mater. Veg.* 17:113 (1969).
44. Clemens, E., *Landwirtsch. Forsch.* 17:202 (1963).
45. Jenkins, F.P., *J. Sci. Food Agric.* 14:773 (1963).
46. Volkjalmsdottier, L., and H. Fisher, *J. Nutr.* 101:1185 (1971).
47. Gardner, H.K., E.L. D'Aquin, S.P. Koltun, E.J. McCourtney, H.L.E. Vix, and E.A. Gastrock, *J. Am. Oil Chem. Soc.* 37:142 (1960).

48. Fuller, G., H.G. Walker, A.C. Mottola, D.D. Kuznicky, G.O. Kohler, and P. Vohra, *Ibid.* *48*:616 (1971).
49. Olsnes, I., and A. Pihl, *Trends Biochem. Sci.* *3*:7 (1978).
50. Ham., T.H., and W.B. Castle, *Trans. Assoc. Am. Physicians* *55*:127 (1940).
51. Dameschek, W., and E.B. Miller, *Arch. Inter. Med.* *72*:1 (1943).
52. Borchers, R., and C.W. Ackerson, *J. Nutr.* *41*:339 (1950).
53. Orru, A., and V.C. Demel, *Quad. Nutr.* *7*:273 (1941).
54. Shone, D.K., *Rhod. Agric. J.* *58*:18 (1961).
55. Dennison, C., R.H. Stead, and G.V. Quicke, *Agroplantae* *3*:27 (1971).
56. Carlini, C.R., and J.A. Guimaraes, *Toxicon* *19*:667 (1950).
57. Huprikar, S.V., and K. Sohonie, *Enzymologia* *28*:333 (1965).
58. Salgarkar, S., and K. Sohonie, *Indian J. Biochem.* *2*:193 (1965).
59. Higuchi, M., I. Tsuchiya, and K. Iwai, *Agric. Biol. Chem. (Tokyo)* *48*:695 (1984).
60. Mager, J., M. Chevion and G. Glaser *Toxic Constituents of Plant Foodstuffs*, edited by I.E. Liener, Academic Press, New York, NY, 1980, pp. 266-294.
61. Sitren, H., E.M. Ahmed and D.E. George, *J. Food Sci.* *50*:418 (1985).
62. Valdebouze, P., E. Bergerno, T. Gaborit, and J. Delort-Laval, *Can. J. Plant Sci.* *60*:695 (1980).
63. Bender, A.E., *Food Chem.* *11*:309 (1983).
64. Etzler, M.E., *Am J. Clin. Nutr.* *32*:133 (1979).
65. Freeman, H.J., M.E. Etzler, A.B. Garrido and Y.S. Kim, *Gastroenterology* *75*:1066 (1978).
66. Ichev, K., and W. Ovtscharoff, *Acta Histochem.* *69*:119 (1981).
67. Jacobs, L.R., D. DeFontes, and K.L. Cox, *Dig. Dis. Sci.* *28*:422 (1983).
68. Peschke, P., W.D. Kuhlmann, and K. Wurster, *Experientia* *39*:286 (1983).
69. Mahmood, A., and R. Torres-Pinedo, *Biochem. Biophys. Res. Commun.* *113*:400 (1983).
70. Ovtscharoff, W., and K. Ichev, *Acta. Histochem.* *74*:21 (1984).
71. Jacobs, L.R., and P.W. Huber, *J. Clin. Invest.* *75*:112 (1985).
72. King, T.P., A. Pusztai, E.W. Clarke, *Histochem. J.* *12*:201 (1980).
73. Banwell, J.G., R. Howard, D. Cooper, and J.W. Costerton, *Appl. Environ. Microbiol.* *50*:68 (1985).
74. Hara, T., Y. Mukunoki, I. Tsukamoto, M. Miyoshi, and K. Hasegawa, *J. Nutr. Sci. Vitaminol.* *30*:381 (1984).
75. Tedeschi, G.G., F. Putelli, and D. Amici, *Ital. J. Biochem.* *14*:237 (1965).
76. Pusztai, A., E.M.W. Clarke, T.P. King, *Proc. Nutr. Soc.* *38*:115 (1979).
77. Pusztai, A., E.M.W. Clarke, T.P. King, J.C. Stewart *J. Sci. Food Agric.* *30*:843 (1979).
78. King. T.P., A. Pusztai, E.M.W. Clarke, *J. Comp. Pathol.* *90*:585 (1980).
79. King. T.P., A. Pusztai, E.M.W. Clarke, *Ibid.* *92*:357 (1982).
80. Rossi, M.A. J.M. Filho and F.M. Lajolo *Br. J. Exp. Pathol* *65*:117 (1984).
81. Sotelo, A., M.E. Arteago, I. Frias and M.T. Gonzalez-Garza *Qual. Plant. Plant Foods Hum. Nutr.* *30*:79 (1980).

82. Kimura, T., C. Santanochte and A. Yoshida, *J. Nutr. Sci. Vitaminol. 28*:27 (1982).
83. Lorenzsonn, V., and W.A. Olsen, *Gastroenterology 82*:838 (1982).
84. Lorenz-Meyer, H., H. Roth, P. Elasser, and U. Hahn *Eur. J. Clin. Invest. 15*:227 (1985).
85. Torres-Pinedo, R., *J. Pediatr. Gastroenterol 2*:588 (1983).
86. Wieser, M.M., *Chronic Diarrhea in Children* edited by E. Lebenthal, Raven Press, New York, NY, 1984, p. 279-287.
87. Pusztai, A., *Lectins*, Vol. V, edited by I.C. Bog-Hansen and E. Van Driessche, Walter de Gruyter and Co., Berlin, W. Germany, 1986, pp. 317-327.
88. Jaffé, W.G., and G. Camejo, *Acta. Cient. Venez. 12*:59 (1961).
89. Donatucci, D.A., I.E. Liener, and C.J. Gross, *J. Nutr. 117*:2154 (1987).
90. Lasheras, B., J. Bolufer, M.N. Cenarruzabeitia, M. Lluch and J. Larralde, *Rev. Exp. Fisiol. 36*:89 (1980).
91. Santidrian, S., *J. Anim. Sci. 53*:414 (1981).
92. Santidrian, S., B. Lasheras, M.N. Cenarruzabeitia, J. Bolufer, and J. Larralde, *Poultry Sci. 60*:887 (1981).
93. Ishiguro, M., M. Mitarai, H. Harada, I. Sekine, L. Nishimori and M. Kikutani *Chem. Pharm. Bull. 31*:3222 (1983).
94. Ishiguro, M., H. Harada, O. Ichiki, I. Sekine, I. Nishmoni, and M. Kikutani *Ibid. 32*:3141 (1984).
95. Kawatra, B.L., and I.S. Bhatia, *Biochem. Physiol. Pflanz 174*:283 (1979).
96. Dhaunsi, G.S. U.C. Garg, G.S. Sidhu and R. Bhatnagar, *IRCS Med. Sci. 13*:469 (1985).
97. Banwell, J.G., D.R. Boldt, J. Meyers, F.L. Weber Jr., B. Miller, and R. Howard, *Gastroenterology 84*:506 (1983).
98. Dobbins, J.W. J.P. Laurenson, F.S. Gorelick, and J.G. Banwell, *Ibid. 90*:1907 (1986).
99. Banwell, J.G., C.R. Abranowsky, F. Weber, and R. Howard, *Dig. Dis. Sci. 29*:921 (1984).
100. Triadou, N., and E. Audran, *Digestion 27*:1 (1983).
101. Erickson, R.H., and Y.S. Kim *Biochim. Biophys. Acta. 743*:37 (1983).
102. Kim, S., E.J. Brophy, and J.A. Nicholson, *J. Biol. Chem. 251*:3206 (1976).
103. Rouanet, J.-M., and P. Besancon, *Ann. Nutr. Aliment, 33*:405 (1979).
104. Lasheras, B., M.N. Cenaruzabeitia, J. Fontan, M. Lluch, and J. Larralde *J. Rev Esp. Fisiol. 36*:331 (1980).
105. Jindal, S., G.L. Soni, and R. Singh, *IRCS Med. Sci. 10*:214 (1982).
106. Rouanet, J.-M., J. Lafont, M. Chalet, A Creppy, and P. Besancon, *Nutr. Rep. Int. 31*:237 (1985).
107. Rouanet, J.-M., P. Besancon, and J. Lafont, *Experientia 39*:1356 (1983).
108. Thompson, L.U. A.V. Tenebaum and H. Hui, *J. Food Sci. 51*:150 (1986).
109. Rea, R.L., L.U., Thompson and D.J.A. Jenkins, *Nutr. Res. 5*:919 (1985).
110. Thompson, L.U., and J.E. Gabon, *J. Food Sci. 52*:1050 (1987).
111. Pusztai, A., G. Grant, and J.T.A. deOliveira, *IRCS Med. Sci. 14*:205 (1988).

112. Liener, I.E., M.L. Kakade, *Toxic Constituents of Plant Foods* edited by I.E. Liener, Academic Press, New York, NY, 1980, pp. 7-71.
113. Singleton, V.L., and F.H. Kratzer, *Toxicants Occurring Naturally in Foods* Natl. Acad. Sci. Washington, DC., 1973, pp. 309-345.
114. Carnovale, E., E. Lugaro, and G. Lombardi-Boccia, *Cereal Chem. 65*:114 (1988).
115. Motliva, M.J., J.A. Martinez, A. Iludain and J. Larralde, *J. Sci. Food Agric. 34*:239 (1983).
116. Barcini, Y., A.J. Alcalde, A. Iludain and J. Larralde, *Ibid. 35*:996 (1984).
117. Mitjavila, S., C. Lacombe, G. Carrera, and R. Derache, *J. Nutr. 107*:2113 (1977).
118. Jayne-Williams, D.J., *Nature New Biol. 243*:150 (1973).
119. Jayne-Williams, D.J., and D. Hewitt, *Ibid. 35*:331 (1972).
120. Jayne-Williams, D.J., and C.D. Burgess, *J. Appl. Bacteriol. 37*:149 (1974).
121. Hewitt, D., M.E. Coates, M.L. Kakade, and I.E. Liener, *Br. J. Nutr. 29*:423 (1973).
122. Rattray, E.A.S., R. Palmer and A. Pusztai, *J. Sci. Food Agric. 25*:1035 (1974).
123. Wilson, A.B., T.P. King, E.M.W. Clarke, and A. Pusztai, *J. Comp. Pathol. 90*:597 (1980).
124. Untawale, G.G., A. Pietraszek and J. McGinnis, *Proc. Soc. Exp. Biol. Med. 159*:276 (1978).
125. Untawale, G.G., and J. McGinnis, *Poultry Sci. 58*:928 (1979).
126. Untawale, G.G., and J. McGinnis, *Ibid. 55*:2101 (1976).
127. Etzler, M.E., in *The Lectins: Properties, Functions, and Applications in Biology and Medicine*, edited by I.E. Liener, N. Sharon, and I.J. Goldstein, Academic Press, New York, NY, 1986, pp. 371-435.
128. Gibbons, R.J., and I. Dankers, *Arch. Oral Biol. 28*:561 (1983).
129. Williams, P.E.V., A. Pusztai, A, Macdearmid and G.M. Igges, *Animal Feed Sci. Technol. 12*:1 (1984).
130. Grant, G., F. Green, N.H. McKenzie, and A. Pusztai, *J. Sci. Food Agric. 36*:409 (1985).
131. Begbie, R., and T.P. King, in *Lectins: Biology-Biochemistry, Clinical Biochemistry*, Vol. 4 edited by T.C. Bog-Hanson and J. Breborowicz, Walter de Gruyter, Berlin, W. Germany, 1985, pp. 15-27.
132. Liener, I.E., and E.G. Hill, *J. Nutr. 49*:609 (1953).
133. Pak, N., A. Matcluna, and H. Araya, *Arch. Latinoam. Nutr. 28*:184 (1978).
134. Bender, A.E., and G.B. Reaidi, *J. Plant Foods 4*:15 (1982).
135. Grant, G., L.J. More, N.H. McKenzie and A. Pusztai, *J. Sci. Food Agric. 33*:1324 (1982).
136. Thompson, L.U., R.L. Rea, and D.J.A. Jenkins, *J. Food Nutr. 48*:235 (1983).
137. Nachbar, M.S., and J.D. Oppenheimer, *Am. J. Clin. Nutr. 33*:2338 (1980).
138. Brady, P.G., A.M. Vannier, and J.G. Banwell, *Gastroenterology 75*:236 (1978).

139. Jaffé, W.G., and C.L. Vega Lette, *J. Nutr. 94*:203 (1968).
140. Pusztai, A., G. Grant, and R. Palmer, *J. Sci. Food Agric. 26*:149 (1975).
141. Pusztai, A., T.P. King, and E.M.W. Clarke, *Toxicon 20*:195 (1982).
142. Nachbar, M.S., J.D. Oppenheimer, and O.J. Thomas, *J. Biol. Chem. 235*:2056 (1980).
143. Kilpatrick, D.C., A. Pusztai, G. Grant, C. Graham, and S.W.B. Ewen, *FEBS Lett. 185*:299 (1985).
144. Faschingbauer, H., and L. Kofler, *Wien. Klin. Wochenschr. 42*: 1069 (1929).
145. Barker, B.E., *In Vitro 4*:64 (1964).
146. Griebel, C., *Z. Lebensm –Unters. Forsch. 90*:191 (1950).
147. Korte, R., *Ecol. Food Nutr. 1*:303 (1972).
148. Noah, N.D., A.E. Bender, G.B. Reaidi, and R.J. Gilbert, *Br. Med. J. 281*:236 (1980).
149. Lowgren, M., and I.E. Liener, *Qual. Plant. Plant Foods Hum. Nutr. 36*:147 (1986).
150. Coffey, D.G., M.A. Uebersax, G.L. Hosfield, and J.R. Brunner, *J. Food Sci. 50*:78 (1985).

Chapter Twenty

α-Amylase Inhibitors of Higher Plants and Microorganisms

John R. Whitaker

> Department of Food Science and Technology
> University of California
> Davis, CA

Naturally-occurring compounds that inhibit α-amylases have been reported in a number of plants and microorganisms, but there appears to be only two reports of their presence in animals (1,2). α-Amylase inhibitors in cereals were reported as early as 1933 (3,4), again in 1943 (5,6), but were not further studied until 1973 (7,8). After 1973, these

TABLE 20.1

Number of Reports of α-Amylase Inhibitors from Different Sources (1967-1987)[a,b]

Source	No. of reports	Source	No. of reports
Higher plants		*Microorganisms*	
Barley	12	Streptomyces	
Wheat	37	Proteinaceous	53
Millet (ragi)	9	Carbohydrate	25
Beans	17	Unknown	6
Sorghum	3	Cladosporium	1
Rye	3	Total	85
Triticale	1	*Animals*	2
Peanut	2		
Lupin	1		
Corn	1		
Bajra	1		
Acorn	1		
Total	86		

[a]Computer search of Chemical Abstracts in February 1988 for years 1967-1987, plus the addition of some references.
[b]Numbers in Tables 20.1 and 20.2 do not add to the same numbers as a α-amaylase inhibitor from more then one source mentioned in same papers.

TABLE 20.2
Reports on α-Amylase Inhibitors by Year[a,b]

Year	Number	
	Higher plants	Microorganisms
Before 1972	6	0
1972-1977	14	13
1978-1982	18	38
1983-1987	58	36

[a]Based on computer search of Chemical Abstracts in February 1988.
[b]Numbers in Tables 20.1 and 20.2 do not add to the same numbers as α-amylase inhibitor from more than one source mentioned in same papers.

inhibitors in cereals have received much attention, particularly by Silano's group in Italy. In 65 published papers (Table 20.1), α-amylase inhibitors were reported in legumes as early as 1945 (9) and again in 1968 and 1973 (10,11). Jaffé et al. (11) reported α-amylase inhibitor in 79 of the 95 legume cultivars tested. Serious study of the α-amylase inhibitors of legumes began with the detailed work of Marshall and Lauda (12). Some 17 publications were found in the literature after 1966. α-Amylase inhibitors have been reported in maize (13), peanuts (14), mangos (15), taro root (16,17) and acorns (18).

α-Amylase inhibitors of various types were reported in microorganisms, particularly the *Streptomyces*, as early as 1972 (19). Since that time 84 additional publications on α-amylase inhibitors of microorganisms have appeared (Table 20.1).

Table 20.2, showing the publications on α-amylase inhibitors by periods, indicates that these inhibitors are receiving more and more attention. Some of the microbial inhibitors have been successfully cloned in other organisms (20-22), and cDNA clones have been proposed for inhibitors of higher plants (23,24). Considerable evidence on their mechanisms of action is accumulating.

Investigation of the naturally-occurring α-amylase inhibitors is important as model systems for protein-protein interaction, possible control of insects in plants or regulators of reactions (25-27), use in patients with diabetes or hyperglycemia (28) and for analytical uses (29-31).

Specificity of Various Amylase Inhibitors

The higher plant and microbial α-amylase inhibitors show marked diversity in the inhibition of glucosidases. None has been shown to inhibit β-amylase. The only reference to a naturally-occurring β-amylase is glycosyldeoxynojirimycin from a *Streptomyces* sp. (32). Wheat contains a complex array of α-amylase inhibitors (Table 20.3). The bifunctional α-amylase inhibitor/subtilisin inhibitor inhibits α-amylases of wheat, barley, rye and oats, but does not inhibit α-amylase of rice, sorghum, *Tenebrio molitor* larvae, hog pancreas, human saliva, *Aspergillus oryzae* or *Bacillus subtilisin* (33). The 0.19-, 0.28-, 0.53- (35) and WAI-65 (59) inhibitors are active against animal and insect α-amylases and also against a 93,000 MW *B. subtilis* α-amylase (34). The barley bifunctional α-amylase inhibitor/subtilisin inhibitor has similar activity as the wheat bifunctional inhibitor against plant enzymes described above (33,36) while the barley CM_a protein has specificity only for insect α-amylases (37).

The α-amylase inhibitor from corn inhibits several insect α-amylases, corn α-amylase and *B. subtilis* α-amylase but not wheat, rye, barley, triticale or *A. oryzae* α-amylase (13). The α-amylase inhibitors from sorghum, rye, millet, bajra and *Echinocloa fruneutacea* grains inhibit animal α-amylases (38-46) and the inhibitors of rye and *E. fruneutacea* grains are reported to inhibit Thermamyl (Novo bacterial α-amylase; 40) and *A. oryzae* α-amylase (45), respectively. A bifunctional α-amylase inhibitor/trypsin inhibitor has been isolated from millet (41). The sorghum inhibitor is active only against one of the two porcine pancreatic α-amylase isoenzymes (38). The millet α-amylase inhibitor is 100 times more effective on human pancreatic α-amylase than on human salivary α-amylase and is inactive on porcine pancreatic α-amylases (43). The α-amylase inhibitor from *E. fruneutacea* grains inhibits human pancreatic, *A. oryzae*, human salivary and porcine pancreatic α-amylases in the ratio 27:6:1:1 (45). The α-amylase inhibitor of red kidney beans inhibits animal and insect α-amylases but not plant or microbial α-amylases (47). The peanut inhibitor shows similar specificity as a red kidney bean inhibitor but was apparently not tested on insect α-amylases (14).

Certain microorganisms produce complex mixtures of α-amylase inhibitors. Tajiri *et al.* (60) reported the separation of 14 kinds of amylase inhibitors from cultures of *Streptomyces* species No. 80. Some were proteins, others were nitrogen-containing carbohydrates. The sizes ranged from 310-1300 daltons by gel filtration. The polypeptide inhibi-

TABLE 20.3
Specificity of Various α-Amylase Inhibitors

Source	Designation (if any)	α-Amylases Animal	Insects	Plant	Microbial	Gluco-amylase	Other gluco-hydrolases	References
Wheat	1. Bifunctional α-amylase/subtilisin inhibitor	-[a]		+	-	-	-	33
	2. α-Amylase inhibitors	+	+	-	+[a]			34,35
Barley	1. Bifunctional α-amylase subtilisin inhibitor		+	+				33,36
	2. CM$_a$		+					37
Corn		-	+	+[c]	+[d]			13
Sorghum		+			-			38,39
Rye		+		-	+[f]			40
Millet	1. Bifunctional α-amylase/tripsin inhibitor	+						41
	2. Amylase inhibitor	+[g]		-[ ]				42-44
	3. *Echinocloa fruneutacea* grain	+[h]		+[h]				45
Bajra		+						46
Beans	1. Red kidney bean	+	+	-	-			47
Peanuts		+		-	-			14

TABLE 20.3 — continued
Specificity of Various α-Amylase Inhibitors

Source	Designation (if any)	α-Amylases Animal	Insects	Plant	Microbial	Gluco-amylase	Other gluco-hydrolases	References
Streptomyces								
	1. Haim	+[i]						48,49
	2. Paim	+[j]						48,50,51
	3. S-AI[k]	+				+	−	33, 52–54
	4. S-GI[k]	+[l]			−[l]	+		33, 52–54
	5. BAYe 4609 and BAYg 5241	+						55,56
Streptomyces calidus								
NRRL 8141		+					+	57
Streptomyces nigrifacien								
NTU-3314		+				+		58

[a] ———, no inhibition found; +, inhibition found; blank, data unknown.
[b] On a *Bacillus subtilis* α-amylase of 93,000 MW.
[c] Inhibitory of corn α-amylase only; not wheat, rye, barley, or triticale.
[d] Inhibitory of *B. subtilis* α-amylase, no *Aspergillus oryzae* α-amylase.
[e] Strong inhibition of human salivary and pancreatic α-amylases, bovine pancreatic α-amylase and porcine pancreatic α-amylase (less anionic form); no inhibition of more anionic porcine pancreatic α-amylase.
[f] Thermamyl (NOVO bacterial α-amylase).
[g] 100 x more effective on human pancreatic α-amylase than human salivary α-amylase; no inhibition of porcine pancreatic α-amylase inhibitor.
[h] 27:6:1:1 ratio for inhibition of human pancreatic, *Aspergillus oryzae*, human salivary and porcine pancreatic α-amylases.
[i] Strongly inhibits α-amylases from vertebrates; weakly inhibits α-amylase from invertebrates.
[j] Does not inhibit human and rabbit salivary and pancreatic α-amylases.
[k] From *Streptomyces diastaticum amylostaticus* and *S. lavendulae*, respectively.
[l] Inhibits animal invertases, not microbial invertases.

TABLE 20.4
Some Chemical and Physical Properties of Microbial Peptide α-Amylase Inhibitors

Name	Source	Molecular weight	Subunits no.	Carbohydrates content	Stoichiometry (amylase:inhibitor)	Reference
Paim[a]	S.[b] *corchorushii*	7420(73)[c]	1	0	1:2	48,61
	Streptomyces sp. No. 291-10	~6000				62
Z-2685	S. *parvullus* FH-1641	8129(76)[c]				63
AI-3688	S. *aureofaciens*	3936				64
Haim II[d]	S. *griseosporus*	~8500(77)[c]	1	0	1:1	49,65,66,67
Hoe-467A	S. *tendae* 4158	7958(74)[c]	1	0	1:1	68,69,70
(Tendamistat)						
AI-B	S. *viridosporus* No. 297-A2	8000				71
X-2	S. *fradiae*	6500		0		72

[a]Paim, pig pancreatic α-amylase inhibitor from microbe.
[b]*Streptomyces*.
[c]Number of amino acid residues per mol.
[d]Haim I has 75 amino acids; Haim, human pancreatic α-amylase inhibitor from microbe.

tors, such as Haim (human α-amylase inhibitor from microbes) and Paim (pig α-amylase inhibitor from microbes), inhibit animal α-amylases while the carbohydrate inhibitors also are active against glucoamylase, sucrase and often phosphorylase a. Haim and Paim are homologous proteins (Table 20.4), yet Paim, unlike Haim, does not inhibit human and rabbit α-amylases. Both inhibit pig, dog, cow and horse α-amylases.

Chemical and Physical Properties of α-Amylase Inhibitors

The α-amylase inhibitors include proteins with molecular weights of 12,000 to 60,000 (Table 20.5), polypeptides ranging from 3936 to 8500 (Table 20.4) and nitrogen-containing carbohydrates ranging from 310-1300 (60).

Higher plant inhibitors.

α-Amylase inhibitors have been reported in wheat, barley, millet, sorghum, corn, *Echinocloa fruneutacea* grain, peanut and bean (see Table 20.5). At least seven different α-amylase inhibitors exist in wheat. At least four of them belong to four main families with molecular weights of 60,000, 44,000, 24,000 and 12,000 (88). These four families of inhibitors apparently derived from common ancestral genes. Proposed interrelationships among them are shown in Figure 20.1 (89). The 60,000 and 44,000 molecular weight components have not yet been studied extensively. The WAI-65, 0.19- and 0.53-inhibitors, all of molecular weight near 24,000, contain two subunits. The 0.19 inhibitor is known to form a 1:1 complex with α-amylase; therefore, there is only one binding site for enzyme. Yet, two forms of the complex could be separated (89). Two molecules of the 0.28-inhibitor, of molecular weight 12,000, bind to α-amylase, presumably (but not shown directly) in the catalytic site and in the glycogen binding site (Fig. 20.1). The bifunctional α-amylase inhibitor/subtilisin inhibitor of 20,500 daltons inhibits α-amylase and subtilisin simultaneously. The N-terminal sequence (45 residues) of the bifunctional inhibitor is homologous with cereal and leguminous inhibitors of the soybean trypsin inhibitor (Kunitz) family (33). The 0.19-, 0.28- and 0.53-inhibitors show considerable homology (90). The complete amino acid sequences of the 0.19-, 0.28- and 0.53-inhibitors are available (74,76,78).

TABLE 20.5
Chemical and Physical Properties of Plant α-Amylase Inhibitors

Source	Molecular weight	Subunits (No.)	Carbohydrate (%)	PI	Stoichiometry (E:I)	Reference
Wheat						
1. Endogenous α-amylase inhibitor	19,641(180)[a]	1				73
2. WAI-65	24,000	2				59
3. 0.19[b]-inhibitor	26,500(248)[a]	2		7.1	1:1	31,74,75
4. 0.28[b]-inhibitor	12,000(13,400)			6.2	1:2	31,75,76
5. 0.53[b]-inhibitor	26,390	2 (identical)				77,78
6. Amylase/subtilisin inhibitor[c]	20,500			7.2		33
Barley						
1. Amylase inhibitor	20,000			7.3		79
2. Amylase/subtilisin inhibitor	19,685(181)[a]	1	0			80
Millet						
1. Foxtail	19,000			10		81
2. Proso	14,000					43
3. α-Amylase inhibitor 1-2	9,333(95)[a]	1				82
4. Amylase/trypsin inhibitor	13,300(122)[a]	1		>10		83
5. Amylase/trypsin inhibitor	12,000					41
6. *Echinocloa fruneutacea* grain (millet)	15,000					45

TABLE 20.5 — continued
Chemical and Physical Properties of Plant α-Amylase Inhibitors

Source	Molecular weight	Subunits (No.)	Carbohydrate (%)	PI	Stoichiometry (E:I)	Reference
Sorghum	21,000	2				39
Corn	29,600	1	0			13
Peanut	25,000		0		1:2	14,84
Bean						
Red kidney	49,000	4	8.6;13.0		1:1	47,85
Black kidney bean I	49,000 (55,000)^d	3	Yes	4.93	1:1	86
II	47,000 (49,000)^d					
White kidney bean	49,000	2 (at least)	Yes	4.86	1:1	86
Pinto	45,000		9–10		1:1	12
			14	4.68		87

^{a}Number of amino acids per mol.
^{b}So named based on their electrophoretic motility relative to bromophenol blue.
^{c}May be same as the wheat endogenous α-amylase inhibitor (see 1. above).
^{d}Unpublished data.

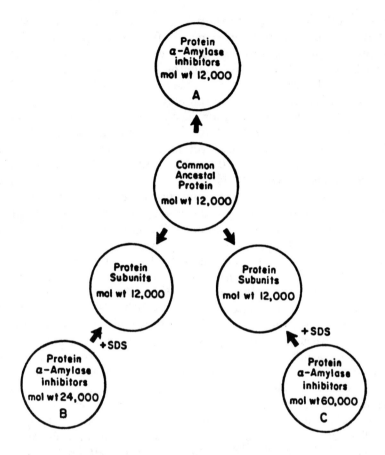

Fig. 20.1. Possible interrelationships among the protein α-amylase isoinhibitor families from hexaploid wheats (89). All three inhibitors may have originated from a common ancestral protein of 12,000 daltons.

Barley contains at least three α-amylase inhibitors. The amylase inhibitors and the bifunctional α-amylase inhibitor/subtilisin inhibitor have molecular weights near 20,000 (Table 20.5). The bifunctional inhibitor binds one molecule of α-amylase and one of subtilisin simultaneously. The third protein, CMa, specifically inhibits *Tenebrio molitor* larvae α-amylase (37). The CM-proteins of barley are a group of five salt-soluble proteins that can be selectively extracted from the endosperm with chloroform/methanol mixtures. The N-terminal sequences of the five proteins are homologous, indicating that they are members of a dispersed multi-gene family. CMc and CMe inhibit trypsin while CMb and CMd do not inhibit trypsin or α-amylase from saliva, pancreas, *A. oryzae*, *Tenebrio molitor*, or barley (37). The complete amino acid sequence of the bifunctional α-amylase inhibitor/subtilisin inhibitor has been determined (80). There is considerable homology with members of the soybean trypsin inhibitor (Kunitz) family.

Millet contains at least five α-amylase inhibitors ranging in molecular weight from 10,700 to 19,000 (Table 20.5). Two of the proteins have pI values at 10 or above (very basic proteins). The amino acid sequences of the α-amylase inhibitor I-2 (10,700 MW) (82) and the bifunctional α-amylase inhibitor/trypsin inhibitor (83) from the same cultivar have been determined. There is no sequence homology between the two proteins. The α-amylase inhibitor I-2 has no sequence homology with known α-amylase inhibitors from other plants. On the other hand, the bifunctional inhibitor has sequence homology with trypsin inhibitors from barley and corn and with α-amylase inhibitor from wheat (83).

α-Amylase inhibitory activity was reported in 79 of the 95 legume cultivars tested (11). Inhibitors have been purified from red, white and black kidney beans and pinto beans (Table 20.5). They are 3–5% of the total protein. The proteins from all four varieties have similar properties. They have molecular weights close to 50,000, contain subunits, appreciable carbohydrate, pI values of 4.7–4.9 and form 1:1 stoichiometic complexes with porcine pancreatic α-amylase. Yet each can be distinguished chromatographically and electrophoretically. Even inhibitors in cultivars of the same variety (red kidney and white kidney beans) can be distinguished chromatographically and electrophoretically. The basis of these differences is not known; however, Powers and Whitaker (47) reported 8.6% carbohydrate in red kidney beans while Wilcox and Whitaker (91) reported 13.0% for a different cultivar of red kidney beans. Methodology used was the same. The red kidney bean α-amylase inhibitor contains four sulfhydryl groups per mol (47) while the black (92) and white kidney (unpublished data) bean α-amylase inhibitors contain one and zero sulfhydryl group per mol respectively. Red kidney bean α-amylase inhibitor probably contains four subunits of three different types based on molecular weight of the subunits, four sulfhydryl groups/mol, and electrophoretic separation of three bands by sodium dodecyl sulfate polyacrylamide gel electrophoresis, with staining of the center band being about twice as intense as the two outer bands (47). Black kidney bean α-amylase inhibitor gives three bands in a similar manner as for red kidney bean, but the authors suggested there are three subunits since three different N-terminal amino acids were

```
Hoe-467A                                                                     Asp-Thr-Thr-
Z-2685                                                                               Ala-
Haim II
Paim I

                                                  10                                  20
Hoe-467A  Val-Ser-Glu-Pro-Ala-Pro-Ser-Cys-Val-Thr-Leu-Tyr-Gln-Ser-Trp-Arg-Tyr-Ser-Gln-Ala-
Z-2685    Thr-Gly-Ser-Pro-Val-Ala-Glu-Cys-Val-Glu-Tyr-Phe-Gln-Ser-Trp-Arg-Tyr-Thr-Asp-Val-
Haim II           Ile-Ala-Ala-Pro-Ala-Cys-Val-His-Phe-Thr-Ala-Asp-Trp-Arg-Tyr-Thr-Phe-Val-
Paim I    Ala-Ser-Glu-Pro-Ala-Pro-Ala-Cys-Val-Val-Met-Tyr-Glu-Ser-Trp-Arg-Tyr-Thr-Thr-Ala-

                                                  30                                  40
Hoe-467A  Asp-Asn-Gly-Cys-Ala-Glu-Thr-Val-Thr-Val-Lys-Val-Val-Tyr-Glu-Asp-Asp-Thr-Glu-Gly-
Z-2685    His-Asn-Gly-Cys-Ala-Asp-Ala-Val-Ser-Val-Thr-Val-Glu-Tyr-Thr-His-Gly-Gln-Trp-Ala-
Haim II   Thr-Asn-Asp-Cys-Ser-Ile-Asp-Tyr-Ser-Val-Thr-Val-Ala-Tyr-Gly-Asp-Gly-Thr-Asp-Val-
Paim I    Ala-Asn-Asn-Cys-Ala-Asp-Thr-Val-Ser-Val-Ser-Val-Ala-Tyr-Gln-Asp-Gly-Ala-Thr-Gly-

                                                  50                                  60
Hoe-467A  Leu-Cys-Tyr-Ala-Val-Ala-Pro-Gly-Gln-Ile-Thr-Thr-Val-Gly-Asp-Gly-Tyr-Ile-Gly-Ser-
Z-2685    Pro-Cys-Arg-Val-Ile-Glu-Pro-Gly-Gly-Trp-Ala-Thr-Phe-Ala----Gly-Tyr-Gly-Thr-Asp-
Haim II   Pro-Cys-Arg-Ser-Ala-Asn-Pro-Gly-Asp-Ile-Leu-Thr-Phe-Pro----Gly-Tyr-Gly-Thr-Arg-
Paim I    Pro-Cys-Ala-Thr-Leu-Pro-Pro-Gly-Ala-Val-Thr-Thr-Val-Gly-Glu-Gly-Tyr-Leu-Gly-Glu-

                                                  70
Hoe-467A  His-Gly-His-Ala-Arg-Tyr-Leu-Ala-Arg-Cys-Leu
Z-2685    Gly-Asn-Tyr-Val-Thr-Gly-Leu-His-Thr-Cys-Asp-Pro-Ala-Thr-Pro-Ser
Haim II   Gly-Asn-Glu-Val-Leu-Gly-Ala-Val-Leu-Cys-Ala-Thr-Asp-Gly-Ser-Ala-Leu-Pro-Val-Asp
Paim I    His-Gly-His-Pro-Asp-His-Leu-Ala-Leu-Cys-Pro-Ser-Ser
```

Fig. 20.2. Amino acid sequence of the *Streptomyces* polypeptide α-amylase inhibitors Hoe-467A, Z-2685, Haim II and Paim I (48).

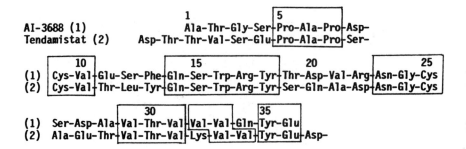

Fig. 20.3. Comparison of the amino acid sequences of AI-3688 and Hoe-467A (64).

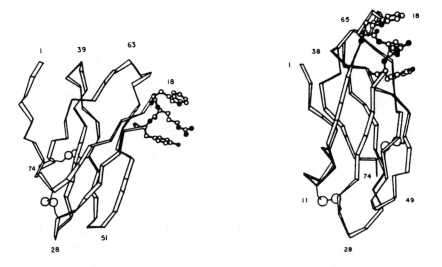

Fig. 20.4. Tertiary structure of α-amylase inhibitor Hoe-467A (94).

kidney bean α-amylase inhibitor gives two bands by sodium dodecyl sulfate polyacrylamide electrophoresis of ≃20,000 daltons each (unpublished data).

Polypeptide α-amylase inhibitors from Streptomyces.

Various species of *Streptomyces* produce polypeptides of 3936–8500 molecular weight which inhibit animal α-amylases (Table 20.4). They all presumably have a single peptide chain and no carbohydrate. Haim II and Hoe-467A form 1:1 stoichiometric complexes with α-amylases while two molecules of Paim I bind to each α-amylase molecule (61).

There is considerable sequence homology among Paim I, Haim II, Hoe-467A, Z-2685 (Fig. 20.2, ref. 48) and AI-3688 (Fig. 20.3, ref. 64). As shown in Figure 20.2, there are four Cys residues at the same aligned position in each molecule. These are residues 8, 24, 42 and 70 in Paim I. Val is in positions 9, 30 and 32, Try-Arg-Tyr at positions 15-17, Asn at position 22, Try at positions 17, 34 and 57, Pro-Gly at positions 47 and 48, Thr at position 52, Gly-Tyr at positions 56 and 57 and Gly at positions 48 and 56 in all four inhibitors. As shown in Figure 20.3, there is considerable homology between AI-3688 and the amino terminal end of Hoe-467A even though the molecular weights are 3936 and 7958, respectively. Note that the numbering of amino acid residues in Figure 20.3 is displaced one position to the left of that in Figure 20.2.

The tertiary structure of Hoe-467A has been determined (94) as shown in Figure 20.4. The polypeptide chain from Cys 11 to Cys 73 is folded into a β-barrel. The two twisted sheets are made up of three antiparallel strands each. The Trp-Arg-Tyr sequence (18-20 in Fig. 20.4, 15-17 in Fig. 20.2) is a prominent feature of the molecule, consistent with the postulation (48,95) that this represents a part of the active site of Hoe-467A. Hirayama et al. (48) suggested that the region Gly58-Val68 (Paim I) and Leu58-Ala68 (Haim II) may also be part of the binding site for α-amylases. This region in Hoe-467A is adjacent to the Trp15-Arg16-Tyr17 sequence (Fig. 20.4).

Nitrogen-containing basic oligosaccharide amylase inhibitors.

Several nitrogen-containing basic oligosaccharides inhibitory of glucosidases are produced by various *Streptomyces* (Fig. 20.5). Oligostatins C, D and E are produced by *S. myxogenes* nov. sp. SF-1130 (96). The three compounds contain a pseudodisaccharide, oligobioamine (Structure A), and variable numbers of α-D-glucose units linked α-1,4. In oligostatin C, the oligobioamine unit is at the nonreducing end followed by three

Structure A

A. OLIGOSTATINS

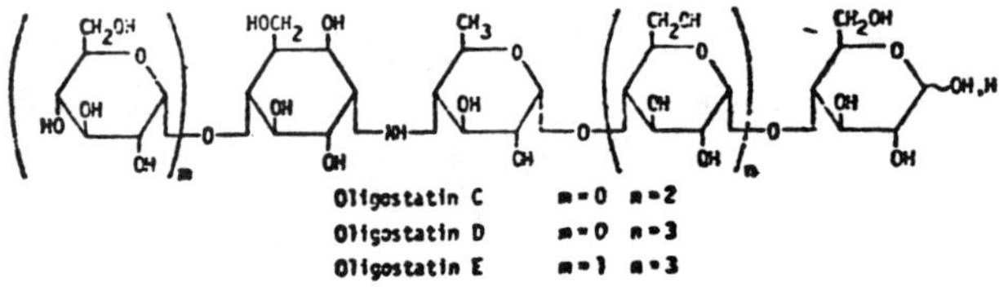

Oligostatin C	m = 0	n = 2
Oligostatin D	m = 0	n = 3
Oligostatin E	m = 1	n = 3

B. AMYLOSTATINS

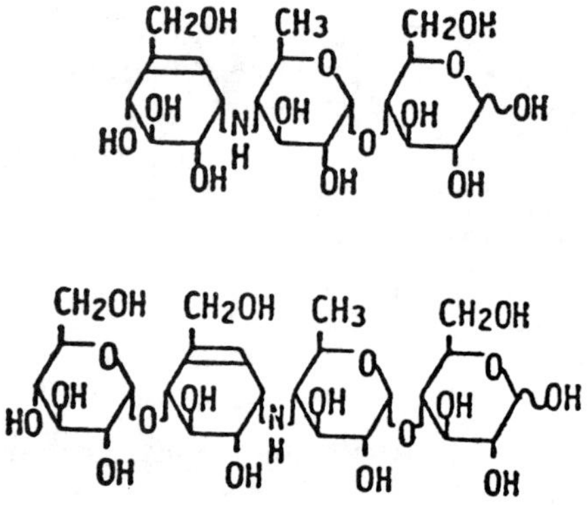

C. TRESTATINS

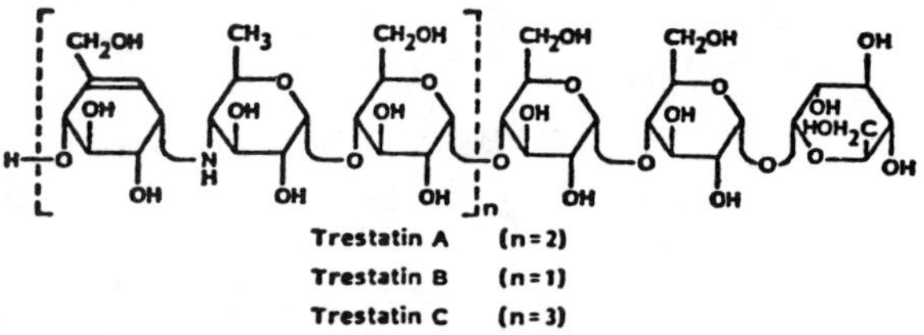

Trestatin A	(n = 2)
Trestatin B	(n = 1)
Trestatin C	(n = 3)

Fig. 20.5. Structures of oligostatins, amylostatins and trestatins.

glucose units. Oligostatin D is similar to oligostatin C, but contains one additional glucose unit (four total) at the reducing end. Oligostatin E has one glucose unit at the nonreducing end and four glucose units at the reducing end. The oligostatins inhibit α-amylases as well as have antibiotic activity against Gram-negative bacteria.

The amylostatins are produced by *D. diastaticus* subsp. *Amylostaticus* No. 9410 (97). They consist of D-glucose plus the pseudodisaccharide, dehydro-oligobioamine (Structure B; unsaturated cyclitol unit (configuration of constituents corresponding to an α-D-glucose unit) bound to the amino sugar, 4,6-dideoxy-4-amino-D-glucopyranose and variable number of α-D-glucose units (Fig. 20.5). There are two types. In one, the dehydro-oligobioamine is at the nonreducing end followed by 1 to 3 α-1,4-D-glucose units at the reducing end. In the second, one D-glucose unit is at the nonreducing end, followed by dehydro-oligobioamine and 1 to 3 α-1,4-D-glucose units at the reducing end. The amylostatins have different inhibitory activities against different amylases and invertase (97). Both the number of D-glucose units and the positions of the dehydro-oligobioamine unit (at nonreducing end vs central) determine the effectiveness of the amylostatins on several amylases and invertase. Endo-type α-amylases such as bacterial liquefying α-amylase, bacterial saccharifying α-amylase and taka-amylase A were more strongly inhibited by GXG, GXGG or GXGGG than by XG, XGG or XGGG (where X = dehydro-oligobioamine, G = glucose). On the other hand, glucoamylase, which is an exo-type amylase, was strongly inhibited by XG or XGG. An exception was porcine α-amylase, an endo-type which was strongly inhibited by XGGG as well as by GXG, GXGG and GXGGG. These compounds were described earlier by Schmidt *et al.* (56). XGG (designated BAYg 5421) was the strongest inhibitor of invertase, while GGXGG and GGGXGG were the strongest inhibitors of animal α-amylases, including the human enzymes.

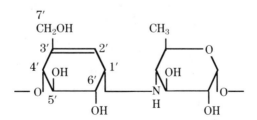

Structure B

The third group of oligosaccharide amylase inhibitors are the trestatins (98). Trestatins A, B and C (Fig. 20.5) contain 2, 1 or 3 repeats of the pseudo disaccharide, dehydro-oligobioamine, plus one D-glucose unit at the nonreducing end, followed by three additional D-glucose units. The first two D-glucose units are linked α-1,4 and the last one α-1,1 (trehalose). Unlike the oligostatins and amyglostatins, the trestatins are nonreducing oligosaccharides. Smaller amounts of three other trestatins have been isolated where two D-glucose units are linked α-1,4 at the nonreducing ends of trestatins A, B and C (preceding the dehydro-oligobioamine units) (96).

Mechanism of Action of the α-Amylase Inhibitors

Several approaches to understanding the mechanism of action of the α-amylase inhibitors have been used. These are: (a) comparison of primary sequence and specificity of the inhibitors for several glucosidases; (b) comparison of tertiary structures and specificity of inhibitors for several glucosidases; (c) chemical modification of inhibitors and/or α-amylases and effect on complex formation; and (d) kinetic approaches. We are a long way from understanding the mechanism of any one of the inhibitors and further away from the answer as to whether there is a common mechanism. This lack of understanding of the mechanism is further complicated by incomplete information of the tertiary structure of α-amylase.

α-Amylase structure.

The complete amino acid sequence of porcine pancreatic α-amylase isoenzyme I (α-1,4-glucan-4-glucanohydrolase; EC 3.2.1.1) is known (99,100). The protein is a single polypeptide chain of 496 amino acids, with a pyrrolidone carboxylic acid residue (cyclization of glutamic acid residue) at the amino-terminal end and Leu at the carboxyl-terminal end. The two cysteine residues are at positions 103 and 119 and the five disulfide bridges involve cysteines 28 and 86, 70 and 115, 141 and 160, 378 and 384, and 450 and 462 (100).

There are several reports on the tertiary structure of α-amylase by X-ray crystallography (101-103). Unfortunately, the 2.9 Å resolution structure has only been published in abstract forms (103). A balsa-wood model of the α-amylase molecule is shown in Figure 20.6 (102). There is a deep cleft, indicated by the arrow, which runs for 30 Å on one side of the molecule (dimensions of molecule, a = 56.3, b = 87.8 and c = 103.4 Å).

The cleft divides the molecule into two domains, a large N-terminal domain (residues 1-410) with "an 8-stranded singly wound β-barrel (successive strands being often connected by parallel helical regions) and a smaller C-terminal domain (residues 410-496) with essentially β-sheet" structure (103). The deep cleft is thought to contain the active site of α-amylase; this is supported by X-ray crystallographic data showing the binding of a substrate analog (Structure C)

Structure C

into the cleft. A second molecule of substrate analog binds on the surface of the α-amylase 20 Å away from the active site. By a completely different method, Loyter and Schramm (104) found two binding sites for glycogen limit dextrin. We have independently verified this by showing that two molecules of maltotriose, a poor substrate, bind to porcine pancreatic α-amylase, as determined by equilibrium dialysis (unpublished data). The two cysteine residues (No. 103 and 119) are not located in the active site cleft but are in a sort of tunnel next to the cleft (102). They are not essential for α-amylase activity. The Cl⁻ ion, which serves as an activator, is located at the center of the large β-barrel (as shown by Br⁻ substitution). The essential Ca^{2+} ion maintains one stretch of the polypeptide chain in the vicinity of the β-barrel, which would explain why it is essential for the integrity of the 3D structure of the enzyme (103).

Comparison of primary sequence and specificity of the inhibitors.

The best comparative data are for the small polypeptide α-amylase inhibitors from *Streptomyces* (Fig. 20.2; ref. 48, 64). The structural similarities of Paim I, Haim II, Hoe-467A, Z-2685 and AI-3688 are described above (CHEMICAL AND PHYSICAL PROPERTIES OF α-AMYLASE INHIBITORS). As aligned, there is complete agreement among the five polypeptides in the location of the four half-cysteine residues and the -Trp15-Arg16-Tyr17-sequence. Based on the difference in α-amylase specificities of Paim I and Haim II, Hirayama et al. (48) suggested that amino acid sequences —Gly58-Val68- in Paim I and Leu58-Ala68 in Haim II might be part of the binding sites for α-amylases. On the other

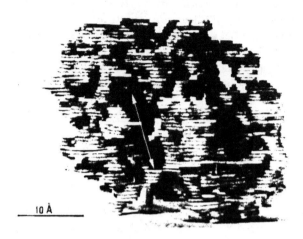

Fig. 20.6. A balsa-wood model of the porcine pancreatic α-amylase molecule. The active site is located in the crevice indicated by the long white arrow (102).

hand, AI-3688 does not contain this region since it ends at residue 36 (64), but it does inhibit α-amylases. The two disulfide bonds are between Cys residues 8 and 24 and 42 and 70 in all four proteins, indicating that the proteins probably have similar tertiary structures (65).

The primary amino acid sequences for several higher plant α-amylase inhibitors are known. These include the bifunctional α-amylase inhibitor/subtilisin inhibitor from barley (80), endogenous α-amylase inhibitor in wheat (73), the 0.19-inhibitor from wheat (74), the 0.53-inhibitor of wheat (78)—including the location of the four disulfide bonds (77), the 0.28-inhibitor of wheat (76), the α-amylase inhibitor I-2 from millet (82) the bifunctional α-amylase inhibitor/trypsin inhibitor of millet (83) and the α-amylase inhibitor of corn (13). There is considerable homology among these inhibitors of higher plants, with the trypsin inhibitors of barley (105) and other cereals and with the Kunitz soybean inhibitor family (80). Bifunctional inhibitors of α-amylase and subtilisin have been isolated from barley (80) and wheat (33) and a bifunctional inhibitor of α-amylase and typsin from millet (83). Comparative examination of the amino acid sequences of the eight inhibitors above do not provide much information on the location of the binding region with α-amylases.

Tertiary structures and specificity of inhibitors.

Little information is available on the tertiary structure of the proteinaceous inhibitors. The crystal structure of the polypeptide Hoe-467A produced by *S. tendae* 4158 has been resolved at 2.5 Å (94) as shown in Figure 20.4. The two twisted sheets are assembled from three antiparallel strands each. There are β-turns at residues 17–20, 37–40 and 49–52. The most noticeable feature of the inhibitor, in relation to its possible mechanism of action, is the prominence of the -Trp18-Arg19-Try20- (Try15-Arg16-Try17 in Fig. 20.2) where the aromatic sidechains are stacked, and the location of the Leu61-Ala71 sequence is adjacent to the -Trp18-Arg19-Try20- sequence. Both segments are postulated to be part of the recognition site for α-amylase (48). This hypothesis is supported by spectroscopic studies of the complex of Haim II and porcine pancreatic α-amylase indicating the involvement of tryptophan and tyrosine residues in complex formation (49), and also by chemical modification (95). Some limited crystallographic data are available for one of the wheat α-amylase inhibitors (106). Two forms of monoclinic crystals were produced. The rod-like crystals had molecular dimensions of 43.5, 64.8 and 32.2 Å, while the plate-like crystals had molecular dimensions of 42.5, 65.2 and 37.2 Å. It is hoped that more complete data will be available soon. X-ray crystallographic data are also needed on one or more of the α-amylase/α-amylase inhibitor complexes.

Chemical modification of inhibitors and of α-amylase.

Chemical modification has been used to a limited extent to determine the essentiality of amino acid residues in complex formations between α-amylase and inhibitors. Chemical modification of Haim gave evidence for the importance of tryptophan, tyrosine and arginine for complex formation with α-amylase (95). When Haim was treated with 1,2-cyclohexanedione, one of the three arginines was modified with major loss of activity. Modification of one of the four tyrosine residues with tetranitromethane caused greater than 90% loss of activity. Modification of tryptophan almost completely eliminated the activity of Haim.

The effect of limited chemical modification of red kidney bean α-amylase inhibitor on its ability to bind with α-amylase has been reported (85). Removal of 70% of the carbohydrate from the inhibitor did not cause loss in activity, nor were the carbohydrate groups alone inhibitory of α-amylase at 3.3×10^5 fold molar excess. Oxidation of one tryptophan residue with N-bromosuccinimide gave 50% loss of inhibitory activity.

Periodate oxidation of the inhibitor led to loss of two tyrosine and one methionine residues per mol; there was not a direct correlation between oxidation of these residues and loss of inhibitory activity. Modification of three of the five histidine residues with diethylpyrocarbonate resulted in about 50% loss in inhibitory activity.

Effect of chemical modification of porcine pancreatic α-amylase on the ability to form a complex with black kidney bean inhibitor has been reported (107). Modification of the histidine residues of α-amylase by photooxidation with rose bengal or by derivatization with diethylpyrocarbonate, tryptophan residues by N-bromosuccinimide and tyrosine residues with N-acetylimidazole reduced α-amylase activity but none of the modifications affected complex formation with the inhibitor. Modification of cysteine with 5,5'-dithio-2-nitrobenzoate had no effect on α-amylase activity or on complex formation with the inhibitor. Whitaker et al. (92) showed that modification of the single cysteine residue of the black kidney bean α-amylase inhibitor or the two cysteine residues of porcine pancreatic α-amylase had no effect on complex formation. Cl$^-$ ions, required for activity of α-amylase at both pH 5.40 and 6.90, were important for inhibitor-enzyme binding at pH 6.90 but not at pH 5.40 (92). Calcium-free α-amylase bound inhibitor at both pH 6.90 and 5.40.

Solvent perturbation studies indicated that the two tryptophan residues on the surface of wheat α-amylase inhibitors alone are no longer accessible to the solvent in the complex (108).

Some limited chemical studies have been done on the basic pseudo oligosaccharides from *Streptomyces*. Treatment of the inhibitors with β-amylase removes maltose units from some of the amylostatins (Fig. 20.5). The inhibitory activity against α-amylases is reduced while activity against invertase is increased (97).

Kinetic studies of α-amylase/α-amylase inhibitor complex formation.

Very few kinetic studies, except for the red and black kidney bean inhibitors, have been done. There is kinetic evidence that the smaller microbial polypeptide and carbohydrate inhibitors do not require preincubation with α-amylases prior to adding the substrate as is required for the larger proteins of higher plants (48). Depending on experimental conditions, complex formation, as measured by loss of α-amylase activity, requires 30-120 minutes of preincubation to reach equilibrium prior to adding the substrate (109). The rate of complex formation is at least 20 times faster at pH 5.40 than at pH 6.90 (92,109,110). Interestingly, the reaction is second order at pH 6.90 and first order at pH 5.40 (92,110). The complex formed with black kidney bean inhibitor and porcine pancreatic α-amylase had a dissociation constant (K_d) of 1.7×10^{-10}M at pH 5.5 and 4.4×10^{-9}M at pH 6.90 (110). For red kidney bean inhibitor and porcine pancreatic α-amylase at pH 6.9, K_d was 3.5×10^{-11}M (109). The Ca^{2+}-free porcine pancreatic α-amylase bound to black kidney bean inhibitor at the same rates at pH 5.40 and 6.90 as with the Ca^{2+}-containing enzyme (92).

Wilcox and Whitaker (91), based on detailed kinetic studies at pH 6.90 in the presence and absence of *p*-nitrophenyl α-maltoside substrate, proposed that red kidney bean inhibitor and bovine pancreatic α-amylase form an initial complex (EI) that is fully active but with time changes to a complex (EI*) with ≈5% of original activity on β-nitrophenyl α-maltoside, but no activity on starch (Equation a).

$$E + I \underset{k_{-1}}{\overset{k_1}{\rightleftarrows}} EI \underset{k_{-2}}{\overset{k_2}{\rightleftarrows}} EI^* \tag{a}$$

The K_{eq} (k_{-1}/k_1) for the first step was determined to be 3.1×10^{-5}M while the K_{eq} (k_{-2}/k_2) for the second step was 3.5×10^{-11}M. k_2 was determined to be 3.05 min^{-1}, while $k_{-2} \simeq 0$. Assuming a value of $\sim 1 \times 10^6$M sec^{-1} for k_1, k_{-1} would be about 31 sec^{-1}. The data indicate that E, EI and EI* bind substrate. With *p*-nitrophenyl α-maltoside as the substrate, ES and EIS form products at the same rates while EI*S hydrolyzes at about 5% the rate of ES and EIS but starch, as substrate, is not hydrolyzed. Starch still binds to the complex (see below).

Mechanism of α-Amylase/α-Amylase Complex Formation and Inhibition

There are few data available that provide information on the mechanism of how α-amylases and the α-amylase inhibitors recognize each other and the α-amylases are inhibited. It is generally assumed that the microbial pseudooligosaccharides bind at the active site, being recognized as substrate-like compounds (97). Wilcox and Whitaker (91) determined that one molecule of acarbose (amylostatin XGGG; Fig. 20.5) binds at the active site of bovine pancreatic α-amylase and that there is considerable nonspecific binding of additional acarbose molecules. K_d was 9.7×10^{-6}M for the dissociation constant of acarbose binding at the active site of the enzyme. The α-amylase/red kidney bean inhibitor complex did not bind acarbose specifically but nonspecific binding was still observed.

The α-amylase/red kidney bean inhibitor complex binds p-nitrophenyl α-maltoside and slowly hydrolyzes it (91); it also binds maltose, starch and Sephadex (109). Because of hydrolysis of p-nitrophenyl α-maltoside by the complex, one may postulate that the red kidney bean inhibitor binds at the second binding site on α-amylase, but not the active site. *Tenebrio molitor* larvae α-amylase binds two molecules of 0.28-inhibitor of wheat, indicating that both the active site and the second site of this enzyme can bind the inhibitor (75,111). From spectral changes on binding the inhibitor, both sites appear to involve tryptophan residues. Porcine pancreatic α-amylase isoenzyme I forms a 1:1 stoichiometric complex with Haim, but the complex still binds S-AI inhibitor forming a triple complex (66), presumably by using both binding sites on the enzyme. Paim forms 2:1 (I:E) complexes with animal α-amylases (61). To further complicate postulation of a mechanism, a 10,000 MW barley seed protein homologous with an α-amylase inhibitor from Indian finger millet was shown by difference UV spectroscopy to combine with porcine pancreatic α-amylase but without inhibiting it (112). Paim was shown to combine to human salivary α-amylase in a ratio of 2:1 (I:E) without inhibiting the enzyme (61).

Effect of changes in solvent ionic strength and dielectric constant have been used to determine the nature of the bonds formed between α-amylase and inhibitor in the complex. Whitaker et al (92) showed that an increase in ionic strength increased the rate of complex formation between porcine pancreatic α-amylase and black kidney bean inhibitor, while a decrease in dielectric constant decreased the rate of complex formation. The equilibrium constant for complex formation increased markedly at temperatures below 30°C (92). These three sets of data indicate complex formation involves primarily hydrophobic bonds. On the other hand, increase in ionic strength was reported to increase dissociation of the barley α-amylase inhibitor/barley α-amylase II complex (36).

Acknowledgments

The author thanks Pam Bains and Clara Robison for typing the manuscript and Fulbright and National Science Foundation International/Brazil for support of some of the research.

References

1. Fossom, K., and J.R. Whitaker, *J. Nutr. 104*:930 (1974).
2. Zvyagintseva, T.N., V.V. Sova, I. Pereva and L.A. Elyakova, *Khim. Prir. Soedin, 3*:343 (1982); *Chem Abstr. 97*:89155c (1982).
3. Chrzaszcz, T., and J. Janicki, *Biochem. Z. 260*:354 (1933).
4. Chrzaszcz, T., and J. Janicki, *Biochem. J. 28*:296 (1934).
5. Kneen, E., and R.M. Sandstedt, *J. Am. Chem. Soc. 65*:1247 (1943).
6. Kneen, E., and R.M. Sandstedt, *Arch. Biochem. Biophys. 9*:235 (1946).
7. Saunders, R.M., and J.A. Lang, *Phytochemistry 12*:1237 (1973).
8. Silano, V., and F. Pocchiari and D.D. Kasarda, *Biochim. Biophys. Acta 317*:139 (1973).
9. Bowman, D.E., *Science, 102*:358 (1945).
10. Hernandez, A., and W.G. Jaffe, *Acta Cient. Venez. 19*:183 (1968).
11. Jaffé, W.G., R. Moreno and V. Wallis, *Nutr. Rep. Int. 7*:169 (1973).
12. Marshall, J.J., and C.M. Lauda, *J. Biol. Chem. 250*:8030 (1975).
13. Blanco-Labra, A., and F.A. Iturbe-Chinas, *J. Food Biochem. 5*:1 (1981).
14. Irshad, M., and C.B. Sharma, *Biochim. Biophys. Acta 659*:326 (1981).
15. Mattoo, A.K., and V.V. Modi, *Enzymologia 39*:237 (1970).
16. Narayana Rao, M., K.S. Shurpalekar and O.E. Sundaravalli, *Indian J. Biochem. 4*:185 (1967).
17. Narayana Rao, M., K.S. Shurpalekar and O.E. Sundaravalli, *Ibid. 7*:241 (1970).
18. Stankovic, S.C., and N.D. Markovic, *Glasnik Hem. Drustva, Beograd 25-26*:519 (1960-61); *Chem Abstr. 59*:3084d (1963).
19. Frommer, W., W. Puls, D. Schaefer and D. Schmidt, Ger. Offen, DE 2,064,092 (1972).
20. Koller, K.P., J. Engels and E. Uhlmann, in *Eur. Congr. Biotechnol.*, Vol. 3, 1984, p. 273.
21. Saito, S., H. Takahashi, H. Saito, M. Arai and S. Murao, *Biochem. Biophys. Res. Commun. 141*:1099 (1986).
22. Koller, K.P., and G.J. Riess, Ger. Offen. DE 3,536,182 (1987).
23. Paz-Ares, J., F. Ponz, P. Rodriguez-Palenzuela, A. Lazaro, C. Hernandez-Lucas, F. Garcia-Olmedo and P. Carbonero, *Theor. Appl. Genet. 71*:842 (1986).
24. Mundy, J., and J.C. Rogers, *Planta 169*:51 (1986).
25. Richardson, M., *Phytochemistry 16*:159 (1977).
26. Yetter, M.A., R.M. Saunders and H.P. Boles, *Cereal Chem. 56*:243 (1979).
27. Powers, J.R., and J.D. Culbertson, *J. Food Protection 45*:655 (1982).
28. Puls, W., and U. Keup, *Diabetologia 9*:97 (1973).
29. Harmoinen, A., H. Jokela, T. Koivula and W. Poppe, *J. Clin. Chem. Clin. Biochem. 24*:903 (1986).
30. Courtois, P., and J.R.M. Franckson, *Ibid. 23*:733 (1985).

31. Buonocore, V., P. Giardina, R. Parlamenti, E. Puerio and V. Silano, *J. Sci. Food Agric. 35*:225 (1984).
32. Arai, M., M. Sumida, S. Nakatani and S. Murao, *Agric. Biol. Chem. 47*:183 (1983).
33. Mundy, J., J. Hejgaard and I. Svendsen, *FEBS Lett. 167*:210 (1984).
34. Orlando, A.R., P. Ade, D. Di Maggio, C. Fanelli and L. Vittozzi, *Biochem. J. 209*:561 (1983).
35. Pace, W., R. Parlamenti, A.U. Rab, V. Silano and L. Vittozzi, *Cereal Chem. 55*:244 (1978).
36. Weselake, R.J., A.W. MacGregor and R.D. Hill, *J. Cereal Sci. 3*:249 (1985).
37. Barber, D., R. Sanchez-Monge, E. Mendez, A. Lazaro, F. Garcia-Olmedo and G. Salcedo, *Biochim. Biophys. Acta 869*:115 (1986).
38. Kutty, A.V.M., and T.N. Pattabiraman, *Biochem. Arch. 2*:203 (1986).
39. Kutty, A.V.M., and T.N. Pattabiraman, *J. Agric. Food Chem. 34*:552 (1986).
40. Dojczew, D., J. Andrzejczuk-Hybel and J. Kaczkowski, *Nahrung 30*:275 (1986).
41. Manjunath, N., P.S. Veerabhadrappa and T.K. Virupaksha, *Phytochemistry 22*:2349 (1983).
42. Nagaraj, R.H., and T.N. Pattabiraman, *Indian. J. Med. Res. 84*:89 (1986).
43. Nagaraj, R.H., and T.N. Pattabiraman, *J. Biosci. 7*:257 (1985).
44. Shivaraj B., and T.N. Pattabiraman, *Indian J. Biochem. Biophys. 17*:181 (1980).
45. Kutty, A.V.M., and T.N. Pattabiraman, *Ibid. 22*:155 (1985).
46. Chandrasekher, G., and T.N. Pattabiraman, *Ibid. 20*:241 (1983).
47. Powers, J.R., and J.R. Whitaker, *J. Food Biochem. 1*:217 (1977).
48. Hirayama, K., R. Takahashi, S. Akashi, K. Fukuhara, N. Oouchi, A. Murai, M. Arai, S. Murao, K. Tanaka and I. Nojima, *Biochemistry 26*:6483 (1987).
49. Goto, A., Y. Matsu, K. Ohyama, M. Arai and S. Murao, *Agric. Biol. Chem. 49*:435 (1985).
50. Murao, S., N. Oouchi, A. Goto and M. Arai, *Ibid. 49*:107 (1985).
51. Murao, S., N. Oouchi, A. Goto and M. Arai, *Ibid. 47*:453 (1983).
52. Murao, S., K. Ohyama, H. Murai, A. Goto, Y. Matsui, K. Fukuhara, S. Miyata, M. Sumida and M. Arai, *Denpun Kagaku 26*:157 (1979); *Chem. Abstr. 94*:188298w (1981).
53. Ohyama, K., and S. Murao, *Agric. Biol. Chem. 41*:2221 (1977).
54. Murao, S., K. Ohyama and S. Ogura, *Ibid. 41*:919 (1977).
55. Hidaka, H., T. Takaya and J.J. Marshall, *Denpun Kagaku 27*:114 (1980); *Chem. Abstr. 95*:37976k (1981).
56. Schmidt, D.D., W. Frommer, B. Junge, L. Mueller, W. Wingender, E. Truscheit and D. Schaefer, *Naturwissenschaften 64*:535 (1977).
57. Belloc, A., J. Florent, J. Lunel, D. Mancy and J.C. Palla, Ger. Offen. DE 2,702,417 (1977).
58. Su, Y.C., R.J. Chiu, N. Yu and W.R. Chang, *Proc. Natl. Sci. Counc., Repub. China, Part B. Life Sci. 8*:292 (1984).

59. Oriental Yeast Co., Ltd., Jpn. Kokai Tokyo Koho JP 60/4132 A2 (1985).
60. Tajiri, T., Y. Koba and S. Ueda, *Agric. Biol. Chem. 47*:671 (1983).
61. Arai, M., N. Oouchi and S. Murao, *Ibid. 49*:987 (1985).
62. Miyagawa, E., T. Hamakado, H. Mimura and Y. Yoshinobu, Jpn. Kokai Tokyo Koho JP 61/74587 A2 (1986); *Chem Abstr. 105*:77591n (1986).
63. Hofmann, O., L. Vertesy and G. Braunitzer, *Hoppe-Seyler's Z. Physiol. Chem. 366*:1161 (1985).
64. Vertesy, L., and D. Tripier, *FEBS Lett. 185*:187 (1985).
65. Murai, H., S. Hara, T. Ikenaka, A. Goto, M. Arai and S. Murao, *J. Biochem. (Tokyo) 97*:1129 (1985).
66. Goto, A., Y. Matsui, K. Ohyama and S. Murao, *Denpun Kagaku 27*:91 (1980); *Chem Abstr. 95*:37977m (1981).
67. Goto, A., Y. Matsui, K. Ohyama, M. Arai and S. Murao, *Agric. Biol. Chem. 47*:83 (1983).
68. Aschauer, H., L. Vertesy, G. Nesemann and G. Braunitzer, *Hoppe-Seyler's Z. Physiol. Chem. 364*:1347 (1983).
69. Aschauer, H., L. Vertesy and G. Braunitzer, *Ibid. 362*:465 (1981).
70. Vertesy, L., V. Oeding, R. Bender, K. Zepf and G. Nesemann, *Eur. J. Biochem. 141*:505 (1984).
71. Fuji Zoki Seiyaku K.K., Jpn. Kokai Tokyo Koho JP 57/2684 A2 (1982); *Chem Abstr. 96*:160822e (1982).
72. Goto, H., T. Inukai and M. Amano, Jpn. Kokai HP 50/77594 (1975); *Chem. Abstr. 83*:176706n. (1975).
73. Maeda, K., *Biochim. Biophys. Acta 871*:250 (1986).
74. Maeda, K., S. Kakabayashi and H. Matsubara, *Ibid. 828*:213 (1985).
75. Buonocore, V., F. Gramenzi, W. Pace, T. Petrucci, E. Poerio and V. Silano, *Biochem. J. 187*:637 (1980).
76. Kashlan, N., and M. Richardson, *Phytochemistry 20*:1781 (1981).
77. Maeda, K., S. Wakabayashi and H. Matsubara, *J. Biochem. (Tokyo) 94*:865 (1983).
78. Maeda, K., T. Hase and H. Matsubara, *Biochim. Biophys. Acta 743*:52 (1983).
79. Weselake, R.J., A.W. MacGregor and R.D. Hill, *Plant Physiol. 73*:1008 (1983).
80. Svendsen, I., J. Hejgaard and J. Mundy, *Carlsberg Res. Commun. 51*:43 (1986).
81. Tashiro, M., and Z. Maki, *Agric. Biol. Chem. 50*:2955 (1986).
82. Campos, F.A.P., and M. Richardson, *FEBS Lett. 167*:221 (1984).
83. Campos, F.A.P., and M. Richardson, *Ibid. 152*:300 (1983).
84. Irshad, M., and C.B. Sharma, *Indian J. Biochem. Biophys. 22*:371 (1985).
85. Wilcox, E.R., and J.R. Whitaker, *J. Food Biochem. 8*:189 (1984).
86. Frels, J.M., and J.H. Rupnow, *Ibid. 8*:281 (1984).
87. Kotaru, M., K. Saito, H. Yoshikawa, T. Ikeuchi and F. Ibuki, *Agric. Biol. Chem. 51*:577 (1987).

88. Vittozzi, L. and V. Silano, *Theor. Appl. Genet.* **48**:279 (1976).
89. DePonte, R., R. Parlamenti, T. Petrucci, V. Silano and M. Tomasi, *Cereal Chem.* **53**:805 (1976).
90. Barber, D., R. Sanchez-Monge, F. Garcia-Olmedo, G. Salcedo and E. Mendez, *Biochim. Biophys. Acta* **873**:147 (1986).
91. Wilcox, E.R., and J.R. Whitaker, *Biochemistry* **23**:1783 (1984).
92. Whitaker, J.R., F. Finardi Filho and F.M. Lajolo, *Biochemie*, **70**:1153 (1988).
93. Lajolo, F.M., and F. Finardi Filho, *J. Agric. Food Chem.* **33**:32 (1985).
94. Pflugrath, J.W., G. Wiegard, R. Huber and L. Vertesy, *J. Mol. Biol.* **189**:383 (1986).
95. Arai, M., N. Oouchi, A. Goto, S. Ogura and S. Murao, *Agric Biol. Chem.* **49**:1523 (1985).
96. Omoto, S., J. Itoh, H. Ogino, K. Iwamatsu, N. Nishizawa and S. Inouye, *J. Antibiotics* **34**:1429 (1981).
97. Fukuhara, K., H. Murai and S. Murao, *Agric. Biol. Chem.* **46**:2021 (1982).
98. Yokose, K., K. Ogawa, Y. Suzuki, I. Umeda and Y. Suhara, *J. Antibiotics* **36**:1166 (1983).
99. Kluh, I., *FEBS Lett.* **136**:231 (1981).
100. Pasero, L., Y. Mazzei-Pierron, B. Abadie, Y. Chicheportiche and G. Marchis-Mouren, *Biochim. Biophys. Acta* **869**:147 (1986).
101. Haser, R., F. Payan, M. Pierrot, G. Buisson and E. Dúee, *J. Chim. Phys. Phys.-Chim. Biol.* **76**:823 (1979).
102. Payan, F., R. Haser, M. Pierrot, M. Frey, J.P. Astier, B. Abadie, E. Duée and G. Buisson, *Acta Cryst. B36*:416 (1980).
103. Buisson, G., E. Dúee, R. Haser and F. Payan, *Ibid.* **A40**:C-38 (Suppl.) (1984).
104. Loyter, A., and M. Schramm, *J. Biol. Chem.* **241**:2611 (1966).
105. Odani, S., T. Koide and T. Ono, *FEBS Lett.* **141**:279 (1982).
106. Maeda, K., M. Sato, Y. Kato, N. Tanaka, Y. Hata, Y. Katsube and H. Matsubara, *J. Mol. Biol.* **193**:825 (1987).
107. Tanizaki, M.M., F.M. Lajolo and F. Finardo Filho, *J. Food Biochem.* **9**:91 (1985).
108. Vittozzi, L., G. De Angelis and V. Silano, *Int. J. Biochem.* **19**:281 (1987).
109. Powers, J.R., and J.R. Whitaker, *J. Food Biochem.* **1**:239 (1977).
110. Tanizaki, M.M., and F.M. Lajolo, *Ibid.* **9**:71 (1985).
111. Silano, V., E. Puerio and V. Buonocore, *Mol. Cell. Biochem.* **18**:87 (1977).
112. Svensson, B., K.J. Asano, I. Jonassen, F.M. Poulsen, J. Mundy and I. Svendsen, *Carlsberg Res. Commun.* **51**:493 (1986).

Chapter Twenty-one

Toxic Compounds in Plant Foodstuffs: Cyanogens

Jonathan E. Poulton

Department of Botany
University of Iowa
Iowa City, Iowa 52242

Chemical Nature of Cyanogens and Their Occurrence in Economically Important Crops

Although cyanogenesis is clearly widespread within the plant kingdom, the source of HCN has been identified in only about 300 species. In certain Sapindaceous seeds, HCN may arise during cyanolipid hydrolysis. The distribution, biosynthesis and chemistry of these natural products have been reviewed by Mikolajczak (1). All known cyanolipids possess the same branched, five-carbon nitrile skeleton, presumably derived from leucine (Fig. 21.1), but variations exist in the position of the double bond and in the number and location of hydroxyl groups. Only two of these (Figs. 21.1A & B) release HCN by enzymic hydrolysis. Endogenous lipases remove the acyl groups, forming an unstable cyanohydrin, which subsequently decomposes to HCN and a carbonyl compound. Cyanolipids may occur in copious amounts in species of economic interest. For example, kusum oil, the seed oil of *Scleichera trijuga*, may be quite toxic if untreated, but after processing it finds usage as an edible and medicinal oil, a hair dressing, and as a raw material for soap production. Extensive studies of the toxic properties of cyanolipids have not been conducted.

More frequently, HCN production in higher plants results from the catabolism of cyanogenic glycosides. The approximately 75 documented cyanogenic glycosides are all β-glycosidic derivatives of α-hydroxynitriles. The great majority are derived by multistep biosynthetic sequences from the five common protein amino acids *L*-phenylalanine, *L*-tyrosine, *L*-leucine, *L*-isoleucine, and *L*-valine, but a further group is based on the cyclopentene ring structure, and presumably originates from *L*-2-cyclopentenyl-1-glycine (2). Mostly, these glycosides are monosaccharides in which the unstable cyanohydrin moiety is stabilized by

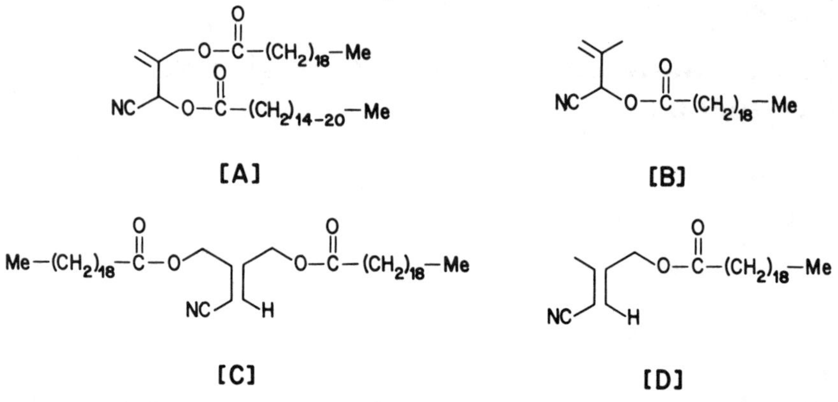

Fig. 21.1. Structures of cyanolipids found in seed oils of the Sapindaceae.

glycosidic linkage to a single sugar residue. Alternatively, in the cyanogenic disaccharides (e.g., (*R*)-amygdalin, (*R*)-vicianin, and linustatin), two sugar residues are involved in such stabilization (Fig. 21.2).

Depending on the species, L-phenylalanine serves as the precursor for several cyanogenic glycosides including the monosaccharides (*R*)-prunasin and (*S*)-sambunigrin, and the disaccharides amygdalin, vicianin, and (*R*)-lucumin. (*R*)-Prunasin is widespread in many families including the Rosaceae, where it occurs in the leaves of numerous members of the domesticated stone fruits (Table 21.1). The epimeric glycoside (*S*)-sambunigrin is present in the fruit and leaves of the European black elder (*Sambucus nigra*). Undoubtedly the best known of all cyanogenic glycosides is amygdalin, which was the first cyanogen to be isolated and fully characterized. A general feature of cyanophoric plants is that the cyanogen(s) may not necessarily be distributed equally throughout all tissues and organs of the plant. A clear illustration is afforded by the fruits of domesticated rosaceous species (e.g., cherry, apple, peach, apricot and pear). The fleshy portion of these fruits, which have long been enjoyed by humans, is noncyanogenic; in contrast, the seeds enclosed within these fruits may be highly cyanogenic (Table 21.1), with amygdalin being the sole or major cyanogen.

It is fortunate that the dry, starch-rich *Sorghum* seed, used as a food source by millions of humans in India and Africa and by cattle in North America, is noncyanogenic. By contrast, young sorghum seedlings may be highly cyanogenic (Table 21.1), and have frequently caused accidental poisoning of livestock (4). The toxic agent responsible was identified

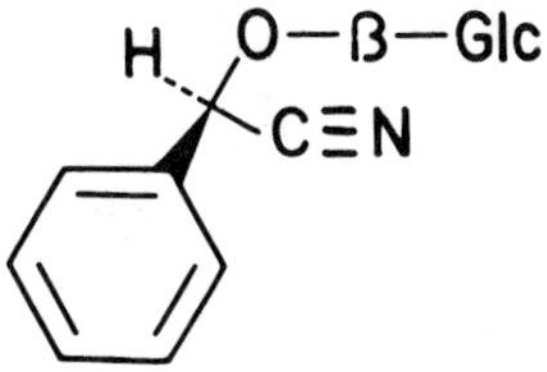

(R)- Prunasin

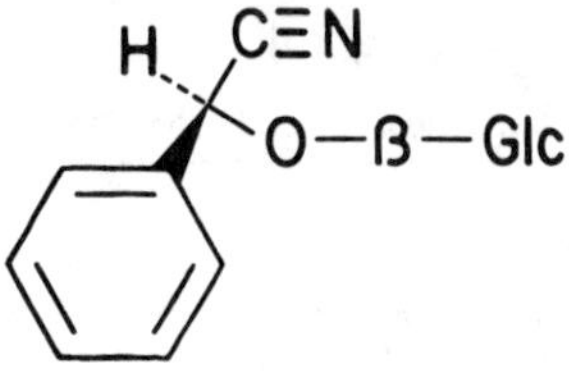

(S)- Sambunigrin

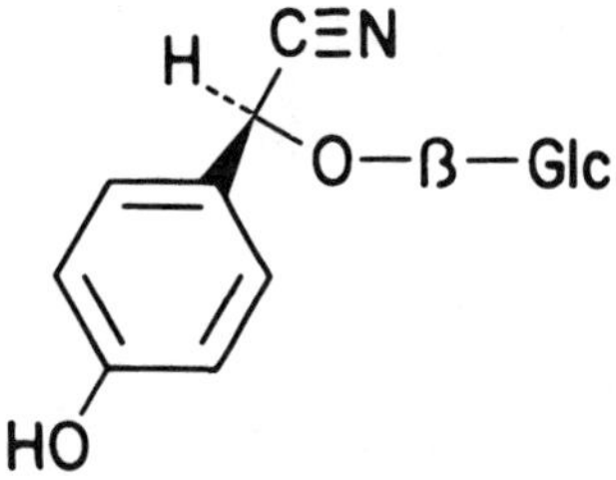

(S)- Dhurrin

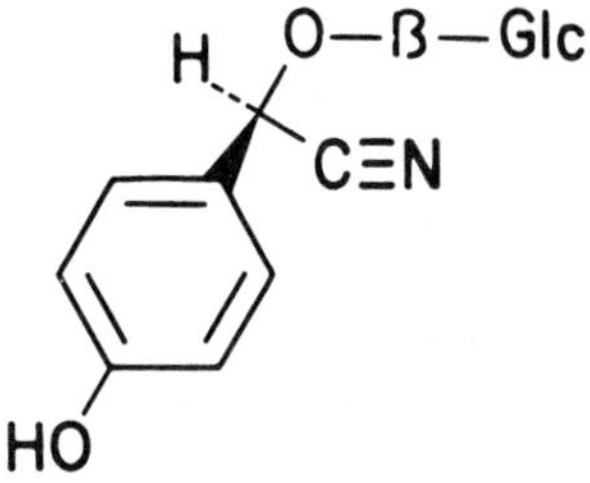

(R)- Taxiphyllin

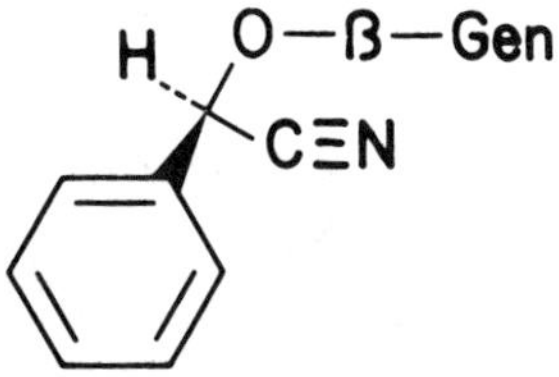

(R)- Amygdalin

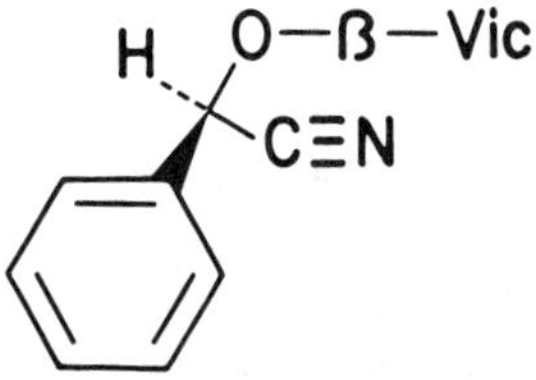

(R)- Vicianin

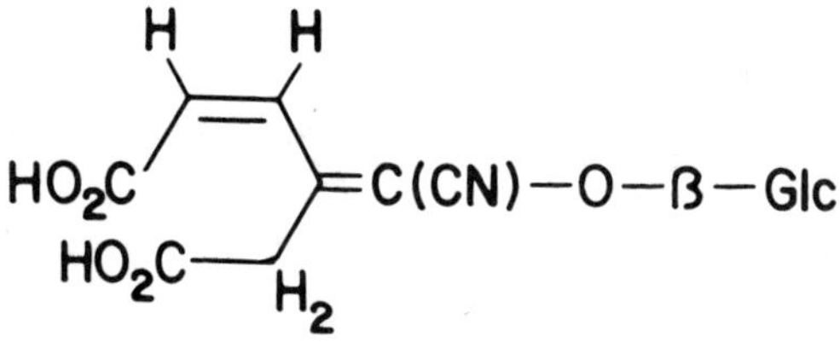

Triglochinin

Fig. 21.2. Structures of common cyanogenic glycosides derived from *L*-phenylalanine and *L*-tyrosine.

TABLE 21.1
Yield of HCN Released from Food Plants[a]

Plant	HCN yield (mg/100 g FW)
Bitter almond	
Seed	290
Young leaves	20
Wild cherry, leaves	90–360
Apricot, seed	60
Peach	
Seed	160
Leaves	125
Sorghum	
Mature seed	0
Etiolated shoot tips	240
Young green leaves	60
Bamboo	
Stem, unripe	300
Tops of unripe sprouts	800
Linen flax	
Seedling tops	91
White clover	
Young leaves	0.3–35
Bitter cassava	
Leaves	104
Bark of tuber	84
Inner part of tuber	33
Lima bean, mature seed	
Puerto Rico, small black	400
Puerto Rico, black	300
Java, colored	312
Burma, white	210
Jamaica, speckled white	17
Arizona, colored	17
American, white	10
Vicia sativa, seed	52

[a] Modified from Poulton (3).

as the cyanogenic glycoside (*S*)-dhurrin (Fig. 21.2) (5). A dramatic synthesis of dhurrin from *L*-tyrosine occurs within the first week after germination of sorghum in either light or darkness. Dhurrin may constitute up to 30% of the dry weight of the leaves and coleoptiles of two-day-old, dark-grown seedlings. The biosynthetic pathway to dhurrin and other cyanogenic glycosides, elucidated by Conn and co-workers (6), involves the rarely encountered *N*-hydroxyamino acids, aldoximes and nitriles as intermediates (Fig. 21.3). (*R*)-Taxiphyllin, the epimer of dhurrin, occurs in young bamboo shoots.

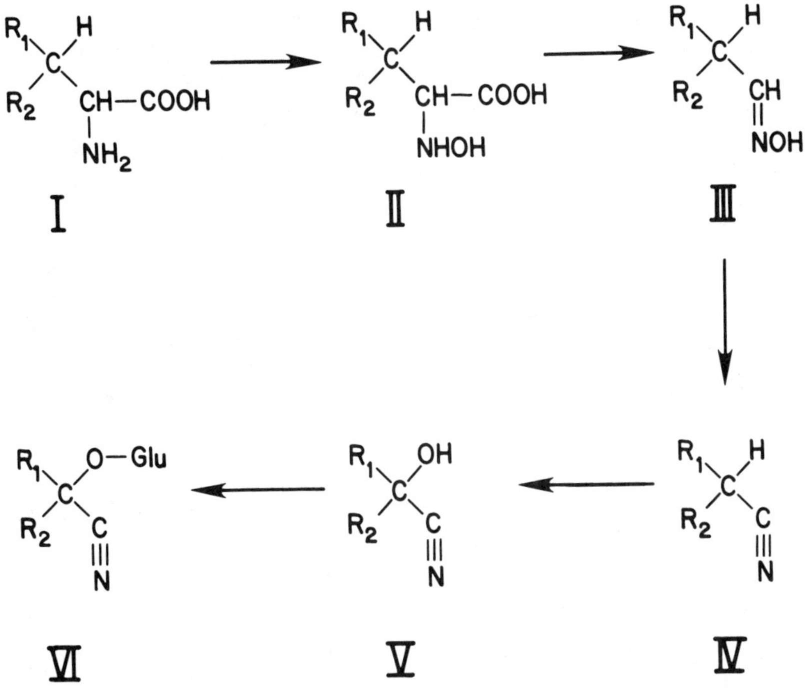

Fig. 21.3. Biosynthetic pathway of cyanogenic glycosides (VI) from *L*-amino acids (I), involving *N*-hydroxyamino acids (II), aldoximes (III), nitriles (IV), and α-hydroxynitriles (V) as intermediates.

In addition to the foregoing aromatic glycosides, significant concentrations of aliphatic cyanogenic glycosides may also be present in foodstuffs. Linamarin and its homolog lotaustralin (Fig. 21.4) are widely distributed among cyanophoric plants including linen flax, white clover, cassava, lima beans and *Hevea brasiliensis*. While both compounds are generally present within the same plant, the observed tissue ratio of linamarin to lotaustralin is species and variety dependent. Linamarin predominates in cassava (ratio 90:10), *Lotus arenarius Brot.* (99:1) and *Linum usitatissimum* (55:43), whereas the same ratio in *Lotus tenuis* L. is 6:94 (7). Studies on the ability of linseed to protect animals against the toxic effects of ingested selenium have revealed two new cyanogenic glycosides in linseed meal (8); these are the disaccharides linustatin and neolinustatin corresponding to the monosaccharides linamarin and lotaustralin, respectively (Fig. 21.4). Although flax seed possesses relatively little cyanogenic glycoside, flax seedling tops may release over 90 mg HCN per 100 g fresh weight (7). White clover (*Trifolium repens*) is used extensively as an important cover crop world-wide due to its ability to fix nitrogen and thereby increase the nitrogen content of the soil. In young leaves of certain varieties, the cyanogenic potential may reach 35 mg of HCN per 100 g of tissue, with lotaustralin predominating over linamarin (7).

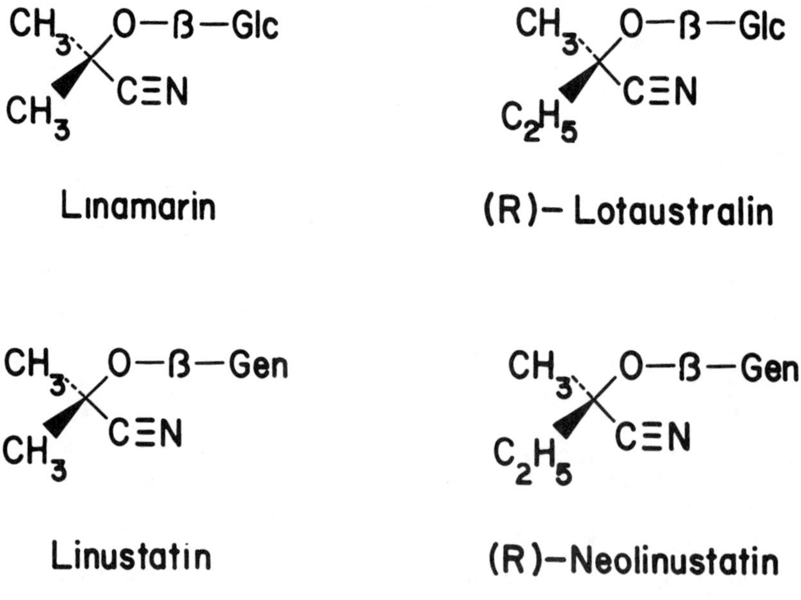

Linamarin

(R)-Lotaustralin

Linustatin

(R)-Neolinustatin

Fig. 21.4. Structures of common cyanogenic glycosides derived from *L*-valine and *L*-isoleucine.

Two agricultural crops which have caused great concern because of their cyanogen contents are cassava and lima beans. Cassava (*Manihot esculenta* Crantz) constitutes a major food staple in developing countries, being the fourth most important dietary source of calories in the tropics after rice, maize and sugar-cane (9). A conservative calculation suggests that 450-500 million people from 26 countries consume approximately 300 kcals/day as cassava. World production in 1980 was estimated at 118 million tons, with 38% raised in Africa, 35% in the Far East and 27% in South America. About two-thirds was processed for human consumption, but cassava also finds increasing usage as livestock feed, in alcohol production and as a source of starch in the textile, paper and food industries (10). Cassava agriculture is promoted by the remarkable growth of this species on poor soils, its resistance to locust attacks and drought, and by its low production costs. The chief limitation to cassava usage is its cyanogenic nature. The cyanogen concentration in edible roots (tubers) varies in the range 1.5-40 mg of HCN per 100 g fresh weight depending on the plant variety and climatic and cultural conditions (11). Leaves are also highly cyanogenic, showing a normal range of 10-110 mg HCN per 100 g fresh weight.

In a comprehensive review of the edible and poisonous beans of the lima type (*Phaseolus lunatus* L.), Viehoever (12) discussed the origin, distribution and toxicity of this important edible legume. All parts of the plant including the developing seed pods are cyanogenic, possessing linamarin predominantly. The cyanogenic potential of the mature bean varies with its color, shape and size. The white lima beans of American origin ("butter beans") yield only low amounts of HCN on analysis and may be considered desirable and safe for consumption. On the other hand, the seed of the small black wild lima bean native to most of Central America is highly toxic, containing up to 400 mg of HCN per 100 g of seeds, and has caused frequent human and livestock poisoning (13). The following legume seeds are also reported as being cyanogenic (levels given in mg HCN per 100 g): kidney bean (*Phaseolus vulgaris*, 2.0 mg), common pea (*Pisum sativum*, 2.3 mg), black-eyed pea (*Vigna sinensis*, 2.1 mg), chick pea (*Cicer arietinum*, 0.8 mg) and common vetch (*Vicia sativa*, 52 mg) (13).

A recent survey has shown that cyanogenic glycosides are more widely distributed among cereal plants than was previously believed (14). Glycosides have been isolated and characterized from *Avena sativa* L. (linamarin), *Eleusine coracana* (L.) Gaertn. (triglochinin), *Hordeum vulgare* L. (epiheterodendrin), *Secale cereale* L. (dhurrin), *Triticum aestivum* L. ssp. *spelta* (dhurrin) and *T. monococcum* (linamarin, lotaustralin and epilotaustralin), but tissue concentrations were not always given.

Toxicology of Cyanogenic Glycosides in Food Plants

It is generally accepted that the toxicity of cyanophoric plants is due not to the cyanogenic glycosides themselves but to the HCN which is released, often at high levels, upon tissue disruption (Table 21.1). Under normal physiological conditions, the tissues of a cyanogenic species contain little or no detectable free HCN. However, when that plant tissue is crushed or otherwise disrupted, HCN may be rapidly liberated from cyanogenic glycosides. Considerable evidence suggests that cyanogenesis may constitute an effective defense mechanism against herbivores (15,16). The catabolism of a cyanogenic glycoside is initiated by cleavage of the carbohydrate moiety by one or more β-glycosidases, yielding the corresponding α-hydroxynitrile (Fig. 21.5). This inter-

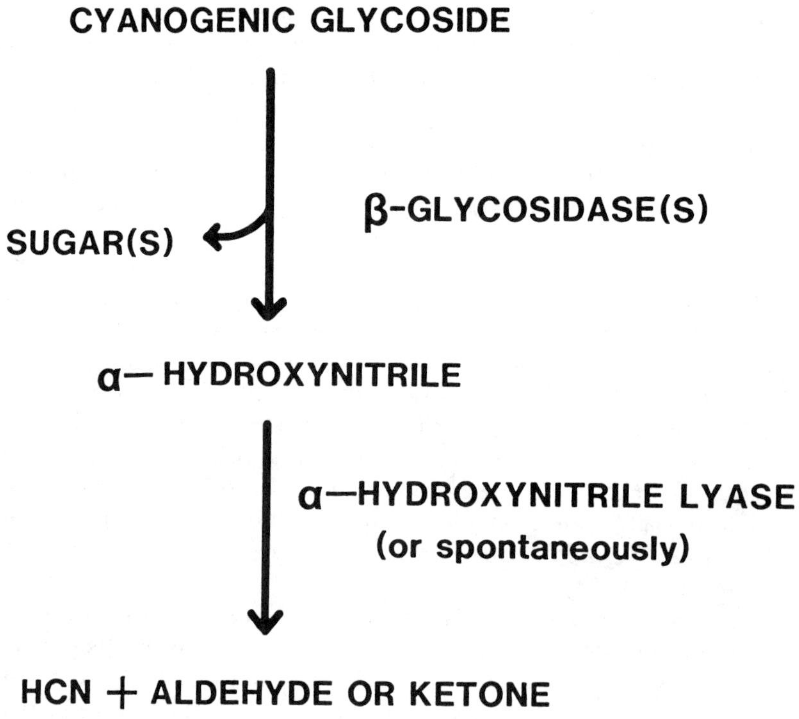

CYANOGENIC GLYCOSIDE

SUGAR(S) ← β-GLYCOSIDASE(S)

↓

α—HYDROXYNITRILE

↓ α—HYDROXYNITRILE LYASE (or spontaneously)

HCN + ALDEHYDE OR KETONE

Fig. 21.5. Catabolism of cyanogenic glycosides by β-glycosidases and α-hydroxynitrile lyases.

mediate may decompose either spontaneously or enzymically in the presence of an α-hydroxynitrile lyase to yield HCN and an aldehyde or ketone. These enzymes show optimum activity in the slightly acidic pH range exhibited by most plant homogenates (17).

Whether an animal actually suffers cyanide poisoning following the ingestion of plant material containing potentially toxic levels of cyanogenic glycosides depends on several factors (18). These include the nature and size of the animal, the level of β-glycosidases in the plant, the length of time between tissue disruption and ingestion, the presence and nature of other components of the meal, and the rate of detoxication of cyanide by the animal. For acute toxicity, enough plant material must be ingested in a sufficiently short time period such as to accumulate lethal cyanide levels. In humans, the minimum lethal dose of HCN taken orally is approximately 0.5-3.5 mg/kg body weight (19-21). Christensen (22) cites the following values for oral lethal doses of HCN (in mg/kg body weight): cat, 2.0; mouse, 3.7; dog, 4.0; and rat, 10.0. The lethal HCN dose for cattle and sheep is 2.0 mg/kg body weight. However, sheep could tolerate 15-20 mg/kg body weight per day in the normal grazing situation when the ingestion of forage is relatively slow. Cyanide exerts its acute toxic effect by combining with metalloporphyrin-containing enzyme systems. Most importantly, it has an extremely strong affinity for cytochrome oxidase. Cyanide concentrations of only 33 μM can completely block electron transfer through the mitochondrial electron transport chain, thus swiftly preventing the utilization of O_2 by the cell. Death ensues from generalized cytotoxic anoxia, with the brain, heart and central nervous system being most rapidly affected.

Given the above figures, it is not surprising that the literature contains numerous examples of acute cyanide poisoning of animals believed to be caused by ingestion of cyanogenic plants. There are reports of human poisoning following ingestion of amygdalin-containing seeds of several rosaceous species, notably bitter almonds, apricot kernels and choke cherry seeds (23). The leaves of these species may also be toxic due to their prunasin content (18). Accidental poisoning of children after drinking tea made from peach leaves is documented. Cattle fed saskatoon serviceberry (*Amelanchier alnifolia*), an important, if not preferred, browse for rangeland livestock and wildlife showed symptoms of cyanide poisoning attributed to prunasin breakdown (24). Young sorghum and arrow grass have been cited for much loss of livestock in the United States (4). The toxicity of arrowgrass (*Triglochin maritima* L.) and small arrowgrass (*T. palustris*) is attributed to the cyanogens triglochin (Fig. 22.2) and taxphyllin (18,25). Majak *et al.* (26), investigating the seasonal variation in the cyanide potential of arrowgrass, noted that cyanogen levels in leaves were substantially elevated during a period of severe moisture deficit; this correlates with the higher toxicity of this plant to sheep during extreme drought (27). Young bamboo shoots, a delicacy in many countries, may also cause cyanide poisoning (28). Cyanophoric species of acacia have been blamed for the death of sheep and cattle in Australia, South Africa and the United States (18,29,30).

In cases where acute cyanide poisoning occurs rapidly, one can probably safely assume that the plant material has supplied both the glycoside and active endogenous β-glycosidases capable of hydrolyzing it. Alternative mechanisms for cyanide release from cyanogenic glycosides have been considered. For example, it has been suggested that acid hydrolysis of cyanogenic glycosides might occur in the stomach of humans and other monogastric organisms, where the pH of the contents may be approximately 2. This appears improbable since these glycosides are stable at 37°C to dilute acid (2,31) and to gastric juices of pH 2.2-4.1 (32). Since plant β-glycosidases exhibit their highest activity in the pH range around pH 5-6, one might expect that the acidity of the monogastric stomach ($\sim$pH 2) would cause inactivation or at least inhibition of ingested β-glycosidases. If this effect were reversible, the breakdown of cyanogens to HCN by plant β-glycosidases might resume when the stomach contents pass to the duodenum and are neutralized by the bile juices. The ability of mammalian digestive enzymes to hydrolyze cyanogenic glycosides has also been questioned, but this too seems unlikely since, being α-glycosidases, they would, in theory, not cleave the β-glycosidic linkage in known cyanogenic glycosides. In contrast to the monogastric stomach, the rumen of ruminants maintains a pH near neutrality, which would favor the continuing hydrolysis of cyanogenic glycosides by ingested β-glycosidases. This may explain, at least partially, the observation that ruminants are more susceptible to poisoning by cyanogenic plants than are nonruminants (33).

Evidence for the release of HCN from cyanogenic glycosides *in vivo* by intestinal microfloral enzymes has been provided by studies with germ-free and antibiotically-treated rats (32,34). Likewise, microbial participation in amygdalin catabolism in humans may be inferred from clear differences in response to oral and intravenous administration (35). Gut microflora capable of hydrolyzing cyanogenic glycosides have been identified. Ten strains of the bacterial genera *Enterobacteria* and

Enterococci, which were isolated from mouse intestine, released HCN from amygdalin *in vitro* (36). Newton *et al.* (32) demonstrated that 53% of the available HCN was released by hydrolysis when amygdalin was incubated with approximately 1 g of human feces. *Bacteroides fragilis*, a major constituent of human stool, efficiently hydrolyzed amygdalin *in vitro*. Several other human fecal bacteria, including *Enterobacter aerogenes*, *Streptococcus fecalis* and *Clostridium perfringens*, but not *Escherichia coli*, showed weaker activity. Moreover, *E. coli* and *S. faecalis*, but not *C. perfringens*, hydrolyzed linamarin when grown on glucose or lactose media (37). Rumen flora are also capable of cyanogenic glycoside hydrolysis (36,38,39).

Whether mammalian β-glycosidases also contribute to cyanogenesis following ingestion of cyanogenic glycosides is difficult to assess, but the sensitivity of certain species to parenteral cyanogen administration suggests their participation (32,40). β-Glycosidases are widely distributed within mammalian tissues (41,42), but until recently little was known about their capacity to hydrolyze cyanogenic glucosides (43). Freese *et al.* (44) detected β-glucosidase activities towards amygdalin and prunasin in extracts from cat, rat and rabbit kidney. These activities were also present in the small intestine and intestinal contents of germ-free rats and in human small intestine (45). Frakes *et al.* (46) reported hydrolysis of linamarin, prunasin and amygdalin by a crude β-glucosidase preparation from hamster caecum and its contents. Enzyme activities were ascribed to intestinal microflora, although the possible role of a mammalian hydrolase apparently was not excluded.

In conclusion, it appears that routes of administration which allow the most direct contact of the cyanogens with the gastrointestinal flora maximize the possibility of HCN release and subsequent toxicity. This danger is increased dramatically if sources of β-glucosidase are simultaneously ingested with the glycoside (35,47).

Factors Affecting Cyanogen Levels in Plants

Knowledge of the tissue distribution of cyanogens and of factors affecting glycoside levels in plants is critical in reducing the potential toxicity of cyanogenic plants used as foodstuffs. It is difficult to generalize regarding the tissue distribution of cyanogenic glycosides within cyanophoric plants. Leaves and seeds frequently contain the highest concentration but, as shown in Table 22.1, these compounds may be found in all plant organs. In general, young shoots are more dangerous than older tissue (e.g., sorghum, wild cherries, white clover).

In a given plant, glycosides may be present in one or more organs. However, it should be borne in mind that the tissues where cyanogens accumulate may not necessarily be the site of synthesis since translocation of glycosides may be possible (48,49). The amount of glycoside present at a particular time may be influenced not only by plant age but also by diurnal, seasonal, environmental, nutritional and genetic factors (50-54).

In certain species, the polymorphic nature of cyanogenesis has been indicated by the existence of both cyanogenic and noncyanogenic plants within natural populations (55,56). This phenomenon has been most extensively studied in *Trifolium repens* and *Lotus corniculatus* (57). In *T. repens*, cyanogenesis is controlled by two independently inherited genes, whose relationship is shown in Figure 21.6. Alleles of the gene *Ac* determine the biosynthesis of the cyanogenic glycosides linamarin and lotaustralin from their respective amino acid precursors valine and isoleucine. The presence or absence of the enzyme linamarase is governed by alleles of the *Li* gene. Only plants possessing at least one dominant functional allele of both genes can liberate HCN when damaged. Of the four homozygous genotypes *AcAcLiLi*, *AcAclili*, *acacLiLi*, and *acaclili* obtained by selective breeding, solely the first genotype is cyanogenic, since it possesses both glycoside and β-glycosidase. The second genotype contains the glycoside but fails to release HCN upon tissue disruption because it lacks the β-glucosidase. It should be realized, however, that even these individuals may be potentially toxic if eaten with other food materials possessing β-glycosidases capable of hydrolyzing linamarin or lotaustralin. Compared with homozygous individuals, plants heterozygous for *Ac* or *Li* have intermediate levels of cyanogenic glucosides or linamarase, respectively.

Reduction of Cyanogenic Potential by Plant Breeding

Decades of selective breeding has yielded low-cyanide varieties of important food plants including almonds (*Prunus amygdalus*), cassava, lima beans and sorghum. Amygdalin is apparently responsible both for the bitterness and toxicity of seeds of many rosaceous species such as bitter almonds (100 μmol/g) and apricots (20-80 μmol/g). Selective breeding has produced the sweet almond, whose kernel is edible, sweet and of commercial value (58). Its amygdalin concentration is far lower than in bitter almonds but still detectable by qualitative tests for HCN. Bitter almonds are grown only as a source of the bitter almond

Fig. 21.6. Genetics of cyanogenesis in *Trifolium repens*.

oil used as a flavoring agent. Several thousand distinct cassava varieties, originating from traditional selection and breeding programs, have been identified. They are commonly subdivided according to their taste into sweet and bitter cultivars. Bitterness has been linked to higher linamarin levels, but this assertion is equivocal (59). Factors other than cyanogen content which contribute to cassava bitterness require further examination (11). It should be noted that the bitter variety may sometimes be preferred because of the generally bland, starchy taste of the sweet form (60). Although breeding and selection efforts continue, acyanogenic cassava plants have yet to be found. Of the 88,510 plants screened by Sadik et al. (61), more than 99% were highly cyanogenic. New "high-yielding low-cyanide" varieties, like those developed at the International Institute of Tropical Research in Nigeria, appear promising but still release 5-10 mg of cyanide per 100 g fresh weight. Since the grass species *Sorghum sudanense* (sudangrass) and *S. bicolor* and their hybrid sorghum-sudangrass find usage in many countries as animal feed, continuing efforts are being made by plant breeders to produce low-cyanide strains. The genetics of cyanogenesis in *Sorghum* appear more complex than exist in *T. repens*, and authors differ as to the number of genes involved and the dominance relationship of these genes (57).

Reduction of Toxicity of Cyanogenic Plant Material by Food Processing

The potential toxicity of cyanogenic plants used as foodstuffs by humans and domestic animals may be significantly reduced during food processing, whereby one attempts either to remove the cyanogenic glycosides, to inactivate the β-glycosidases, or to effect both goals. Several techniques are discussed here with special reference to cassava, which provides a high-carbohydrate, low-protein staple food in many tropical countries. Similar arguments apply to the detoxication of other cyanogenic plants such as lima beans. Until recently, the amount of analytical data available on the efficacy of these processes was limited and often unreliable (11,62). Furthermore, most studies reported total cyanide levels and drew no distinction between free cyanide (nonglycosidic) and bound cyanide (cyanogenic glycosides), which may differ in their response to processing techniques. In 1978, Cooke (63) optimized an enzymic assay for both quantities in cassava and cassava products. This was later automated to allow screening of 300 samples per day (64). As analyzed by these procedures, cyanide losses during processing are generally lower than published in previous studies (65,66). Methods used to analyze residual cyanide in cassava foods have been critically reviewed (67).

As described earlier, cyanogen levels in plants may be affected by genetic and environmental factors. Where climatic and soil conditions permit their cultivation, low-cyanide varieties are obviously desirable. Initial processing steps are facilitated if the tissue distributions of cyanogens and their catabolic enzymes are known. With cassava, Kojima et al. (68) confirmed the large variations in cyanide content and linamarase activity among roots, even from the same plant. The cyanide content showed both radial and longitudinal gradients. In most varieties, the cyanogen concentration is appreciably higher (5–10 to 1) in the cortex (cassava peel) than in the parenchyma tissue (peeled root). Linamarase showed a similar radial gradient with greater concentrations existing in the peel than in parenchyma tissue (65,68). Thus, the common practice of peeling cassava root as an initial processing step serves to eliminate the most toxic tissues.

Few effective means are available for removal of cyanogenic glycosides from food material. In general, these glycosides are heat stable. Thus cooking cassava or other cyanophoric plants by boiling, roasting, frying or steaming before consumption would tend to inactivate the endogenous β-glycosidases but have little or no effect on the cyanogen itself (e.g., 69). Cooke and Maduagwu (70) showed that free cyanide, which constituted only 8–12% of the total cyanide content, was rapidly removed from fresh cassava chips by boiling, leaching in cold water, or air-drying. By contrast, boiling removed only 55% of the bound cyanide, which was recovered in the water used. Nambisan and Sundaresan (71) made similar observations but noted that the extent of glycoside retention (between 25–75%) was dependent upon chip size and the ratio of tissue weight to volume of boiling water. Less than 5% of the bound cyanide was lost by leaching cassava chips in cold water for four hours (70). This increased to 50% after 18 hours, but was attributed to the onset of fermentation. The effect of oven- or air-drying on cyanide content received much attention over several decades but no clear picture emerged (70). More recent findings show that cyanogen retention is determined both by chip size and temperature. Air-drying at 46.5°C removed 29% of the bound cyanide; lower losses occurred at higher temperatures (70). Comparing the efficacies of oven-drying at two different temperatures (50°C and 70°C) with sun-drying, Nambisan and Sundaresan (71) reported that 36%, 26% and 47%, respectively, of the cyanogen content was lost with chips of 3 mm thickness. The corresponding figures for thicker chips (10 mm) were 54%, 40% and 72%. Taken together, these data indicate that drying thicker slices at milder temperatures provides temperature and moisture conditions which prolong linamarase action and thus achieve highest linamarin losses. The thermostability of linamarin may explain the cases of people poisoned by eating colored lima beans after boiling and draining them (12). Taxiphyllin, being unusually thermolabile (72), constitutes an exception. This glycoside is present in young bamboo shoots but should be destroyed by boiling the shoots in water for 35-40 min; boiling should also drive off the liberated HCN if the cooking pot remains uncovered.

At least in theory, the release of HCN from cyanophoric plants could be prevented by inactivation or removal of the endogenous β-glycosidases which catabolize the cyanogenic glycosides upon tissue disruption. This supposition may be false however, since, if treatment of the food does not simultaneously remove or destroy the glycosides, toxicity may still result from cyanogen catabolism in the gastrointestinal tract by β-glycosidases of microfloral or mammalian origin. Similar arguments would apply following ingestion of plants which due to their genetic character, possess cyanogen but lack the β-glycosidase. Noting these hazards, Seddon and King (73) performed model experiments using sweet almonds to release HCN from foliage of *Acacia glaucescens* and *Eremophila maculata*, which contained a cyanogenic glycoside, but no β-glucosidase. One should also be aware that some noncyanogenic plants can hydrolyze cyanogenic glycosides. Extracts derived from common components of a typical salad (e.g., lettuce, celery and mushrooms) catalyzed amygdalin hydrolysis (60), so clearly this calls for some caution in consuming these plants and a source of cyanogenic glycosides simultaneously.

Alternative food processing techniques which have shown the greatest success in reducing the acute toxicity of cyanogenic plants are exemplified by the wide variety of traditional ways devised to detoxicate the more poisonous cassava lines. The interested reader is directed to several detailed articles covering the production of cassava-based food products in various parts of the world (74,75). These procedures are designed to disrupt normal cellular compartmentation, thereby allowing cyanogens to come into contact with endogenous β-glycosidases which effect their hydrolysis. Coursey (11) describes how cassava tubers are grated or ground to initiate HCN release. The liberated HCN may be subsequently removed by solution in running or static water, by cooking or air-drying, or during later fermentation. Nambisan and Sundaresan (71) showed that >95% of the cyanogen content of eight cassava lines (ranging from 5-110 mg HCN per 100 g fresh weight) was lost by crushing tubers followed by sun-drying for eight hours. Fermentation, widely used in cassava processing to improve palatibility, also effects cyanide detoxication. For example, in *gari* production in west Africa, washed roots are peeled and grated. The pulp undergoes fermentation by microorganisms including *Cornybacterium manihot* and *Geotrichum candida* for 3-10 days during which pressure is maintained by simple means to help express juices. After fermentation, heating the pulp with stirring yields a free-flowing food product while driving off much of the liberated cyanide. The original belief that organic acids produced early in fermentation chemically hydrolyze cyanogens has been dispelled by the demonstrated stability of linamarin to acids. It now appears that the endogenous linamarase plays the major role in cyanide detoxication of pulped cassava roots and that fermentation processes contribute only minimally (76,77).

Although the tubers are considered more palatable, cassava leaves provide an inexpensive and rich source of supplementary protein, minerals and vitamins. In addition to use as animal feed, these leaves constitute an important part of the human diet in areas of Sierra Leone, Tanzania, Zaire and Gabon, being used as a vegetable or sauce constituent. Despite being highly cyanogenic, they may be successfully detoxified by similar processing techniques as used for roots, but concomitant losses in nutritional value are often observed (78,79).

The fact that millions of people world-wide consume cassava as a daily staple suggests that the foregoing detoxication techniques are generally effective. The incidence of disease that can be related to cassava consumption is quite low and is seemingly restricted to a few specific locations. If insufficiently processed, bitter clones can cause acute cyanide poisoning, but such cases are rare because cassava-eating populations are usually aware of this danger. International concern is currently directed more towards those localities where long-term ingestion of cassava products containing low but significant levels of HCN aggravates preexisting dietary deficiencies. In mammals, the principal detoxication mechanism for cyanide involves combination with thiosulfate (derived from dietary sulfur amino acids) in a reaction catalyzed by rhodanese [EC 2.8.1.1.] which yields thiocyanate. It has become increasingly clear that cyanide exposure from insufficiently processed cassava contributes to endemic goiter and cretinism in areas of Africa where iodine deficiency prevails (62). This is due to the interference of iodine uptake into the thyroid gland by thiocyanate. Depending on the extent of the problem, it can be alleviated by iodine supplementation and/or improvement of cassava processing practices. While the contribution of cassava toxicity to tropical diabetes remains unclear, there is mounting evidence to link the high cyanide and low sulfur content of cassava-dominated diets with two neurological disorders, namely tropical ataxic neuropathy recorded in Nigeria and epidemic spastic paraparesis described in Mozambique, Zaire and Tanzania. The nature and occurrence of these diseases have been extensively reviewed elsewhere (3,13,62,80).

Conclusions

The problem of acute poisoning by HCN of plant origin has been largely overcome in many countries by education, plant selection and by government regulation of plant importation. However, the possible dangers of long-term exposure to low cyanide levels that in single doses do not produce clinical signs of poisoning are not well understood. The use of cassava as a food source is projected to increase in future years (9,10). One should expect that the chronic toxicity observed in humans and animals on high-cassava diets may become a problem of increasing significance unless effective means are taken to reduce its toxicity. Screening for acyanogenesis and low cyanide levels in the extensive collections of cultivated lines of cassava and also of uncultivated "wild" material is undoubtedly desirable (81). Newly developed assay methods, which rapidly assess both bound and free cyanide levels, will greatly facilitate this search and be beneficial in developing improved detoxification procedures. In addition, preventive work should strive to improve the nutritional balance among the exposed population, while reducing exposure to cyanogen.

Acknowledgments

The assistance of Linda Donohoe and Peggy Schroder in typing the manuscript is gratefully acknowledged.

References

1. Mikolajczak, K.L., *Prog. Chem. Fats Other Lipids* 15:97 (1977).
2. Seigler, D.S., in *Cyanide in Biology*, edited by B. Vennesland, E.E. Conn, C.J. Knowles, J. Westley and F. Wissing, Academic Press, New York, NY, 1981, p. 133.
3. Poulton, J.E., in *Handbook of Natural Toxins*, Vol. I, edited by R.F. Keeler and A.T. Tu, Marcel Dekker, New York, NY, 1983, p. 117.

4. Gibb, M.C., J.T. Carbery, R.G. Carter and S. Catalinac, *New Zealand Vet. J.* *22*:127 (1974).
5. Dunstan, W.R., and T.A. Henry, *Philos. Trans. R. Soc. London, Ser. A.* *199*:399 (1902).
6. Conn, E.E., in *Cyanide in Biology*, edited by B. Vennesland, E.E. Conn, C.J. Knowles, J. Westley and F. Wissing, Academic Press, New York, NY, 1981, p. 183.
7. Butler, G.W., *Phytochemistry* *4*:127 (1965).
8. Smith, C.R., D. Weisleder, R.W. Miller, I.S. Palmer and O.E. Olson, *J. Org. Chem.* *45*:507 (1980).
9. Cock, J.H., *Science* *218*:755 (1982).
10. Nestel, B., in *Chronic Cassava Toxicity*, edited by B. Nestel and R. MacIntyre, International Development Research Center, Ottawa, Canada, 1973, p. 11.
11. Coursey, D.G., *Ibid.*, p. 27.
12. Viehoever, A., *Thai Sci. Bull.* *2*:1 (1940).
13. Montgomery, R.D., in *Toxic Constituents of Plant Foodstuffs*, edited by I.E. Liener, Academic Press, New York, NY, 1969, p. 143.
14. Pitsch, Ch., M. Keller, H.D. Zinsmeister and A. Nahrstedt, *Planta Medica* *50*:388 (1984).
15. Jones, D.A., in *Biochemical Aspects of Plant and Animal Coevolution*, edited by J.B. Harborne, Academic Press, New York, NY, 1978, Chap. 2.
16. Jones, D.A., in *Cyanide in Biology*, edited by B. Vennesland, E.E. Conn, C.J. Knowles, J. Westley and F. Wissing, Academic Press, New York, NY, 1981, p. 509.
17. Poulton, J.E., in *Cyanide Compounds in Biology*, edited by S. Harnett, Ciba Foundation Symposium 140, John Wiley and Sons, Chicester, UK, 1988, p. 67.
18. Kingsbury, J.M., *Poisonous Plants of the United States and Canada*, Prentice-Hall, Englewood Cliffs, 1964, p. 364.
19. Chen, K.K., C.L. Rose and G.H.A. Clowes, *Am. J. Med. Sci.* *188*:767 (1934).
20. Gettler, A.O., and J.O. Baine, *Ibid.* *195*:182 (1938).
21. Halstrom, F., and K.D. Moller, *Acta Pharmacol. Toxicol.* *1*:18 (1945).
22. Christensen, H.E., *Registry of Toxic Effects of Chemical Substances*, DHEW Publ. (NIOSH), U.S. Public Health Service, Rockville, MD, 1976, p. 76.
23. Lewis, J., *West J. Med.* *127*:55 (1977).
24. Majak, W., T. Udenberg, L.J. Clark and A. McLean, *Can. Vet. J.* *21*:74 (1980).
25. Muenscher, W.C., *Poisonous Plants of the United States*, Macmillan, New York, NY, 1945, p. 32.
26. Majak, W., R.E. McDiarmid, J.W. Hall and A.L. Van Ryswyk, *Can. J. Plant Sci.* *60*:1235 (1980).
27. Clawson, A.B., and E.A. Moran, *U.S. Dept. Agric. Tech. Bull. No. 580*, Washington, D.C. 1937.
28. Baggchi, K.N., and H.D. Ganguli, *Indian Med. Gaz.* *78*:40 (1943).

29. Hurst, E., *The Poison Plants of New South Wales*, Snelling Printing Works, Sydney, Australia, 1942.
30. Steyn, D.G., and C. Rimington, *Onderstepoort J. Vet. Sci. Anim. Ind.* 4:51 (1935).
31. Caldwell, R.J., and S.L. Cortauld, *J. Chem. Soc.* 91:666 (1907).
32. Newton, G.W., E.S. Schmidt, J.P. Lewis, E.E. Conn and R. Lawrence, *West J. Med.* 134:97 (1981).
33. Moran, E.A., *Am. J. Vet. Res.* 15:171 (1954).
34. Carter, J.H., M.A. McLafferty and P. Goldman, *Biochem. Pharmacol.* 29:301 (1980).
35. Moertel, C.G., M.M. Ames, J.S. Kovach, T.P. Moyer, J.R. Rubin and J.H. Tinker, *J. Am. Med. Assoc.* 245:591 (1981).
36. Smith, R.L., in *A Symposium on Mechanisms of Toxicity*, edited by W.N. Aldridge, Macmillan, New York, NY, 1971, p. 229.
37. Fomunyam, R.T., A.A. Adegbola and O.L. Oke, *Can. J. Microbiol.* 30:1530 (1984).
38. Coop, I.E., and R.L. Blakley, *New Zealand J. Sci. Technol. Sect. A.* 30:277 (1949).
39. Majak, W., and K.-J. Cheng, *J. Anim. Sci.* 59:784 (1984).
40. Khandekar, J.D., and H. Edelman, *J. Am. Med. Assoc.* 242:169 (1979).
41. Cohen, R.B., S.H. Rutenberg, K-C. Tsou, M.A. Woodbury and A.M. Seligman, *J. Biol. Chem.* 195:607 (1952).
42. Neufeld, E.F., T.L. Lim and L.J. Shapiro, *Annu. Rev. Biochem.* 44:357 (1975).
43. Ng, S.J., M.S. Thesis, University of California, Davis, CA, 1975.
44. Freese, A., R.O. Brady and A.E. Gal, *Arch. Biochem. Biophys.* 201:363 (1980).
45. Newmark, J., R.O. Brady, P.M. Grimley, A.E. Gal, S.G. Waller and J.R. Thistlethwaite, *Proc. Natl. Acad. Sci. U.S.A.* 78:6513 (1981).
46. Frakes, R.A., R.P. Sharma and C.C. Willhite, *Food Chem. Toxic.* 24:417 (1986).
47. Schmidt, E.S., G.W. Newton, S.M. Saunders, J.P. Lewis and E.E. Conn, *J. Am. Med. Assoc.* 239:943 (1978).
48. Clegg, D.O., E.E. Conn and D.H. Janzen, *Nature (London)* 278:343 (1979).
49. Ramanujam, T., and P. Indira, *Indian J. Plant Physiol.* 27:355 (1984).
50. De Waal, D., Ph.D. Thesis, Agricultural University, Wageningen, The Netherlands, 1942.
51. Dement, W.A., and H.A. Mooney, *Oecologia* 15:65 (1974).
52. Nelson, C.E., *Agron. J.* 45:615 (1953).
53. Michely, D., H.D. Zinsmeister, E. Roth and A. Nahrstedt, *Z. Lebensm. Unters Forsch.* 177:350 (1983).
54. Butler, G.W., P.F. Reay and B.A. Tapper, in *Chronic Cassava Toxicity*, edited by B. Nestel and R. MacIntyre, International Development Research Center, Ottawa, Canada, 1973, p. 65.

55. Jones, D.A., in *Phytochemical Ecology*, edited by J.B. Harborne, Academic Press, New York, NY, 1972, p. 103.
56. Fikenscher, L.H., and R. Hegnauer, *Pharm. Weekbl. 112*:11 (1977).
57. Hughes, M.A., in *Cyanide in Biology*, edited by B. Vennesland, E.E. Conn, C.J. Knowles, J. Westley and F. Wissing, Academic Press, New York, NY, 1981, p. 495.
58. Heppner, M.J., *Genetics 11*:605 (1926).
59. Sundaresan, S., B. Nambisan and C.S. Easwari Amma, *Indian J. Agric. Sci. 57*:37 (1987).
60. Conn, E.E., *Int. Rev. Biochem. 27*:21 (1979).
61. Sadik, S., O.U. Okereke and S.K. Hahn, in *Screening for Acyanogenesis in Cassava*, Int. Inst. Trop. Agric., Ibadan, Nigeria, Tech. Bull. No. 4, undated.
62. Rosling, H., in *Cassava Toxicity and Food Security*, Tryck kontact, Uppsala, Sweden, 1987, p. 16.
63. Cooke, R.D., *J. Sci. Food Agric. 29*:345 (1978).
64. Rao, P.V., and S.K. Hahn, *Ibid. 35*:426 (1984).
65. DeBruijn, G.H., in *Chronic Cassava Toxicity*, edited by B. Nestel and R. MacIntyre, International Development Research Center, Ottawa, Canada, 1973, p. 43.
66. Raymond, W.D., W. Jojo and Z. Nicodemus, *E. Afr. Agric. J. 6*:154 (1941).
67. Cooke, R.D., *Food Laboratory Newsletter* No. 9, p. 18 (1987).
68. Kojima, M., N. Iwatsuki, E.S. Data, C.D.V. Villegas and I. Uritani, *Plant Physiol. 72*:186 (1983).
69. Joachim, A.W.R., and D.G. Pandittesekere, *Trop. Agric. 100*:150 (1944).
70. Cooke, R.D., and E.N. Maduagwu, *J. Food Technol. 13*:299 (1978).
71. Nambisan, B., and S. Sundaresan, *J. Sci. Food Agric. 36*:1197 (1985).
72. Schwarzmaier, U., *Chem. Ber. 109*:3379 (1976).
73. Seddon, H.R., and R.O.C. King, *J. Counc. Sci. Ind. Res. Aust. 3*:14 (1930).
74. Lancaster, P.A., J.S. Ingram, M.Y. Lim and D.G. Coursey, *Econ. Bot. 36*:12 (1982).
75. Lancaster, P.A., and J.E. Brooks, *Ibid. 37*:331 (1983).
76. Ikediobi, C.O., and E. Onyike, *Agric. Biol. Chem. 46*:1667 (1982).
77. Maduagwu, E.N., *Toxicol. Lett. 15*:335 (1983).
78. Maduagwu, E.N., and I.B. Umoh, *Ibid. 10*:245 (1982).
79. Ravindran, V., E.T. Kornegay and A.S.B. Rajaguru, *Anim. Feed Sci. Technol. 17*:227 (1987).
80. Montgomery, R.D., in *Handbook of Clinical Neurology*, Vol. 36, edited by J.N. Vinken and W. Bruyn, North-Holland, Amsterdam, The Netherlands, 1979, p. 515.
81. Jennings, D.L., and C.H. Hershey, in *Progress in Plant Breeding*, Vol. 1, edited by G.E. Russell, Butterworths, Cambridge, U.K., 1985, p. 89.

Chapter Twenty-two

New Perspectives on the Antinutritional Effects of Tannins

Larry G. Butler

Department of Biochemistry
Purdue University
West Lafayette, IN 47907

Tannins are phenolic polymers produced in many plants as secondary metabolites. The two major types are the condensed tannins (polymers of flavonoids joined by carbon-carbon bonds) and the hydrolyzable tannins (polygalloyl glucose). At least in some plants, tannins provide protection against biological challenges such as herbivory (1), although they may not have been selected primarily as defense chemicals (2). Tannins are present in many plants used as human food (3), where they may contribute a pleasing astringency, as in red wines or fresh fruits. But tannins also have undesirable nutritional effects (4), which would appear to be inevitable for materials which function as deterrents to herbivory. The nutritional consequences of the occurrence of tannins in dietary staples—such as beans and cereals (sorghum and barley)—as well as the human proclivity for consuming tannin-rich (5) but nutrient-poor (3) foodstuffs direct attention to these common plant components. New perspectives on the nature and severity of the effects of dietary tannins are briefly and selectively reviewed here.

Literature Review

Nature of the antinutritional effects.

Evidence of the antinutritional effects of dietary tannins has been obtained using laboratory animals (6,7), livestock (8,9), wild animals (10,11), and humans (12). Antinutritional effects of tannins are often manifest as diminished growth rates and/or feed conversion (8,9), but lowered fecundity (13) and egg production (14) or increased mortality (15) are also reported. Hydrolyzable tannins tend to depress food intake (16). These effects are more difficult to demonstrate for ruminants (17), in which there are reports of beneficial effects of dietary tannin due to inhibition of bloating (18) or prevention of protein loss in the rumen (19).

Molecular basis of the antinutritional effects: Protein binding?

The most widely recognized characteristic of tannins is their propensity for binding proteins (20). This binding involves both hydrogen bonding and hydrophobic associations (21,22). Several common but seldom acknowledged assumptions about the nature and effect of protein binding by tannins have recently been addressed (20). These assumptions include: tannins nonselectively bind all proteins, the binding is irreversible, tannin-protein complexes are invariably insoluble, the antinutritional effects of tannins are due to their protein binding (precipitating?) capacity, and the predominant basis for tannins' antinutritional effects is the inhibition of digestion of dietary protein. The validity of these assumptions is questionable in most instances (20,23), some of which are briefly reviewed below.

Specificity of protein binding by tannins.

Although Zucker (24) predicted on theoretical grounds that tannin-protein interactions should be highly specific, we were somewhat surprised to observe that tannin-protein binding can be quite highly specific for both the protein (25) and tannin (26) components. Using a competitive binding assay to compare the relative affinity of a series of well-characterized proteins for a single tannin preparation, it was found that proteins differ in their relative affinity for tannin by more than four orders of magnitude (Table 22.1). A protein which has a high affinity for tannin may be selectively bound out of a large excess of proteins with lesser affinity. Likewise, condensed tannin preparations from several different sources were found to vary widely with respect to their affinity for different purified proteins (26). The carbohydrate moieties of glycoproteins affect both the strength and specificity of the tannin-protein interaction and the solubility of the resulting complex (22,27,28). Tannins and proteins are often associated as soluble complexes under conditions in which precipitation does not occur (29). Proteins which strongly bind most tannins are rich in the imino acid proline, are relatively large having conformationally loose structures, and are relatively hydrophobic (22,25,30). Abundant proteins which generally conform to these characteristics are collagen from animal tissues, and prolamines, the alcohol-soluble storage proteins of cereal seeds. These proteins, which are preferentially bound by tannins out of mixtures (25,30), are relatively deficient in essential amino acids. In the diet, selective binding of these proteins by tannins could result in less harmful nutritional consequences than would result from random non-specific binding of all proteins, including those with a better balance of essential amino acids.

TABLE 22.1
Relative Affinity for Sorghum Tannin

Gelatin	14.0
Proline-rich salivary protein	6.8
Pepsin	1.1
Bovine serum albumin	1.0
Bovine hemoglobin	0.068
Ovalbumin	0.016
β-Lactoglobulin	0.0087
Lysozyme	0.0048
Soybean trypsin inhibitor	<0.001

Inhibition of digestion.

Tannins are widely regarded as digestibility-reducing substances (31). Because tannins strongly bind to many proteins, it is not unexpected that when they are brought together with enzyme proteins under appropriate conditions, tannins would bind and inhibit the enzymes. Tannins are recognized for their interference with *in vitro* assays of plant enzymes (32). Indeed, tannins inhibit a broad spectrum of enzymes in *in vitro* assays (33,34). Under these conditions the enzyme protein may be the only material present capable of binding the tannin. *In vitro* assays of proteolytic activity in which tannin may also bind the substrate present in larger amounts than the enzyme can demonstrate an enhancement rather than inhibition of digestive activity (35,36), possibly by partially denaturing the substrate protein to make it more digestible.

The evidence that tannins are effective inhibitors of digestive enzymes *in vivo* is equivocal (36,37). Under *in vivo* conditions digestive enzymes may retain full activity in the presence of tannins (34,37), possibly due to the presence of detergents and unfavorable pH conditions which prevent tannin from binding protein (37). The levels of digestive enzymes of animals on tannin-containing diets appear to be normal or in some cases were moderately increased as well as depressed (39,40). Some animals prefer tannin-rich diets and thrive on them (41,42), suggesting their digestion is unimpaired. Increase in fecal nitrogen observed on tannin-containing diets and attributed to inhibition of digestion of dietary proteins (43) could instead be due to increased demands for nonprotein nutrients as sources of metabolic energy for phenol detoxication (40). There is evidence that the increased fecal nitrogen is not dietary protein, but is of endogenous origin (44). Induction by dietary tannin of phenol detoxifying enzymes (16,45) or of the secretion of tannin-binding proteins (46) would seem to require absorption into the body of some factors, presumably polyphenols, from tannin-containing foods. Preliminary studies have shown significant uptake of radiolabeled tannin from rat diets (1,47). Recent investigations described below have provided support for multiple mechanisms of antinutritional effects of dietary tannins. Inhibition of digestion apparently makes only a minor contribution to these effects.

Inhibition of Post-Digestive Metabolism of Absorbed Nutrients

In studies of food utilization by insects, the efficiency with which ingested food is converted to biomass (ECI) is considered to be the product of (a) the approximate digestibility (AD) of the food and (b) the efficiency with which digested food is then converted to biomass (ECD) (48). Evaluation of ECD involves changes in biomass expressed as dry weight so this analysis has seldom been applied to vertebrates. We have now completed a study of the effects of dietary tannin in rats using this method to distinguish between effects on digestion (AD) and on metabolism of digested and absorbed nutrients (ECD). Results were determined on the basis of both dry weight and N content in order to distinguish between digestion/metabolism of proteins and total dry matter (40). The results show that dietary tannins cause a moderate inhibition of AD (protein digestion is inhibited to a greater degree than digestion of dry matter), but that this minor inhibition cannot account for the antinutritional effects of the tannin. For example, under conditions where tannin almost completely blocked rat growth, AD was 80% compared to 88% for a tannin-free diet. There was a similar small depression of feed consumption by tannin, but again not sufficient to account for the failure to gain weight. The major effect of dietary tannin was a strong depression of ECD to a value less than 10% of the ECD on the comparable tannin-free diet. These results suggest that dietary tannin is more effective as an inhibitor of the utilization of digested and absorbed nutrients than it is as an inhibitor of digestion. Recoveries of tannin in the feces were quite low, indicating that most of the ingested tannin was either absorbed from the intestine, metabolized to nontannin materials, or inextricably bound to fecal material. We are pursuing these possibilities.

Metabolic Defenses Against Dietary Tannins

Animals which normally consume tannin-containing plant materials presumably have survived by developing at least partially effective defenses against these potentially harmful conponents of their diet. When antinutritional effects are observed, it should be recognized that these effects do not represent the total toxic capacity of the plant material, but only that part of it which is not neutralized by the animal's defenses. Defenses which develop only after induction by the presence of tannin in the diet are the most readily detected because the introduction of tannin into a previously tannin-free diet results in short-term acute effects before the defensive response is completely induced.

We have recently shown, with our collaborators, that mammalian herbivores (but not carnivores) produce unique proline-rich (up to 45%) salivary proteins which have a very high affinity for tannin. In rats (46) and mice (49), these tannin-binding proteins are virtually absent until induced by dietary tannin, and in hamsters, which are extraordinarily vulnerable to dietary tannin, these proteins are not inducible by tannin although they are induced by treatment with isoproterenol, a beta adrenergic agonist (50). In all other animals we have examined including humans (51) and ruminants (52) these proteins appear to be constitutive and are present in amounts which reflect the approximate level of tannins and related polyphenols in their normal diets. We have proposed that these proline-rich tannin-binding salivary proteins constitute the first line of defense against tannins in the digestive tract (1,50,53). The enhanced vulnerability to dietary tannin of hamsters (50) and of rats in which induction of these salivary proteins is blocked by beta-1-specific blockers (54) is consistent with this proposed defensive mechanism. Supplementation of the tannin-containing diet with additional protein (to bind the tannin) or detoxification of the tannin with dilute aqueous ammonia (55) eliminated both the induction of salivary proline-rich protein and the diminished weight gains due to tannin (20). Without the defense provided by salivary proline-rich proteins, the nutritional consequences of consuming tannin would undoubtedly be more severe. The interrelations of dietary tannins and salivary proline-rich proteins have recently been reviewed (3).

The strategy through which other organisms defend themselves against noxious polyphenolic defense chemicals of plants may similarily include utilization of specialized tannin-binding proteins. The pathogenic fungus which causes the cereal disease anthracnose protects its spores against toxic phenolics produced by the host plant by making a glycoprotein which has an extremely high affinity for tannins (56). Insects (37) and birds (57) apparently defend themselves against dietary tannin by mechanisms which do not involve proline-rich proteins. Supplementation of tannin-rich diets for poultry with free methionine overcomes the antinutritional effect of tannin (57); the mechanism is not yet known.

Conclusions

The antinutritional effects of tannins are more complex than formerly recognized, and tannin-consuming animals actively defend themselves against these effects and thus diminish their severity. The next level of understanding the basis of these effects will be the identification of the site(s) within the body, not just within the intestine, where tannins inhibit utilization of digested and absorbed nutrients.

Acknowledgment

Research from this laboratory was supported by INTSORMIL (UnSAID Title XII Sorghum-Millet CRSP).

References

1. Butler, L.G., J.C. Rogler, H. Mehansho, and D.M. Carlson, *Plant Flavonoids in Biology and Medicine: Biochemical, Pharmacological, and Structure-Activity Relationships*, edited by V. Cody, J.B. Harborne and E. Middleton, Alan R. Liss, New York, NY, 1986, pp. 141-157.
2. Beart, J.E., T.H. Lilley, and E. Haslam, *Phytochemistry 24*:33 (1985).
3. Mehansho, H., L.G. Butler, and D.M. Carlson, *Annu. Rev. Nutr. 7*:423 (1987).
4. Pierpoint, W.S., *Biochemistry of Plant Phenolics*, edited by C.F. van Sumere and P.J. Lea, Academic Press, Oxford, England, 1985, pp. 427-451.
5. Morton, J.F., *J. Crude Drug Res. 12*:1829 (1972).
6. Glick, Z., and M.A. Joslyn, *J. Nutr. 100*:509 (1970).
7. Freeland, W.J., P.H. Calcott, and D.P. Geiss, *Biochem. Systematics Ecol. 13*:195 (1985).
8. Myer, R.O., and D.W. Gorbet, *Anim. Feed Sci. Technol. 12*:179 (1985).

9. Rostagno, H.S., W.R. Featherston, and J.C. Rogler, *Poultry Sci. 52*:765 (1973).
10. Cooper, S.M., and N. Owen-Smith, *Oecologia 67*:142 (1985).
11. Robbins, C.T., T.A. Hanley, A.E. Hagerman, O. Hjeljord, D.L. Baker, C.C. Schwarts, and W.W. Mautz, *Ecology 68*:98 (1987).
12. Hussein, L., and H. Abbas, *Nutr. Rep. Int. 31*:67 (1985).
13. Peaslee, M.H., and F.A. Einhellig, *Comp. Gen. Pharmac. 4*:393 (1973).
14. Sell, D.R., J.C. Rogler, and W.R. Featherston, *Poultry Sci. 62*:2420 (1983).
15. Joslyn, M.A., and Z. Glick, *J. Nutr. 98*:119 (1968).
16. Freeland, W.J., P.H. Calcott, and L.R. Anderson, *Biochem. Systematics Ecol. 13*:189 (1985).
17. Kumar, R., and M. Singh, *J. Agric. Food Chem. 32*:447 (1984).
18. Reid, C.S.W., M.J. Ulyatt, and J.M. Wilson, *Proc. New Zealand Soc. Anim. Prod. 34*:82 (1974).
19. Zelter, S.Z., F. Leroy, and J.P. Tissier, *Ann. Biol. Anim. Biochim. Biophys. 10*:111 (1970).
20. Butler, L.G., *Toxicants of Plant Origin*, Vol. IV, *Phenolics* edited by P. Cheeke, CRC Press, Boca Raton, FL, 1989, pp. 95-121.
21. Butler, L.G., D.J. Riedl, D.G. Lebryk, and H.J. Blytt, *J. Am. Oil Chem. Soc. 61*:916 (1984).
22. McManus, J.P., K.G. Davis, J.E. Beart, S.H. Gaffney, T.H. Lilley, and E. Haslam, *J. Chem. Soc. Perkin Trans. II* 1429 (1985).
23. Mole, S., and P.G. Waterman, *Allelochemicals in Agriculture, Forestry and Ecology*, Am Chem. Soc. Symposium series, Washington, DC, 1987, pp. 572-587.
24. Zucker, W.V., *Am. Naturalist 121*:335 (1983).
25. Hagerman, A.E., and L.G. Butler, *J. Biol. Chem. 256*:4494 (1981).
26. Asquith, T.N., and L.G. Butler, *Phytochemistry 25*:1591 (1986).
27. Asquith, T.N., J. Uhlig, H. Mehansho, L.J. Putman, D.M. Carlson, and L.G. Butler, *J. Agric. Food Chem. 35*:331 (1987).
28. Strumeyer, D.H., and M.J. Malin, *Biochem. J. 118*:899 (1970).
29. Hagerman, A.E., and C.T. Robbins, *J. Chem. Ecol. 13*:1243 (1987).
30. Hagerman, A.E., and L.G. Butler, *J. Agric. Food Chem. 28*:947 (1980).
31. Rhoades, D.F., *Herbivores: Their Interaction with Plant Secondary Compounds*, edited by G.A. Rosenthal and D.H. Janzen, Academic Press, New York, NY, 1979, pp. 3-55.
32. Loomis, W.D., and J. Battaile, *Phytochemistry 5*:423 (1966).
33. Kakiuchi, N., M. Hattori, T. Namba, M. Nishizawa, T. Yamagishi, and T. Okuda, *J. Nat. Prod. 48*:614 (1985).
34. Blytt, H.J., T.K. Guscar, and L.G. Butler, *J. Chem. Ecol. 14*:1455 (1988).
35. Oh, H.I., and J.E. Hoff, *J. Food Sci. 51*:577 (1986).
36. Mole, S., and P.G. Waterman, *Phytochemistry 26*:99 (1987).
37. Martin, J.S., M.M. Martin, and E.A. Bernays, *J. Chem. Ecol. 13*:605 (1987).
38. M. Berenbaum, *Am. Naturalist 115*:138 (1980).

39. Griffiths, D.W., and G. Moseley, *J. Sci. Food Agric. 31*:255 (1980).
40. Mole, S., J.C. Rogler, C.J. Morell, and L.G. Butler, in review.
41. Bernays, E.A., and S. Woodhead, *Science 216*:201 (1982).
42. Atsatt, P.R., and T. Ingram, *Oecologia 60*:135 (1983).
43. Moseley, G., and D.W. Griffiths, *J. Sci. Food Agric. 30*:772 (1979).
44. Glick, Z., and M.A. Joslyn, *J. Nutr. 100*:516 (1970).
45. Sell, D.R., and J.C. Rogler, *Proc. Soc. Exp. Biol. Med. 174*:93 (1983).
46. Mehansho, H., A.E. Hagerman, L.S. Clements, L.G. Butler, J. Rogler, and D.M. Carlson, *Proc. Natl. Acad. Sci. U.S.A. 80*:3948 (1983).
47. Wagner, H., in *Recent Advances in Phytochemistry and Biochemistry of Plant Phenolics*, edited by T. Swain, J. Harborne, and C.F. van Sumere, Plenum, New York, NY, 1979, p. 589.
48. Waldbauer, G.P., *Rec. Adv. Insect Physiol. 5*:229 (1964).
49. Mehansho, H., S. Clements, B.T. Sheares, S. Smith, and D.M. Carlson, *J. Biol. Chem. 260*:4418 (1985).
50. Mehansho, H., D.K. Ann, L.G. Butler, J. Rogler, and D.M. Carlson, *Ibid. 262*:12344 (1987).
51. Mole, S., K.E. Glander, and L.G. Butler, *J. Chem. Ecol.*, in press.
52. Robbins, C.T., S. Mole, A.E. Hagerman, and T.A. Hanley, *Ecology 68*:1606 (1987).
53. Hagerman, A.E., P.A. Austin, L.A. Suchar, and C.T. Robbins, *Fed. Proc. 46*:2025 (1987).
54. Subramanian, V.N., H. Mehansho, and D.M. Carlson, *Fed. Am. Soc. Exp. Biol. J. 2*:A779 (1988).
55. Price, M.L., L.G. Butler, J.C. Rogler, and W.R. Featherston, *J. Agric. Food Chem. 27*:441 (1979).
56. Nicholson, R.L., L.G. Butler, and T.N. Asquith, *Phytopathology 76*:1315 (1986).
57. Armstrong, W.D., and W.R. Featherston, *Poultry Sci. 52*:1592 (1973).

Chapter Twenty-three

Nutritional and Physiological Effects of Phytic Acid

Lilian U. Thompson

Department of Nutritional Sciences
Faculty of Medicine
University of Toronto
Toronto, Ontario, Canada M5S1A8

A number of health agencies recommend an increased intake of complex carbohydrates for better control of chronic diseases such as cardiovascular disease, diabetes, and cancer. With this thrust, an increase in the intake of phytic acid (PA) may be expected in the future since it is rich in fiber-rich starchy foods such as legumes and cereals. Consumption of plant protein products as substitutes for animal protein sources may likewise contribute to the elevated intakes of PA.

Concern for high intake of PA relates to its possible antinutrient effect. PA has been considered as an antinutrient because it can bind with nutrients such as minerals and proteins and render them less available for utilization by the body. However, the same binding capability appears to provide PA with some healthful benefits. This paper discusses some of the health implications of the interactions of PA. Although not an exhaustive review, it provides some current thoughts regarding the nutritional and physiological effects of PA particularly as they relate to the digestibility and bioavailability of zinc, proteins and starch, and to colonic health.

Phytic Acid in the Food and Its Reactions

PA, myoinositol 1,2,3,4,5,6 hexakis-dihydrogen phosphate, is usually found in foods at <0.1-7% levels (1,2). It is the primary phosphate reserve in most seeds and typically accounts for 60-90% of the total phosphorus (1,3). It is largely seen in the globoids of aleurone grains in the aleurone layer of monocotyledenous seeds such as wheat and rice (4), in the crystalloid-type globoids of the dicotyledenous seeds such as legumes, nuts and oilseeds (3,5) and in the germ of corn (6). The discrete location of PA in the plant seed allows reduction or concentration of PA by the separation of the aleurone layer (bran of cereals) or germ of corn through the milling process. The removal of PA from the legumes and oilseeds, however, is more difficult because of its close association with the proteins; PA is often coisolated or coconcentrated with the protein in these foods.

The reactions of PA have been extensively reviewed (1,7,8). PA is strongly negatively charged at pH's normally encountered in foods, thus making it very reactive with other positively charged groups such as cations and proteins. Cations can bind to a phosphate group or two phosphate groups of a PA molecule (Fig. 23.1A) or between phosphate groups of different PA molecules (10). Several cations may complex within the same molecule of PA producing a mixed salt of PA. The presence of more than one cation may synergistically increase the precipitation of the phytate salts as has been observed *in vitro* with Zn and Ca and Zn and Cu (11,12). The binding of PA with mineral ions is pH dependent and complexes of varying solubilities and stabilities are formed (7,13,14).

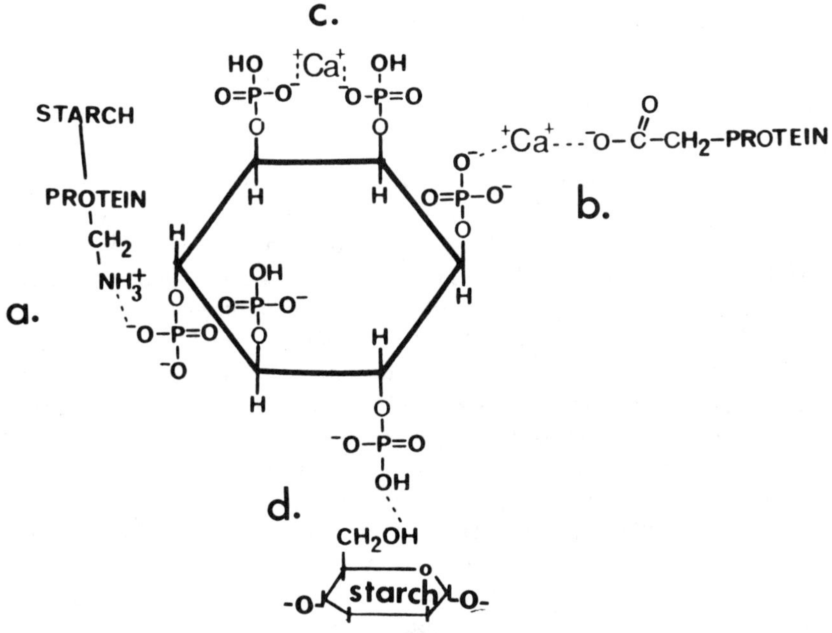

Fig. 23.1. Possible interactions of phytic acid with proteins, minerals and starch (9).

The reaction of PA with proteins also depends on the pH. At low pH below the isoelectric point of the proteins, PA binds directly with the positively charged proteins as a result of electrostatic attraction (Fig. 23.1B). At an intermediate pH above the isoelectric pH, both the PA and proteins are negatively charged. Therefore, electrostatic attraction and direct binding of PA with the proteins do not take place except perhaps in minor quantities with a few terminal amino and epsilon groups of lysine since these groups are still protonated at pH below 10. Primarily, PA binds with proteins mediated by multivalent cations such as Ca and Mg (Fig. 23.1C). At pH >10, the ternary complex may dissociate, the PA may precipitate as cation-PA, while the protein may solubilize as sodium proteinate (7).

PA appears to be structurally capable of binding with starch through phosphate linkages (Fig. 23.1D). Its binding with minerals, proteins or starch, directly or indirectly, may alter the solubility, functionality, digestibility and absorption of these nutrients (8). The nutrient digestibility may also be affected by the binding of PA with the digestive enzymes, which are themselves proteins. These, in turn, may be responsible for both the adverse and healthful effects of PA in the food as it passes through the digestive system.

Mineral Bioavailability

The effect of PA on mineral bioavailability was the focus of many studies and the subject of numerous past reviews (1,7,8,15-20). Of the minerals in the diet, Zn appears to be one of the most affected by high PA concentrations probably because it forms the most stable, insoluble complex with PA (7,14). Therefore, in this paper, Zn is discussed to illustrate the complexity surrounding the PA effect on mineral bioavailability.

The early evidence relating PA to Zn deficiency came about from problems encountered during animal production in the 1950's (21). Growth depression and skin lesions in pigs and productive losses in poultry were seen upon feeding soy and corn as sources of dietary protein. O'Dell and Savage (22) later reported that the Zn requirement was higher with cereal-soya diet than with casein as the protein source and attributed this to the PA content of the diet. Since then, more evidence has been reported on the adverse effects of PA on the availability of Zn in both animals and humans (23-30). Zn deficiency in young men in the Middle East was associated with hypogonadism and dwarfism (31). Marginal Zn deficiency in children produced symptons such as low hair Zn, impaired taste acuity, poor appetite and suboptimal growth (32).

Oberleas (33) first suggested that the availability of Zn in the diet might be better predicted by the PA and Zn molar ratio than by the PA content. This was later supported by studies in rats (34-36) and also in humans (26,37,38), which generally showed that a PA to Zn molar ratio exceeding 10 can significantly inhibit Zn utilization and restrict growth. However, other studies showed that the PA to Zn ratio is a poor indicator of Zn availability, especially when the dietary Ca level is high. High Ca concentration aggravates the effect of PA on zinc absorption (7,11,23,33-35,39-47), and this has been attributed to the formation in the gastrointestinal tract of a Ca-Zn-PA complex which is more insoluble and unavailable for absorption than the complex of PA with either mineral alone (48). Hence it was suggested that the effect of PA on Zn availability may be better predicted by calculating the [PA][Ca]/[Zn] molar ratio (49-51).

Fordyce et al. (44) recently examined the use of the [PA][Ca]/[Zn] molar ratio as a predictor of Zn bioavailability in rats by doing retrospective calculations from several of their published works (27,46,47,52-55). They showed that a ratio greater than 3.51 significantly depressed weight gain. This was in agreement with Davies et al. (49) who estimated a critical ratio of 3.5 from their own published data.

In humans, retrospective calculations from the data of Cossack and Prasad (56) suggest that a [PA][Ca]/[Zn] ratio greater than 0.5M/kg dry diet (49) or 200mM/day (51) may induce negative Zn balance. Since humans have variable energy intake per day, Davies and Mills (57) suggested the ratio [PA][Ca]/[Zn][Energy] as a better predictor of Zn availability. From several studies (26,38,56,58,59) they estimated that 100mM/1000 kcal is the critical ratio. Interestingly, Bindra et al. (51) observed a less than optimum Zn status in 32% of Asian immigrants in Canada eating lacto-ovo vegetarian diets with a molar ratio of 0.9M/kg diet, 475 mM/day or 250mM/1000 kcal as compared with the omnivorous controls taking diets with molar ratio of 0.3M/kg dry diet, 131 mM/day or 90 mM/1000 kcal. The critical molar ratio estimates for humans were lower than for rats, suggesting the greater ability of rats to break down the PA-Ca-Zn complex. Coprophagy in rats may also have a role as bacterial breakdown of PA is known to occur in the colon (44).

Is the current dietary PA intake of humans of concern in relation to Zn availability? Ellis et al. (45) estimated that the regular, ovo-lactovegetarian and soy meat-substituted hospital daily diets contain 378, 440 and 1144 mg PA, while Oberleas and Harland (40) found 2,575 and 290 mg in typical vegetarian and nonvegetarian menus, respectively. However, the mean daily PA intake in self-selected diets has been reported to be 750 mg in average Americans (60), 1275 mg in vegetarian Americans (40), 600-800 mg/day in average British (18), 1480 mg in lacto-ovo vegetarian Asian immigrants and 800 mg in omnivorous Canadians (51). More recently, Ellis et al. (61) reported that the intake of omnivorous Americans, vegetarian Americans and Asian vegetarians to be in general agreement with these data, except that there was a tendency for the female to have a lower intake than the male omnivors. To avoid zinc deficiency due to PA, Harland and Peterson (60) suggested a daily intake of 750 mg PA as reasonable for a 70 kg man. Most omnivorous daily diets are below, while most vegetarian diets are above this limit.

The [PA]/[Zn] and [PA][Ca]/[Zn] molar ratios in typical diets (40,45) and in self selected daily diets of various population groups (51,61) are shown in Figures 23.2 and 23.3. Using both ratios as indices, PA will generally affect the Zn bioavailability in vegetarian diets but not in most American and Canadian omnivorous diets. However, considering the range of intake within a population group, a greater number of omnivors will likely be affected by PA with the [PA][Ca]/[Zn] molar ratio as an index than with the [PA]/[Zn] molar ratio (61).

While useful, the use of molar ratios for estimating Zn deficiency risk, due to PA, is not perfect and should be used with caution since there are many other factors which may also influence Zn availability including the following:

(a) Method of processing.

Food processing which increases the insolubility of the Zn complex with PA will tend to decrease the Zn availability. This was evident in studies (53,55) which showed that acid precipitated soy proteins have higher Zn bioavailability than the neutralized proteins prepared under identical conditions and with similar [PA][Ca]/[Zn] molar ratios. Neutralized soy proteins are thought to have formed tightly bound PA-Zn-protein complexes which are resistant to digestion while the acid precipitated proteins, which are protonated, are less likely to react with the PA and minerals.

(b) Added vs endogenous PA.

Added PA, usually in the form of the sodium salt, is very soluble, and therefore is very reactive with other dietary components such as the minerals. On the other hand, endogenous PA may already be strongly

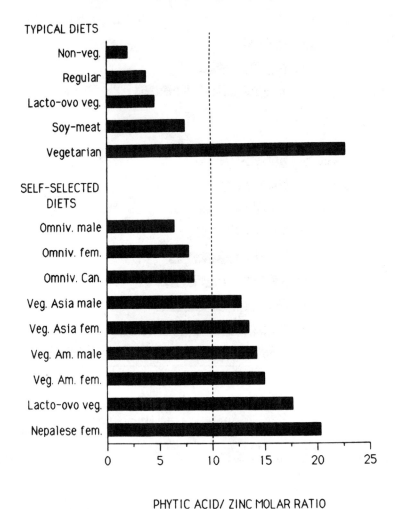

Fig. 23.2. Phytic acid/Zn molar ratio in some typical and self-selected diets (40,45,51,61).

complexed with other food components and may remain insoluble and unavailable for interaction with other dietary minerals. This point is especially important when projecting results of animal studies to humans since animal studies often involve feeding purified PA while humans normally eat endogenous PA in the food.

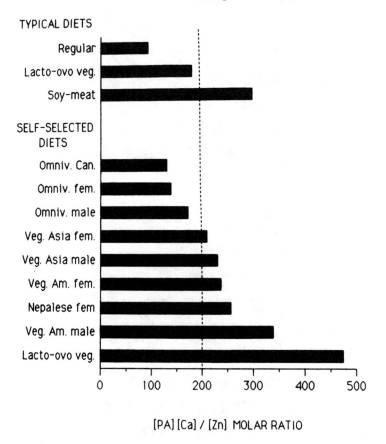

Fig. 23.3. [Phytic acid][Ca]/[Zn] molar ratio in some typical and self-selected diets (45,51,61).

(c) The presence of other mineral binding components such as dietary fiber, oxalic acid and tannins which may compete with PA for binding with minerals.

Dietary fibers are capable of binding minerals, and controversy still exists regarding whether it is the fiber or the PA which is largely responsible for the lowered mineral availability in high fiber diets (18,62-64). However, a strong case has been made for the greater role of PA than dietary fiber in cereal products (38,62,65). There were significant improvements in Zn absorption in meals based on whole wheat flour and/or bran when PA was removed either by fermentation of the bread or acid washing of the bran, and also when low PA bran was used instead of ordinary bran. Nonetheless, Kelsay (63) reported that intakes of 2.0 g PA and as much as 32 g dietary fiber per day do not generally affect mineral balance provided the mineral intake was sufficient.

(d) Dietary protein concentration.

As seen in studies with soy protein meals, the use of the [PA][Ca]/[Zn] molar ratios is a good predictor of Zn availability only in low protein-high Ca meals but not in meals with a high protein content (64). The effect of PA on Zn absorption has been reported to decrease with increased dietary protein level (18,64).

(e) Presence of dietary, intestinal or bacterial phytase.

Many legumes and cereals contain endogenous phytases which, unless inactivated by cooking or processing before intake, can hydrolyze PA under the conditions of the gastrointestinal tract. In studies with rats, pigs and chicks, feeds containing phytase had higher phosphorous availability than those without phytase (66-68). In humans, the breakdown of PA by cereal phytases has been shown to improve the Zn absorption in high phytate meals (69), indicating the significance of dietary phytase in reducing any adverse effect attributed to PA. Although evidence exists for the presence of phytases in the small intestinal mucosa of rats, pigs, chickens and humans (68,70-72), studies with germ-free mice suggest that phytase activity can only occur in the presence of bacteria in the colon (73,74). However, we observed a 30-97% recovery of PA in the ileal effluent of ileostomate volunteers fed various legumes and cereals (likely phytase-free because of the heat treatment involved), suggesting the presence of small intestinal phytases in humans (75). The breakdown of PA by phytases is important since the hydrolysis products of PA(inositol di-, tri-, tetra- or penta-phosphate) have been shown to have lower mineral binding capability compared with the PA (76,77).

(f) Whether the PA is taken in the same meal as the mineral source or in separate meals.

Obviously, PA will have a greater effect when taken in the same meal as the mineral-rich food.

(g) Presence of other minerals.

Cu (78,79) and Fe (80-82) in high concentrations have been shown to affect Zn absorption.

(h) Metabolic adaptation to a high PA diet.

While studies suggest that humans are capable of adapting to PA-containing diets as indicated by a negative mineral balance at the early weeks of feeding and a positive balance after several weeks (83,84), there are also studies showing the contrary (26,85). Discrepancy has been attributed to the amount of PA taken, with a loss of adaptability with very large PA intakes (26).

In addition to the above additional factors which may affect Zn absorption, there is still the question of whether results obtained with rats apply to humans. Note that most of the work establishing the validity of the [PA]/[Zn] and [PA][Ca]/[Zn] molar ratios was done in rats whose mineral requirements and intake differ from those of humans. Certainly, more human studies interrelating all the variables identified here need to be done in the future for a more accurate prediction of Zn bioavailability in foods.

Protein Digestion and Absorption

Limited and conflicting studies have been done on the effect of PA on protein digestibility and amino acid availability.

In vitro studies showed that the addition of PA significantly lowered the rate of pepsin hydrolysis of gluten, casein or bovine serum albumin (86-88); this was attributed to the complexing of PA with the protein substrate. A similar reduction in casein hydrolysis by trypsin in the presence of PA was reported (89), although in this case, the reduction in hydrolysis was primarily attributed to the complexing of PA with the trypsin Ca rather than to the casein substrate. The inhibition of pepsin was inversely correlated with the degree of PA hydrolysis, although beyond 65% hydrolysis, PA appeared to have lost its inhibitory effect (88). The effect of added PA differs with the type of protein with, for example, zein being most affected followed by serum albumin, casein, soy protein isolate and lactalbumin (90).

The above observations are in contrast with the findings of Serraino *et al.* (91) which showed that removal of endogenous PA from rapeseed flour had no significant effect on the *in vitro* rate of protein digestion by pepsin and pancreatin and of Iacobucci *et al.* (92) and Reddy *et al.* (93) who showed no effect of endogenous PA on the digestibility of soybean and Northern bean protein isolates, respectively. The discrepancies in the *in vitro* data may be attributed partly to the differences in the activity of the added vs endogenous PA, the added soluble PA being more reactive towards the substrate or digestive enzymes. Of interest,

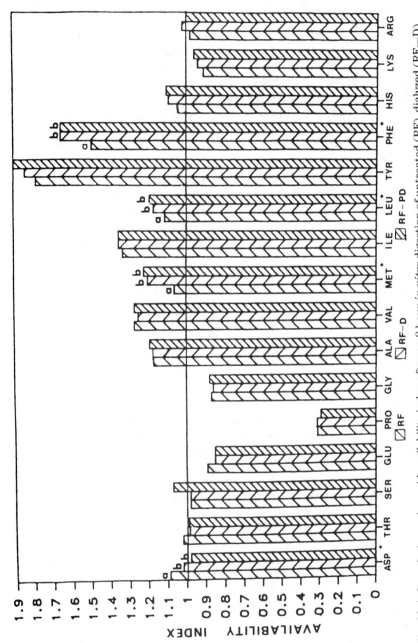

Fig. 23.4. Cumulative amino acid availability index after an 8 hour *in vitro* digestion of untreated (RF), dialyzed (RF–D) and phytase-treated and dialyzed (RF-PD) rapeseed flour containing 5.79, 2.81 and 0.63% phytic acid, respectively. Means with different notations are significantly different (p <0.01) (91).

however, is the observation, *in vitro*, that PA removal from the protein can change the pattern of release of amino acids from the proteins, with some amino acids released faster than the others (91) (Fig. 23.4). This may have some physiological significance since protein synthesis and utilization depend on the type and amount of amino acids which get absorbed at the same period of time.

Atwal *et al.* (94) found that the growth rate, diet consumption and efficiency of protein utilization were lower for rats fed rapeseed protein which provided 1.24% PA in the diet than for those fed casein, but improved as the PA content of the diet was reduced to 0.41%. In contrast, Thompson and Serraino (95) reported that removal of PA from rapeseed flour did not change the true protein digestibility and amino acid absorption in rats, in general agreement with McDonald *et al.* (96) who showed no effect of PA on rapeseed protein quality. Protein digestibility, PER and biological value did not change when the PA content of soy protein isolate was varied but PA reduction in high protein wheat bran flour significantly improved the nutritive value of the proteins (97). The addition of purified PA to levels as high as 2% in the casein diet of rats (98) and 5% in the diet of mice (99) had no effect on protein digestibility. In humans, Reinhold (26) showed that with diets containing 2.5 g PA, the nitrogen balance was not affected when PA was provided as purified sodium phytate but was less positive or even negative at times when PA was provided as PA-rich unleavened whole wheat bread (tanok). However, this has been attributed to the high dietary fiber intake during the tanok period.

Why *in vivo* and some *in vitro* digestibility data do not agree is not clear. However, this could relate partly to the *in vivo* action of intestinal and/or bacterial phytase which is not accounted for in the *in vitro* measurements. Also, most *in vivo* digestibility studies are based on the analysis of fecal nitrogen output; this method may be subject to error since unabsorbed nitrogen in the small intestines can be broken down by the bacterial flora and eventually gets absorbed or used in bacterial synthesis.

Ileostomates are otherwise healthy individuals whose colons have been removed due to some problems in the past. Their undigested products of digestion are collected in a bag attached to their terminal ileum and should provide better information regarding the actual amounts of nutrients absorbed in the small intestines. Using an ileostomate in a study of 21 different leguminous and cereal foods, we observed a low but significant relationship ($r = 0.518$, $p > 0.025$) between the PA/protein intake and the protein output (Thompson *et al.*, unpublished data).

When a basal diet with unleavened bread was fed so that the total PA intake was 0.57 g/day, the ileal protein recovery was 8%. This increased to 11% when the unleavened bread was supplemented with PA so that the total intake was 2.0 g/day, suggesting a small effect of PA on protein digestion and absorption. Although these results may be considered preliminary because only one ileostomate has been tested, it is also noted that this ileostomate has been shown to give nutrient and PA outputs which are similar to three other ileostomates (75,100).

If the small intestinal protein output is indeed increased by PA, this may be due to the slower rate of digestion and absorption of proteins as observed in *in vitro* studies. A slow rate of digestion and absorption has been suggested as more beneficial to health by reducing metabolic stress (101). However, whether the change in digestion rate and small intestinal output of protein due to PA has physiological and nutritional significance remains to be established.

Starch Digestion and Absorption

There is very little work done in relation to the effect of PA on starch digestion and absorption but current information suggests that it may have a role. Several studies revealed the ability of PA to inhibit the activity of amylases from wheat (102,103), maize, chick peas, barley, *Bacillus subtilis* (104), bovine pancreas (105) and human saliva (106,107), although one study showed an insignificant effect of PA on amyloglucosidase and maltase activities and a significant effect on bovine pancreatic amylase activity only at the early stage of digestion (108).

Significant negative relationships were observed between PA concentrations or intake and the *in vitro* rate of starch digestion and the blood glucose response in normal and diabetic volunteers (107,108). The legumes which contained the highest concentrations of PA were digested the slowest and produced the flattest blood glucose response. When PA was added to wheat flour in an unleavened bread preparation to make the concentration approximately equal to that present in legumes (i.e., 2% starch basis), a significant reduction in the blood glucose response was observed (107). Conversely, removal of PA from navy bean flour increased the *in vitro* rate of starch digestion and the blood glucose response while readdition of the PA back to the dephytinized bean flour produced the opposite effect (9) (Fig. 23.5).

The effect of PA on starch digestion can be modified by the addition of Ca (9,102,104,105,107), with small concentrations of Ca reversing the

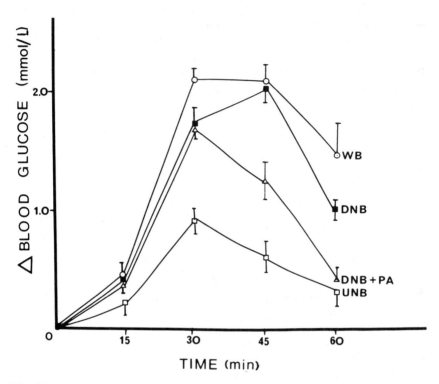

Fig. 23.5. Mean blood glucose response to unleavened breads prepared from white wheat flour (WB), undephytinized navy bean flour (UNB), dephytinized navy bean flour with 1% phytic acid (DNB + PA) (108).

inhibitory effect caused by PA both *in vitro* and *in vivo*. Ca probably precipitated the PA rendering it inactive towards the digestive enzymes.

The significant change in the rate of starch digestion and blood glucose response caused by PA was accompanied by an increase in malabsorbed carbohydrate as seen by breath hydrogen or ileostomate methods (9,108,109). However, the reduction in blood glucose response due to PA was much greater in comparison with the amount of carbohydrate malabsorbed. For instance, when navy bean flour was dephytinized, the blood glucose response increased by 54% while malabsorbed carbohydrate decreased by 7.8%. On the other hand, the addition of PA to the dephytinized flour decreased the blood glucose response by 29% while the malabsorbed carbohydrate increased only by 5% (107,108). Therefore, PA appears to slow down the rate of starch digestion but a high percentage of the starch is digested and absorbed before it reaches the colon.

The mechanism of the PA effect on starch digestion and blood glucose response has been suggested to be due to any or all of the following: (a) its binding with proteins which are closely associated with starch, (b) its association with the digestive enzymes which are themselves proteins, (c) its chelation of Ca required for the activity of amylase, (d) its direct binding with starch, (e) its effect on starch gelatinization during cooking or processing, and (f) *in vivo*, through its influence on gastric emptying (1,7,9,103,104,106-112).

Work on the role of PA on starch digestion and blood glucose response became of interest when it was suggested that a slow rate of starch digestion is more beneficial to health and in the management of diabetes and hyperlipidemia (113-116). High blood glucose rise from rapidly digested starchy foods necessitates greater insulin response (117,118). This, in turn, is believed to lead to reduced sensitivity of insulin receptors and to insulin resistance, a factor which may be crucial in the development of diabetes and obesity (117-119). It has also been related to raised blood triglyceride considered to be an independent risk factor for cardiovascular disease in susceptible individuals (116). Since PA is present in high concentrations in carbohydrate containing foods which are slowly digested, it was suggested to be one of the factors responsible for their slow digestion rate and, probably, their beneficial effects. Many evidences which were recently reviewed (9,107) and mentioned above tended to support these suggestions.

Phytic Acid and Colonic Health

A large proportion of PA can survive the small intestines and enter the colon. The amount may range from 0.30-3.7 g/day, representing 30-97% of the intake (75). Despite this, very little work has been done regarding its implications on colonic health.

In the colon, PA may remain intact and continue to interact with proteins and minerals, or it may be hydrolyzed by the bacterial phytase. Hydrolyzed phytate complexes may either release minerals and proteins into the colonic lumen or continue to bind proteins and minerals in the colon, depending on the stage of PA breakdown and the number of available binding sites. Any of these events may have profound effects on activities occurring in the colon which govern the stage of colonic health. Such activities include epithelial cell proliferation and bacterial enzyme (e.g., β-glucuronidase, mucinase etc.) actions; their increase has been implicated with increased risk for carcinogenesis (120,121) and hence has been suggested as marker systems in assessing the risk of colon cancer.

Some possible mechanisms of PA effect on the colon are summarized in Figure 23.6. If PA remains intact and binds minerals and proteins in the colon, enzyme activities and cell proliferation may be reduced in a number of different ways: (a) Complex formation of PA with the bacterial enzymes, which are proteins, may render them inactive. Furthermore, these enzymes may require minerals as cofactors, which may be unavailable if bound to PA. (b) PA may act as an antioxidant by virtue of the fact that it is able to bind Fe. PA occupies all available Fe coordination sites thus preventing Fe binding to H_2O_2 and catalysis of hydroxy-radical formation catalyzed by Fe (122). This in turn can bring about a reduction in lipid peroxidation and tissue damage — of significance since lipid peroxidation has been suggested as a risk factor in carcinogenesis (123). (c) PA can continue to bind minerals such as Zn, depriving the colonic epithelium of minerals needed for DNA synthesis (124) and hence reducing the rate of cell proliferation. (d) PA may bind Ca—a potentially adverse effect, since Ca will no longer be able to bind to bile acids and free fatty acids and prevent their damaging effects in the colon.

Conversely, if PA is hydrolyzed in the colon, freed minerals such as Ca may bind phosphates, bile acids and fatty acids, which tend to increase cell proliferation (125). Increases in luminal Ca levels have been found

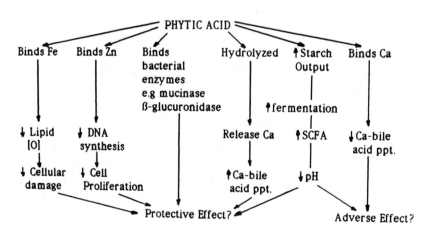

Fig. 23.6. Some possible mechanisms of phytic acid effect on the colon.

to decrease colonic cell proliferation (126). PA has been shown to increase the amount of starch which reaches the colon (9,75,108). Fermentation of this additional substrate by colonic bacteria provides short chain fatty acids which may lower the pH to a more protective level and, more importantly, provide butyric acid which has been shown to have antitumor effects (127,128). However, excessive fermentation and lowering of the pH have also been observed increasing the risk for tumorigenesis (129).

Our studies and those of others tend to support some of the above hypotheses. A significant increase in short chain fatty acid production and a lowering of the fecal pH were observed when diets high in PA were eaten (75,98). Also, a reduction in cell proliferation (Fig. 23.7) and β-glucuronidase and mucinase activities, as a result of the addition of PA to semisynthetic diets, have been observed (98,130,131). That these activities may translate to lower tumor incidence was supported by reports of a 34.7% reduction in tumors in azoxymethane-treated rats fed 1% PA in their drinking water (132,133). Furthermore, an augmentation

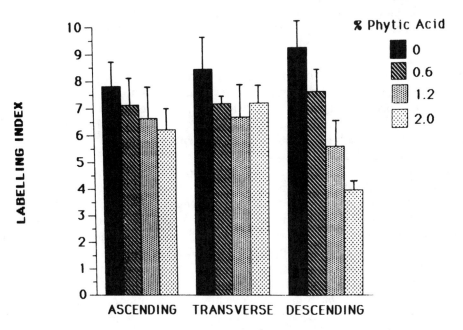

Fig. 23.7. Effect of phytic acid on labeling indices in the ascending, transverse and descending colon.

of dimethylhydrazine colo-rectal tumor induction in animals supplemented with iron compared with the controls has been observed (134); this augmentation was reduced when PA was also included in the diet (Nelson, personal communication).

Evidently, PA and undigested nutrients that reach the colon have some profound effects on colonic health and this could be protective or adverse depending on the concentration and the state of breakdown of the PA.

Conclusion

There is much evidence to show that PA has an adverse effect on the bioavailability of minerals, particularly Zn. However, using the {PA]/[Zn] or [PA][Ca]/[Zn} molar ratios as guides, it appears unlikely that zinc bioavailability will be significantly affected by the PA in most American, Canadian and British omnivorous diets. On the other hand, bioavailability in vegetarian diets is likely to be at risk, especially when the Ca intake is high.

PA can slow down the rate of digestion and absorption and cause a small increase in unabsorbed protein and carbohydrates in the small intestines. The lowering of the blood glucose response caused by PA may be considered potentially healthful since low glycemic index foods have been shown to be beneficial in the management of hyperlipidemic and diabetic individuals. The physiological impact of the slow protein digestion and the change in the pattern of release of amino acids still need to be explored. However, following the general principle that a very high nutrient influx can cause increased metabolic stress, the slowing of protein digestion without extensive desorption may as well be considered as a desirable effect.

The implication of the large amounts of PA, which enter the colon together with the additional starch, proteins and minerals which also reach the colon because of PA, has not been fully studied. Mechanisms exist for both protective and adverse effects on the colon, and they need to be investigated. However, limited available data suggest a possible protective effect, at least in relation to colon cancer risk.

PA has both adverse and potential health benefits and its "antinutrient" label is probably inappropriate. Perhaps it should be called a "physiologically" or "pharmacologically" active minor constituent of foods whose time has come for further reevaluation. It is a matter of dose and this should be established in the future for optimum health benefits.

References

1. Reddy, N.R., S.K. Sathe and D.K. Salunkhe, *Adv. Food Res. 28*:1 (1982).
2. Yoon, J., L.U. Thompson and D.J.A. Jenkins, *Am. J. Clin. Nutr. 38*:835 (1983).
3. Lui, N.S., and A. Altschul, *Arch. Biochem. Biophys. 121*:678 (1967).
4. Tanaka, K., T. Yoshida and Z. Kasai *Plant Cell Physiol. 15*:147 (1974).
5. Lott, J., and M. Buttrose, *Aust. J. Plant Physiol. 5*:89 (1978).
6. O'Dell, B., A. deBoland and S. Koirtyohann, *J. Agric. Food Chem. 20*:718 (1972).
7. Cheryan, M., *CRC Crit. Rev. Food Sci. Nutr. 13*:297 (1980).
8. Cosgrove, D.J., *Inositol Phosphates: Their Chemistry, Biochemistry and Physiology*, Elsevier Publ. Co., New York, NY, 1980.
9. Thompson, L.U., in *Phytic Acid Chemistry and Applications*, edited by E. Graf, Pilatus Press, Minneapolis, MN, 1986, pp. 173-194.
10. Gosselin, R.E., and E.R. Coghlan, *Arch. Biochem. Biophys. 45*:301 (1953).
11. O'Dell, B.L., *Am. J. Clin. Nutr. 22*:1315 (1969).
12. Oberleas, D., in *Toxicants Occurring Naturally in Foods*, edited by F. Strong, Natl. Acad. of Sci., Washington, DC, 1973, pp. 363-371.
13. Vohra, P., G.A. Gray and F.H. Kratzer, *Proc. Soc. Exp. Biol. Med. 120*:447 (1965).
14. Maddaih, V.T., A.A. Kurnick and B.L. Reid, *Ibid. 115*:391 (1964).
15. Erdman, J.W., *J. Am. Oil Chem. Soc. 56*:736 (1979).
16. O'Dell, B., in *Soy Protein and Human Nutrition*, edited by H.C. Wilke, D.T. Hopkins and D.H. Waggle, Plenum Press, New York, NY, 1979, pp. 187-204.
17. Maga, J.A., *J. Agric. Food Chem. 30*:1 (1982).
18. Davis, N.T., in *Dietary Fiber in Health and Disease*, edited by G. Vahouny and D. Kritchevsky, Plenum Press, New York, NY, 1982, pp. 105-116.
19. Morris, E., in *Phytic Acid Chemistry and Applications*, edited by E. Graf, Pilatus Press, Minneapolis, MN, 1986, pp. 57-76.
20. Hallberg, L., *Scan. J. Gastroenterol. 22* (Suppl. 129):73 (1987).
21. Tucker, H.F., and W.D. Salmon, *Proc. Soc. Exp. Biol. Med. 88*:613 (1955).
22. O'Dell, B.L., and J.E. Savage, *Fed. Proc. 16*:394 (1957).
23. Likuski, H.J., and R.M. Forbes, *J. Nutr. 85*:230 (1965).
24. Forbes, R.M., *Ibid. 83*:225 (1964).
25. Forbes, R.M., and M. Yohe, *Ibid. 70*:53 (1960).
26. Reinhold, J.L., K. Nasr, A. Lahimgarzadeh and H. Hedayati, *Lancet 1*:283 (1973).
27. Forbes, R.M., and H. Parker, *Nutr. Rep. Int. 15*:681 (1977).
28. Erdman, J.W., K. Weingartner, H.M. Parker and R.M. Forbes, *Fed. Proc. 37*:891 (1978).
29. Momcilovic, B., and B. Shah, *Nutr. Rep. Int. 13*:135 (1976).
30. Momcilovic, B., and B. Shah, *Ibid. 14*:717 (1976).

31. Prasad, A., in *Zinc Metabolism*, edited by A. Prasad and C.C. Thomas, Springfield, IL, 1966, pp. 250-303.
32. Hambidge, K.M., C. Hambidge, M. Jacobs and J.D. Baum, *Pediatr. Res. 6*:868 (1972).
33. Oberleas, D., in *Proc. Western Hemisphere Nutrition Congress IV*, edited by P.L. White and N. Selby, Publ. Sci. Group, Acton, MA, 1975, pp. 156-161.
34. Davies, N.T., and S.E. Olpin, *Br. J. Nutr. 41*:591 (1979).
35. Morris, E., and R. Ellis, *J. Nutr. 110*:1037 (1980).
36. Lo,G., S.L. Settle, F. Steinke and D.T. Hopkins, *Ibid. 111*:2223 (1981).
37. Morris, E., R. Ellis, P. Steele and P. Moger, *Fed. Proc. 39*:787 (1980).
38. Turnlund, J.R., J.C. King, W.R. Keyies, R. Gong and M.C. Michel, *Am. J. Clin. Nutr. 40*:1071 (1984).
39. Oberleas, D., M.E. Muher and B.L. O'Dell, *J. Nutr. 90*:56 (1966).
40. Oberleas, D., and B. Harland, *J. Am. Diet Assoc. 79*:433 (1981).
41. O'Dell, B.L., *Fed. Proc. 43*:2821 (1984).
42. Davies, N.T., and H. Reid, *Br. J. Nutr. 41*:579 (1979).
43. Forbes, R., and J. Erdman, *Annu. Rev. Nutr. 3*:213 (1983).
44. Fordyce, E.J., R. Forbes, K. Robbibs and J. Erdman, *J. Food Sci. 52*:440 (1987).
45. Ellis, R., E. Morris, A.D. Hill and J.C. Smith, Jr., *J. Am. Diet Assoc. 81*:26 (1982).
46. Forbes, R., H.M. Parker and J. Erdman, *J. Nutr. 114*:1421 (1984).
47. Bafundo, K.W., D.H. Baker and P.R. Fitzgerald, *Poultry Sci. 63*:2430 (1984).
48. Oberleas, D., in *Nutritional Bioavailability of Zinc*, edited by G.E. Inglett, American Chemical Society, Washington, DC, 1983, pp. 145-158.
49. Davies, N.T., A. Carswell and C. Mills, in *Trace Elements in Man and Animals-TEMA 5*, edited by C.F. Mills, I. Bremmer and J.K. Chesters, Aberdeen, Scotland, 1986, pp. 456-457.
50. Ellis, R., R. Reynolds, J. Kelsay and E. Morris, *Fed. Proc. 44*:1506 (1985).
51. Bindra, G.S., R. Gibson and L.U. Thompson, *Nutr. Res. 6*:475 (1986).
52. Weingartner, K.E., J. Erdman, H.M. Parker and R.M. Forbes, *Nutr. Rep. Int. 19*:223 (1984).
53. Erdman, J., K.E. Weingartner, G.C. Mustakas, R.D. Schmutz, H.M. Parker and R.M. Forbes, *J. Food Sci. 45*:1193 (1980).
54. Forbes, R.M., J. Erdman, J. Parker, H. Kondo and S. Ketelsen, *J. Nutr. 113*:205 (1983).
55. Ketelsen, S.M., M.A. Stuart, C.M. Weaver, R.M. Forbes and J. Erdman, *Ibid.114*:536 (1984).
56. Cossack, Z.T., and A.S. Prasad, *Nutr. Res. 3*:23 (1983).
57. Davies, N.T., and C. Mills, *Human Nutr.: Appl. Nutr.*, in press.
58. Sandberg, A., C. Hasselblad, K. Hasselblad and H. Leif, *Br. J. Nutr. 48*:185 (1982).
59. Navert, B., and B. Sandstrom, *Ibid. 53*:47 (1985).
60. Harland, B., and M. Peterson, *J. Am Dietet. Assoc. 72*:259 (1978).

61. Ellis, R., J. Kelsay, R.D. Reynolds, E. Morris, P.B. Moser and C.W. Frazier, *Ibid. 87*:1043 (1987).
62. Reinhold, J.G., B. Faradji, P. Abadi and F. Ismail-Beiji, *J. Nutr. 106*:493 (1976).
63. Kelsay, J., *Am. J. Gastroenterol. 82*:983 (1987).
64. Sandstrom, B., *Scan. J. Gastroenterol. 22* (Suppl 129):*80* (1987).
65. Andersson, H., B. Navert, S.A. Bingham, H.N. Englyst and J.H. Cummings, *Brit. J. Nutr. 50*:503 (1983).
66. Cromwell, G.L., *Feedstuffs 52*:38 (1980).
67. Nelson, T.S., in *Proc. Florida Nutr. Conference*, St. Petersburg, FL, 1980, pp. 59-84.
68. Pointillart, A., A. Fourdin and N. Fontaine, *J. Nutr. 117*:907 (1987).
69. Kivisto, B., H. Andersson, G. Cederblad, A. Sandberg and B. Sandstrom, *Brit. J. Nutr. 55*:255 (1986).
70. Bitar, K., and J. Reinhold, *Biochim. Biophys. Acta 268*:442 (1982).
71. Cooper, J., and H. Gowing, *Br. J. Nutr. 50*:673 (1983).
72. Rao, R., and C. Ramakrishnan, *Biol. Neonate 50*:165 (1986).
73. Wise, A., and D. Gilburt, *Appl. Environ. Microbiol. 43*:753 (1982).
74. Wise, A., C. Richards and M. Trimble, *Ibid. 45*:313 (1983).
75. Thompson, L.U., M. McBurney and D. Jenkins, in *Proc. Annual Meeting Can. Fed. Biol. Soc.*, 1988, p. 86.
76. Tao, S.H., M. Fox, B. Phillippy, B.E. Fry, M.L. Johnson and M.R. Johnston, *Fed. Proc. 45*:819 (1987).
77. Lonnerdahl, B., C. Kunz, A. Sandberg and B. Sandstrom, *Ibid. 46*:599 (1987).
78. Van Campen, D.R., *J. Nutr. 97*:104 (1969).
79. Oestreicher, P., and R.J. Cousins, *J. Nutr. 115*:159 (1985).
80. Solomons, N., O. Pineda, F. Viteri and H. Sandstead, *Ibid. 113*:337 (1983).
81. Sandstrom, B., L. Davidson, A. Cederblad and B. Lonnerdal, *Ibid. 115*:411 (1985).
82. Valberg, L., P. Flanagan and M. Chamberlain, *Am. J. Clin. Nutr. 40*:536 (1984).
83. Walker, A.R.P., F.W. Fox and J.T. Irving, *Biochem. J. 42*:452 (1948).
84. Cullumbine, H., V. Basnayake and J. Lemottee, *Br. J. Nutr. 4*:101 (1950).
85. McCance, R., and E. Widdowson, *Ibid. 2*:401 (1949).
86. Barre, K., and N. Van Huot, *Bull. Soc. Chim. Biol. 47*:1419 (1965).
87. Camus, M.C., and J.C. Laporte, *Annu. Biol. Biochem. Biophys. 16*:719 (1976).
88. Knuckles, B.E., D. Kuzmicky and A. Betschart, *J. Food Sci. 50*:1080 (1985).
89. Singh, M., and A. Krikorian, *J. Agric. Food Chem. 30*:799 (1982).
90. Carnovale, E., E. Lugaro and G. Lombardi-Boccia, *Cereal Chem. 65*:114 (1988).
91. Serraino, M., L.U. Thompson, L. Savoie and G. Parent, *J. Food Sci. 50*:1689 (1985).

92. Iacobucci, G.A., D.V. Meyers and K. Okubo, U.S. Patent 3,736,147 (1973).
93. Reddy, N.R., S. Sathe and M. Pierson, *J. Food Sci. 53*:107 (1988).
94. Atwal, A.S., N. Eskin, B. McDonald and M. Vaisey-Genser, *Nutr. Rep. Int. 21*:257 (1980).
95. Thompson, L.U., and M. Serraino, *J. Agric. Food Chem. 34*:468 (1986).
96. McDonald, B., S. Lieden and L. Hambraeus, *Nutr. Rep. Int. 17*:49 (1978).
97. Satterlee, L., and R. Abdul-Kadir, *Lebensm. Wiss. Technol. 16*:8 (1983).
98. Nielsen, B., M.S. Thesis, University of Toronto, Toronto, Canada, 1987.
99. Yoshida, T., S. Shinoda, T. Matsumoto and S. Watarai, *J. Nutr. Sci. Vitaminol. 28*:401 (1982).
100. Jenkins, D.J.A., D. Cuff, T. Wolever, D. Knowland, L.U. Thompson, Z. Cohen and E. Prokipchuk, *Am. J. Gastroentrol. 82*:709 (1987).
101. *Delaying Absorption as a Therapeutic Principle in Metabolic Diseases*, edited by W. Creutzfeldt and U.R. Folsch, George Thieme Verlag, New York, NY, 1983, p. 159.
102. Cawley, R.W., and T.A. Mitchell, *J. Sci. Food Agric. 19*:106 (1968).
103. Sharma, C.B., M. Goel and M. Irshad, *Phytochemistry 17*:201 (1978).
104. Deshpande, S.S., and M. Cheryan, *J. Food Sci. 49*:516 (1984).
105. Thompson, L.U., and J. Yoon, *J. Food Sci. 49*:1228 (1984).
106. Yoon, J., L.U. Thompson and D.J.A. Jenkins, *Am. J. Clin. Nutr. 38*:835 (1983).
107. Thompson, L.U., *Food Technol. 42*:123 (1988).
108. Thompson, L.U., C. Button and D.J.A. Jenkins, *Am. J. Clin. Nutr. 46*:467 (1987).
109. Buonocore, V., E. Poerio, V. Silano and M. Tomasi, *Biochem. J. 153*:621 (1976).
110. Caldwell, M.L., and J.T. King, *J. Am. Chem. Soc. 75*:3132 (1953).
111. Sognen, E., *Acta Pharmacol. Toxicol. 22*:8 (1965).
112. Sognen, E., *Ibid. 22*:31 (1965).
113. Jenkins, D.J.A., T.M.S. Wolever, J. Kalmusky, S. Giudici, C. Giordano, G.S. Wong, J.N. Bird, R. Patten, M. Hall, G. Buckley and A. Little, *Am. J. Clin. Nutr. 42*:604 (1985).
114. Anderson, J.W., and J. Ward, *Ibid. 32*:2312 (1979).
115. Simpsom, H., R. Simpson, S. Lousley, R. Carter, M. Geekie, T. Hockaday and J. Mann, *Lancet 1*:1 (1981).
116. Jenkins, D.J.A., and A.L. Jenkins, *J. Am. Coll. Nutr. 6*:11 (1987).
117. Jenkins, D.J.A., and T. Wolever, *Proc. Nutr. Soc. 40*:227 (1981).
118. Wolever, T.M.S., D.J.A. Jenkins, A.L. Jenkins, M.J. Thorne, L.U. Thompson and R. Taylor, in *Unconventional Sources of Dietary Fiber*, edited by I. Furda, Washington, DC, 1983, pp. 17-31.
119. Read, N.W., and I. Welch, in *Food and the Gut*, edited by J.O. Hunter and V. Alun Jones, Bailliers, Tindall, London, 1985, pp. 45-56
120. Lipkin, M., in *Experimental Colon Carcinogenesis*, edited by H. Autrup and G. Williams, CRC Press, Boca Raton, FL, 1983, p. 139.

121. Reddy, B., S. Mangat, J. Weisburger and E. Wynder, *Cancer Res. 37*:3533 (1977).
122. Graf, E., in *Phytic Acid Chemistry and Applications*, edited by E. Graf, Pilatus Press, Minneapolis, MN, 1986, pp. 1–21.
123. Draper, H., and R. Bird, *J. Agric. Food Chem. 32*:433 (1984).
124. Prasad, A., and D. Oberleas, *J. Lab. Clin. Med. 83*:634 (1974).
125. Deschner, E., B. Cohen and R. Raicht, *Digestion 21*:290 (1981).
126. Bird, R., *Lipids 21*:289 (1986).
127. Kim, S., E.J. Brophy and J.A. Nicholson, *J. Biol. Chem. 251*:3206 (1976).
128. Kim, Y.S., D. Tsao, A. Morita and A. Bella, in *Colonic Carcinogenesis*, edited by R.A. Malt and C.N. Williamson, MTP Press, Lancaster, PA, 1981, pp. 317–323.
129. Jacobs, L., in *Dietary Fiber: Basic and Clinical Aspects*, edited by G. Vahouny and D. Kritchevsky, Plenum Press, New York, NY, 1986, pp. 211–228.
130. Thompson, L.U., and B. Nielsen, Paper presented at FASEB Annual Meeting, 1988.
131. Nielsen, B., L.U. Thompson and R. Bird, *Cancer Lett. 37*:317 (1987).
132. Shamsuddin, A., A. Elsayed and A. Ullah, *Carcinogenesis 9*:577 (1988).
133. Elsayed, A., A. Chakravarthy and A. Shamsuddin, *Lab. Invest. 56*:21A (1987).
134. Nelson, R.L., *AntiCancer Res. 7*:259 (1987).